Pearson Education
AP* Test Prep Series

AP* BIOLOGY

Fred W. Holtzclaw
Theresa Knapp Holtzclaw

To accompany:

BIOLOGY

Campbell • Reece

8e AP Edition*

*AP is a registered trademark of the College Board, which was not involved in the production of, and does not endorse, this product.

PEARSON

Benjamin Cummings

San Francisco • Boston • New York
Cape Town • Hong Kong • London • Madrid • Mexico City
Montreal • Munich • Paris • Singapore • Sydney • Tokyo • Toronto

Vice President/Editor-in-Chief: Beth Wilbur
Senior Editorial Manager: Ginnie Simione Jutson
Senior Supplements Project Editor: Susan Berge
Project Editor: Elizabeth Campbell
Executive Marketing Manager: Lauren Harp
Managing Editor: Michael Early
Production Supervisor: Jane Brundage
Manufacturing Buyer: Michael Penne
Production Service: S4Carlisle Publishing Services
Cover Design: Yvo Riezebos Design
Text and Cover Printer: Courier, Stoughton
Cover Photo Credit: Magnolia Flower-Corbis
Photographer: Chris Fox

Figures taken from:

Biology, Seventh Edition
by Neil A. Campbell and Jane B. Reece
Copyright © 2005 by Pearson Education, publishing as Benjamin Cummings
San Francisco, California 94111

Biology, Eighth Edition
by Neil A. Campbell and Jane B. Reece
Copyright © 2008 by Pearson Education, publishing as Benjamin Cummings
San Francisco, California 94111

Biology: Concepts & Connections, Fifth Edition
by Neil A. Campbell, Jane B. Reece, Martha R. Taylor, and Eric J. Simon
Copyright © 2006 by Pearson Education, publishing as Benjamin Cummings
San Francisco, California 94111

Biology: Exploring Life, First Edition
By Neil A. Campbell, Brad Williamson, and Robin J. Heyden
Copyright © 2004 by Pearson Education, publishing as Prentice Hall
Upper Saddle River, New Jersey 07458

Essential Biology with Physiology, Second Edition
by Neil A. Campbell, Jane B. Reece and Eric J. Simon
Copyright © 2007 by Pearson Education, publishing as Benjamin Cummings
San Francisco, California 94111

Student Study Guide for Biology, Seventh Edition
by Martha R. Taylor
Copyright © 2005 by Pearson Education, publishing as Benjamin Cummings

PEARSON
Benjamin Cummings

7 8 9 10—CRS—11 10
www.aw-bc.com.

Brief Contents

Part I: *Introduction to the AP Biology Examination* 1

Part II: *Topical Review with Sample Questions and Answers and Explanations* 25

TOPIC 1
The Chemistry of Life 27

TOPIC 2
The Cell 43

TOPIC 3
Respiration and Photosynthesis 75

TOPIC 4
Mendelian Genetics 97

TOPIC 5
Molecular Genetics 117

TOPIC 6
Mechanisms of Evolution 151

TOPIC 7
The Evolutionary History of Biological Diversity 171

TOPIC 8
Plant Form and Function 207

TOPIC 9
Animal Form and Function 231

TOPIC 10
Ecology 291

Part III: *The Laboratory* 311

Part IV: *Sample Tests with Answers and Explanations* 365

Practice Test One 367
Practice Test Two 397

About Your Pearson AP* Guide

Pearson Education is the leading publisher of textbooks worldwide. With operations on every continent, we make it our business to understand the changing needs of students at every level, from Kindergarten to college. We think that makes us especially qualified to offer this series of AP test prep books, tied to some of our best-selling textbooks.

Our reasoning is that as you study for your course, you're preparing along the way for the AP test. If you can tie the material in the book directly to the test you're taking, it makes the material that much more relevant, and enables you to focus your time most efficiently. And that's a good thing!

The AP exam is an important milestone in your education. A high score means you're in a better position for college acceptance, and possibly puts you a step ahead with college credits. Our goal is to provide you with the tools you need to excel on the exam . . . the rest is up to you.

Good luck!

Part I

Introduction to the AP Biology Examination

This section gives an overview of the Advanced Placement* program and the AP Biology Examination. Part I introduces the types of questions you will encounter on the exam, explains the procedures used to grade the exam, and provides helpful test-taking strategies. A correlation chart shows where in Campbell and Reece, *BIOLOGY*, Eighth Edition, you will find key information that commonly appears on the AP Biology Examination. Review Part I carefully before trying the sample test items in Part II and Part III.

The Advanced Placement* Program

Probably you are reading this book for a couple of reasons. You may be a student in an Advanced Placement (AP) Biology class, and you have some questions about how the whole AP Program works and how it can benefit you. Also, perhaps, you will be taking an AP Biology Examination, and you want to find out more about it. This book will help you in several important ways. The first part of this book introduces you to the AP Biology course and the AP Biology Exam. You'll learn helpful details about the different question formats—multiple-choice and free-response—that you'll encounter on the exam. In addition, you'll find many test-taking strategies that will help you prepare for the exam. A correlation chart at the end of Part I shows how to use your textbook, Campbell and Reece *BIOLOGY*, to find the information you'll need to know to score well on the AP Biology Exam. By the way, this chart is useful, too, in helping you to identify any extraneous material that won't be tested. Part II of this book provides an extensive content review correlated to each unit of your textbook, along with sample multiple-choice and free-response questions. Finally, in Part III, you will find two full-length sample tests. These will help you practice taking the exam under real-life testing conditions. The more familiar you are with the AP Biology Exam ahead of time, the more comfortable you'll be on testing day.

The AP Program is sponsored by the College Board, a nonprofit organization that oversees college admissions examinations. (The College Board is composed of college and high school teachers and administrators.) The AP Program offers thirty-four college-level courses to qualified high school students. If you receive a grade of 3 or higher on an AP exam, you may be eligible for college credit, depending on the policies of the institution you plan to attend. Over 3,000 colleges and universities around the world grant credit to students who have performed well on AP exams. Some institutions grant sophomore status to incoming first-year students who have demonstrated mastery of several AP subjects. You can check the policies of specific institutions on the College Board's website (www.collegeboard.com). In addition, the College Board confers a number of AP Scholar Awards on students who score 3 or higher on three or more AP exams. Additional awards are available to students who receive very high grades on four or five AP exams.

Why Take an AP Course?

You may be taking an AP course simply because you like challenging yourself and you are thirsty for knowledge. Another reason may be that you know that colleges look favorably on applicants who have AP courses on their secondary school transcripts. AP classes involve rigorous, detailed lessons, a lot of homework, and numerous tests. College admissions officers may see your willingness to take these courses as evidence of your work ethic and commitment to your education. Because AP course work is more difficult than average high school work, many admissions officers evaluate AP grades on a higher academic level. For example, if you receive a B in an AP class, it might carry the same weight as an A in a regular-level high school class.

Your AP Biology course prepares you for many of the skills you will need in college. For example, your teacher may assign a major research paper or require you to perform several challenging laboratory exercises using proper scientific protocol. AP Biology teachers routinely give substantial reading assignments, and students learn how to take detailed lecture notes and participate vigorously in class discussions. The AP Biology course will challenge you to gather and consider information in new—and sometimes unfamiliar—ways. You can feel good knowing that your ability to use these methods and skills will give you a leg up as you enter college.

Each college or university decides whether or not to grant college credit for an AP course, and each bases this decision on what it considers satisfactory grades on AP exams. Depending on what college you attend and what area of study you pursue, your decision to take the AP Biology Exam could save you tuition money. You can contact schools directly to find out their guidelines for accepting AP credits, or use the College Board's online feature, "AP Credit Policy Info."

Taking an AP Examination

The AP Biology Exam is given annually in May. Your AP teacher or school guidance counselor can give you information on how to register for an AP exam. Remember, the deadline for registration and payment of exam fees is usually in March, two months before the actual exam date in May. The cost of the exam is subject to change and can differ depending on the number of exams taken. However, in 2008 a single exam costs $84. For students who can show financial need, the College Board will reduce the price by $22, and your school might also waive its regular rebate of $8, so the lowest possible total price is $54. Moreover, schools in some states are willing to pay the exam fee for the student. If you feel you may qualify for reduced rates, ask your school administrators for more information.

The exams are scored in June. In mid-July the results will be sent to you, your high school, and any colleges or universities you indicated on your answer sheet. If you want to know your score as early as possible, you can get it (for an additional charge of $25) beginning July 1 by calling the College Board at (888) 308-0013. On the phone, you'll be asked to give your AP number or social security number, your birth date, and a credit card number.

If you decide that you want your score sent to additional colleges and universities, you can fill out the appropriate information on your AP Grade Report (which you will receive by mail in July) and return it to the College Board. There is an additional charge of $15 for each additional school that will receive your AP score.

On the other hand, if a feeling of disaster prevents you from sleeping on the nights following the exam, you could choose to withhold or cancel your grade. (Withholding is temporary, whereas canceling is permanent.) Grade withholding carries a $10 charge per college or university, whereas canceling carries no fee. You'll need to write to (or email) the College Board and include your name, address, gender, birth date, AP number, the date of the exam, the name of the exam, a check for the exact amount due, and the name, city, and state of the college(s) from which you want the score withheld. You should check the College Board website for the deadline for withholding your score, but it's usually in mid-July. It is strongly suggested that you *do not* cancel your scores, since you won't know your score until mid-July. Instead, relax and try to assume that the glass is half full. At this point, you have nothing to lose and a lot to gain.

If you would like to get back your free-response booklet for a post-exam review, you can send another check for $7 to the College Board. You'll need to do this by mid-September. Finally, if you have serious doubts about the accuracy of your score for the multiple-choice section, the College Board will re-score it for an additional $25.

AP Biology: Course Goals

The two central goals of the AP Program in Biology are to help students develop a conceptual framework for modern biology and gain an appreciation of science as a process. AP Biology courses are built around topics, concepts, and themes. The College Board defines topics as the subject areas of biology. A concept is an important idea or principle that forms or enhances our current understanding of a particular topic. Themes are the overarching features of biology that recur, connect, and unify our understanding of topics. The College Board lists eight themes that should be stressed in AP Biology courses. The themes are from "Explanation of the Major Themes", page 6, *Biology Course Description*, published by the College Board.

- Science as a process;
- Evolution;
- Energy transfer;
- Continuity and change;
- Relationship of structure to function;
- Regulation;
- Interdependence in nature;
- Science, technology, and society.

During the year, AP Biology students will be applying these themes to a wide range of topics. In fact, the College Board has created a topic outline to illustrate the topics that make up a typical college biology course—and so should form the basis for AP Biology courses. The percentages in parentheses show how much of the course should be spent on particular topics.

    I. Molecules and Cells (25% of the AP Biology course)
       A. Chemistry of Life (7%)
          Water
          Organic molecules in organisms
          Free energy changes
          Enzymes
       B. Cells (10%)
          Prokaryotic and eukaryotic cells
          Membranes

Subcellular organization
Cell cycle and its regulation
C. Cellular Energetics (8%)
Coupled reactions
Fermentation and cellular respiration
Photosynthesis

II. Heredity and Evolution (25% of the AP Biology course)
A. Heredity (8%)
Meiosis and gametogenesis
Eukaryotic chromosomes
Inheritance patterns
B. Molecular Genetics (9%)
RNA and DNA structure and function
Gene regulation
Mutation
Viral structure and replication
Nucleic acid technology and applications
C. Evolutionary Biology (8%)
Early evolution of life
Evidence for evolution
Mechanisms of evolution

III. Organisms and Populations (50% of the AP Biology course)
A. Diversity of Organisms (8%)
Evolutionary patterns
Survey of the diversity of life
Phylogenetic classification
Evolutionary relationships
B. Structure and Function of Plants and Animals (32%)
Reproduction, growth, and development
Structural, physiological, and behavioral adaptations
Response to the environment
C. Ecology (10%)
Population dynamics
Communities and ecosystems
Global issues

No doubt, AP Biology courses vary somewhat from teacher to teacher and from school to school. For the most part, many high school teachers study the AP Biology course outline each year and meticulously customize their curriculum to fit it. In short, you may wish to consult the AP Biology course outline in order to take note of topics and concepts that might require further study.

Understanding the AP Biology Examination

You are probably aware that in general AP exams are long. The AP Biology Exam takes three hours. The exam probably looks like many other tests you've taken. It is made up of a multiple-choice section and a free-response (essay) section. At the core of the examination are questions designed to measure your knowledge and understanding of modern biology. You should be prepared to recall basic facts and concepts, to apply scientific facts and concepts to particular problems, to synthesize facts and concepts, and to demonstrate reasoning and analytical skills by organizing written answers to broad questions.

The AP Biology Exam is very challenging. When you sit down to take the test, exam administrators expect you not only to be fluent in the areas of biology that you find fascinating (the ones that probably inspired you to take a special interest in the subject originally), but also to have an intimate knowledge of topics you don't find interesting at all. Whatever those topics might be—DNA replication, the dizzying details of animal and plant classification, or cell organization—you need to be comfortable with and knowledgeable about all of the AP Biology topics.

Section I: Multiple-Choice Questions

Section I contains 100 multiple-choice questions that test both scientific facts and their applications. You will have 80 minutes to complete Section I. This portion of the exam is followed by a 5–10 minute break—the only official break during the examination. The directions for the multiple-choice section of the test are straightforward and similar to the following:

> **Directions:** Each of the questions or incomplete statements below is followed by five suggested answers or completions. Select the choice that best answers the question or completes the statement.

It will probably not surprise you to know that not all multiple-choice questions are the same. In fact, the AP Biology Exam will contain the different types of multiple-choice questions listed and described below.

Factual Questions

The first type of multiple-choice question is your basic factual recall question, which will test whether you've mastered certain facts, processes, cycles, systems, etc. Here's an example of one of these:

1. In plants, which hormone is responsible for fruit ripening?
   (A) Auxin
   (B) Ethylene
   (C) Cytokinin
   (D) Phytochrome
   (E) Gibberellin

With a question like this, you either know the answer or you don't. It's what the College Board calls a "factual" question: They ask a question, and you're expected to know the answer. The best way to approach factual questions is to read the question and every one of the five choices carefully. If you are certain you know the answer, fill in the corresponding oval on the answer sheet. However, what if you're not certain? The next step is to see if you can eliminate one or more of the choices.

Let's look again at the question. Imagine that you don't recall that ethylene (choice *B*) is responsible for fruit ripening. You might realize that you can eliminate answer choice *D*, because the prefix *phyto-* suggests that the hormone phytochrome probably has something to do with plants' response to light and therefore phytochrome probably doesn't contribute to ripening. Now you have a one-in-four chance of making a correct guess. If you remember what any of the other three plant hormones do, your odds of guessing correctly increase even more. As a general rule on AP exams, if you can eliminate at least one answer choice, you are better off making an educated guess than you are leaving the question unanswered. We'll discuss this idea further in the section "Grading Procedures for the AP Biology Examination."

"Reverse" Multiple-Choice Questions

Sometimes the College Board question developers modify the format for factual questions slightly, and the result is a slightly more difficult type of multiple-choice question. This type of question features four answers that are correct and only one that is incorrect; you are asked to find the incorrect choice. Generally these questions contain the word *not* or the word *except*. Pay attention to the capitalization of these words in Section I of the exam to avoid making a careless mistake. Here is an example of a question you could expect to see on the AP Biology Exam.

2. All of the following statements about photosynthesis are true EXCEPT
    (A) the light reactions convert solar energy to chemical energy in the form of ATP and NADPH
    (B) the Calvin cycle uses ATP and NADPH to convert CO_2 to sugar
    (C) photosystem I contains P700 chlorophyll a molecules at the reaction center; photosystem II contains P680 molecules
    (D) in chemiosmosis, electron transport chains pump protons (H^+) across a membrane from a region of high H^+ concentration to a region of low H^+ concentration
    (E) the steps of the Calvin cycle are sometimes referred to as the dark reactions because they do not require light in order to take place

This question asks you not only to remember one simple fact (like the function of a plant hormone), but also to consider the results of cellular processes and how systems in the cell compare. The correct answer here is *D*, because the statement in choice *D* is not true (in fact, the electron transport chains pump protons across membranes from regions of low H^+ concentrations to regions of high H^+ concentrations). Of course, the statements in choices *A*, *B*, *C*, and *E* are true. You may recall that this proton pumping occurs in both mitochondria and chloroplasts, and that the protons then diffuse—down the concentration gradient—back across the membrane (through ATP synthase), and that this drives the synthesis of ATP. However, you can also apply common sense to see that D doesn't look right. Why would a pump be needed to transport H^+ down its concentration gradient?

Conceptual-Thematic Questions

Now let's look at another type of multiple-choice question that you'll encounter on the exam known as "conceptual-thematic" questions.

3. Which of the following groups is characterized by having a gastrovascular cavity, with a single opening acting as both mouth and anus, and existing in either polyp or medusa form?
   (A) Sponges
   (B) Cnidarians
   (C) Ctenophores
   (D) Platyhelminthes
   (E) Rotifers

Though it may be hard to see the difference between this question and a factual recall question, the difference is that this one asks you to use logic and synthesis to glean the sum of the organismal characteristics listed above. In other words, you're required to synthesize information rather than merely to recall a fact. In this case, the correct choice is B. The cnidarians you might encounter are hydras, jellies, sea anemones, and coral animals. They are simple, sac-like creatures with a gastrovascular cavity that contains a single opening.

Matching Questions

Another type of multiple-choice question you'll see in Section I of the AP Biology Exam is the matching question. Below is an example of how matching questions are presented.

Questions 4–8
   (A) Savanna
   (B) Chaparral
   (C) Rain forest
   (D) Coniferous forest
   (E) Tundra

4. Characterized by epiphytes, closed canopy, and pronounced vertical stratification

5. Characterized by cone-bearing trees, dominated by one or two species of trees, and receiving heavy snowfall in the winter

6. Dominated by spiny evergreen shrubs, which are dependent on seasonal shrub fires for growth

7. Characterized by having insects as the dominant herbivores, predominant grass growth, and large grazing mammals

8. Characterized by permafrost, very low temperatures, and low annual rainfall

(To satisfy your curiosity, the correct answers above are 4. *C*; 5. *D*; 6. *B*; 7. *A*; and 8. *E*.)

Lab-Based or Experimental Questions

The last type of multiple-choice question you'll see in Section I is the lab-based or experimental question. These questions either present you with a set of data in graph (or other) form, or they describe an experiment and ask you to make educated guesses and to form hypotheses. The graph below shows the results of a study to determine the effect of soil air spaces on plant growth.

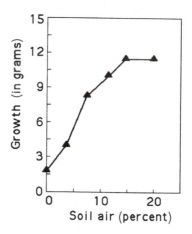

9. The data from the above graph show that the plant
   (A) grows fastest when the soil is 5–10% air
   (B) grows fastest when the soil is 15–20% air
   (C) grows at the same rate regardless of the soil air percentage
   (D) grows most slowly when the soil is 5–10% air
   (E) does not grow at all when the soil is 0–3% air

The correct choice is *A*. The graph shows the line with the greatest slope (the highest degree of change over the shortest amount of time) between the percentages 5 and 10. During this time, the plant grows by about $9 - 5 = 4$ grams. Just to be sure, check the amount this plant grows when the soil is 15–20% air. At the start, when the air was 15% air, the plant weighed 12 grams. At the end, when the soil is 20% air, the plant weight is the same—12 grams. Virtually no growth occurred during this time.

Clearly this question requires you to be able to interpret a graph, but at this point in your biology education you should be quite capable of doing that. In order to brush up on the various ways that graphs present information, you might spend some time looking over Chapter 53 and Chapter 54 of Campbell and Reece, *BIOLOGY*, 8th Edition.

Section II: Free-Response Questions

Section II of the AP Biology Exam is made up of four free-response questions relating to the following topics: molecules and cells, heredity and evolution, and organisms and populations (two questions). One or more of the four free-response questions will be lab-based, because the test writers expect you to understand laboratory concepts and techniques. Many of the questions you encounter in Section II of the AP Biology Exam will require you to integrate material from across the topic outline. The free-response questions are often broken down into parts, and the parts vary in difficulty. Presenting free-response questions in this form is the College Board's way of making sure that you really understand the underlying concepts of biology—and that you aren't just a really lucky guesser.

At the beginning of Section II of the AP Biology Exam, you'll be given a 10-minute reading period to examine the questions and outline your response. After the reading period, you will be given a response booklet in which to write your essays. Following are some sample free-response questions.

Biology

Section II

Time—1 hour and 30 minutes

Directions: Answer all questions.

Answers must be in essay form. Outline form is NOT acceptable. Labeled diagrams may be used to supplement discussion, but in no case will a diagram alone suffice. It is important that you read each question completely before you begin to write.

1. Water comprises roughly 70% of the human body; cells are roughly 70–95% water, and water covers about three-quarters of the Earth's surface.
   (a) **Describe** the major physical properties of water that make it unique from other liquids.
   (b) **Explain** the properties of water that enable it to travel up through the roots and stems of plants to reach the leaves.
   (c) **Explain** why the temperature of the oceans can remain relatively stable and support vast quantities of both plant and animal life, when air temperature fluctuates so significantly throughout the year.

Like many free-response questions on the AP Biology Exam, this sample is broken into three distinct parts. Each contains a clear directive. In fact, they are printed in boldface to help you focus on exactly how you should answer the question. First you will need to explain the uniqueness of water by describing its major physical properties. (In your response to this first part of the question, you might wish to include a labeled diagram of the structure of water, complete with electrons and bonds.) Then you must explain the properties of water that allow it to travel from root to leaf. Finally, you should explain the reason(s) why ocean water temperature remains stable and supports plant and animal life—even in the face of great air temperature variations. Of course, limiting your answer by addressing exactly what the question asks will make writing the essay easier for you and earn you a higher score. Always take the time to determine precisely what is being asked before you begin to formulate a concrete thesis and focus on writing your relevant supporting paragraphs.

Grading Procedures for the AP Biology Examination

The raw scores of the AP Biology Examination are converted to the following 5-point scale:

5—Extremely Well Qualified
4—Well Qualified
3—Qualified
2—Possibly Qualified
1—No Recommendation

Some colleges give undergraduate course credit to students who achieve scores of 3 or better on AP exams. Other colleges require students to achieve scores of 4 or 5. You may check the policy for individual colleges on the College Board website (www.collegeboard.com). Below is a breakdown of how the grading of the AP Biology Exam works.

Section I: Multiple-Choice Questions

The multiple-choice section of the exam is worth 60% of your total score. The raw score of Section I is determined by crediting one point for each correctly answered question and by deducting ¼ point for each question answered incorrectly. No points are gained or lost for unanswered questions. Consequently, if you are able to eliminate at least one choice as incorrect, it is to your advantage to make an educated guess rather than to leave the answer blank. As you take practice exams, determine your score with this equation (number right − ¼ number wrong). Some students find they are timid and leave too many blank, while others should be more cautious. Practice tests will reveal if you are too timid or too bold and allow you to adjust your test-taking strategy accordingly.

Section II: Free-Response Questions

Section II counts for 40% of your examination grade. Within Section II, each of the four essay responses is weighted equally. The free-response section is scored by several hundred faculty consultants, including high school teachers and college instructors from all over the country who work in a central location to grade the essays. This period of scoring exams is called the "Reading." To ensure that scoring of all exams is consistent, grading rubrics, or standards, are developed and then faculty consultants are trained in their application. Because of this intense training, group discussion, and supervision, your essay should receive the same score regardless of who reads it. Ongoing internal checks during the Reading ensure this. Each of your four essays will be evaluated by a faculty consultant trained to score that single response.

Your answers to the free-response questions must be presented in essay form. Outlines or unlabeled and unexplained diagrams are not given credit. Each of the questions is scored on a scale from 0 to 10 points, and your performance on any single essay is evaluated independently of the other essays. Do not assume that information provided in one question will be considered during the grading of another essay. You should repeat information from question to question if it is necessary to illustrate your point.

Test-Taking Strategies for the AP Biology Examination

Here are a few tips for preparing yourself in the weeks leading up to the examination.

▪ The earlier you start studying for the AP Biology Exam, the better. Some students use this AP Biology prep book along with their textbook throughout the course, taking notes in the margin to supplement their teacher's lectures. You should definitely begin serious preparation for the test at least one month in advance.

▪ Each chapter in Part II is correlated to *BIOLOGY,* 8th Edition. For each topic, review your lecture notes, study the figures in your text that explain key concepts, and then make your way through the corresponding section of Part II of this book. If possible, retake your unit test on the topic, and also answer the questions in this guide for each unit. This will help you identify topics that will require further study. Do not try to reread your text; use it as a tool for those topics that need further study. You can use the correlation guide at the end of Part I of this book to link AP Biology topics to your textbook. Pace yourself!

AP Review: Lab Essays

The College Board suggests the following twelve labs for AP Biology courses, so you should be familiar with them. You can find more detailed descriptions of these labs in the *Course Description for AP Biology* or on the College Board's website.

1. Diffusion and Osmosis
2. Enzyme Catalysis
3. Mitosis and Meiosis
4. Plant Pigments and Photosynthesis
5. Cell Respiration
6. Molecular Biology
7. Genetics of Organisms
8. Population Genetics and Evolution
9. Transpiration
10. Physiology of the Circulatory System
11. Animal Behavior
12. Dissolved Oxygen and Aquatic Primary Productivity

▪ At least one essay on the exam will be based on an AP laboratory. To prepare for this question, review the objectives for all twelve laboratories. The College Board does not expect that you have done *the* lab, but that you have performed a lab that meets the same objectives. It is the objectives that you will be tested on. In the laboratory section of this guide, we have reviewed all twelve labs and included sample test questions to help you review. The essay may ask you to "design an experiment to determine . . .". In this case it is not necessary for you to create a new lab! If you performed a lab in your AP class that would answer this question, it is fine to describe this lab. Here are the items that are generally required for a good response:

- **State a hypothesis** as an "**If** . . . (conditions), **then** . . . (results)" statement. Your hypothesis must be testable.
- **Identify the variable factor** for the experiment (e.g., temperature).
- **Identify a control.** You must explain the control for the experiment.
- **Hold all other variables constant.** Explain how you would do this.
- **Manipulate the variable** (e.g., one group at 10°C, one at 20°C, and one at 30°C).
- **Measure the results** (e.g., cm grown, grams increased in mass).
- **Discuss results expected** as related to hypothesis.
- **Replication or verification.** The experiment must be repeated or large sample sizes must be used.

If appropriate, you could also consider using statistical analysis of data (see Chi-square analysis, Lab 7) and review of the literature.

Graphing Data

Several recent exams have asked students to graph results. You will need to consider the type of graph that is appropriate for your data. Bar graphs are used when data points are discrete, that is, not related to each other, such as the number of girls in AP Biology vs. the number of boys in AP Biology. Line graphs are used when the data are continuous, such as the change in an individual's height at each birthday. Consider if there is a data point at 0 on the graph. Be sure to extend your line to 0 if there is, but do not take the line to 0 if there is no measurement for that data point. Also,

- Label the graph with a descriptive title.
- Label the x- and y-axes. Be sure you know which variable is independent and which is dependent.
- Keep all measurement units constant. Each division on the graph must be a unit equal to all the others.

Sample Tests

When you are ready to check your preparation, take the first sample exam in Part III of this book. Keep track of your time, and try to simulate test conditions. Score your responses as "number right − ¼ number wrong." Circle the items you get wrong, or could not answer, and keep a list of the subject matter of those questions. Then analyze the list to look for patterns—are you having a hard time with questions on animal physiology or the process of photosynthesis specifically? Spend the next week or so studying the topics in which you are weak. Then, perhaps a week or two before the AP Biology Exam, take the second sample exam. Once again, grade this exam and determine your weak areas. Then spend the final days before the test looking through this guide, your class notes, and textbook to fill in any remaining gaps.

The Day of the Exam

If you have followed this suggested study plan, you should feel well prepared by test day. Plan your schedule so that you get two very good nights of uninterrupted sleep before exam day. The night before the exam, relax, think positive thoughts, and focus on getting a good night's rest. Below is a brief list of basic tips and strategies to think about before you arrive at the exam site.

1. **Arrive early!** It's a good idea to arrive at the exam site 30 minutes before the start time. On the day of the exam, make sure that you eat a good, nutritious meal. These tips may sound corny or obvious, but your body must be in peak form in order for your brain to perform well. Remember, you are going to need ATP to fuel brain cells at peak efficiency for more than three hours.

2. **Bring a photo ID.** (It's essential if you are taking the exam at a school other than your own.) Carrying a driver's license or a student ID card will allow you to prove your identity.

3. **Bring at least two sharpened #2 pencils** for the multiple-choice section. Also, bring a clean pencil eraser with you. Many pencils today have cheap erasers that smudge. Invest in a good eraser. The machine that scores Section I of the exam recognizes only marks made by a #2 pencil. Poorly erased responses are often misscored.

4. **Bring two black ballpoint pens** for the free-response portion of the test. Felt-tip pens run and pencils and inks of other colors are harder to read.

5. **Bring a watch** with you to the exam. Most testing rooms do have clocks. Still, having your own watch makes it easy to keep close track of your own pace. Watches with calculators or alarms are not permitted in the exam room.

Several other items that are forbidden from the testing room are books, notes, laptops, beepers, cameras, and portable listening or recording devices. If you must bring a cellular phone with you, be prepared to turn it off and to give it to the test proctor until you are finished with your exam. For a complete list of what not to bring, see the College Board website.

Educational Testing Service prohibits the objects listed above in the interest of fairness to all test-takers. Similarly, the test administrators are very clear and very serious about what types of conduct are not allowed during the examination. Below is a list of actions to avoid at all costs, since each can result in your immediate dismissal from the exam room.

- Do not consult any outside materials during the three hours of the exam period. Remember, the break is technically part of the exam—you are not free to review any materials at that time either.
- Do not speak during the exam. If you have a question for the test proctor, raise your hand to get the proctor's attention.
- When you are told to stop working on a section of the exam, you must stop immediately.
- Do not open your exam booklet before the test begins.

- Never tear a page out of your test booklet or try to remove the exam from the test room.
- Do not behave disruptively—even if you're distressed about a difficult test question or because you've run out of time. Stay calm and make no unnecessary noise.

Section I: Strategies for Multiple-Choice Questions

Obviously, having a firm grasp of biology is, of course, the key to doing well on the AP Biology Examination. In addition, being well-informed about the exam itself increases your chances of achieving a high score. Below is a list of strategies that you can use to increase your comfort, your confidence, and your chances of excelling on the multiple-choice section of the exam.

- Become as familiar as possible with the format of Section I. The more comfortable you are with the multiple-choice format and with the kinds of questions you'll encounter, the easier the exam will be. Remember, Part II and Part III of this book provide you with invaluable practice on the kinds of multiple-choice questions you will encounter on the AP Biology Exam.
- Every question you answer correctly is a point, so pacing is important. If you have done all the suggested practice tests, you should have a good sense of how to pace yourself. You will have 80 minutes to answer 100 questions (about 48 seconds per question). Keep track of time!
- Some of the questions will require calculations. If you encounter a question that will require extra time, leave it blank and make a note. Your goal should be to reach the end of the test, picking up all the points from questions you can answer easily.
- The test is organized with three types of questions: standard multiple choice, matching, and lab sets. Lab sets are generally the most tedious. When a data table or graph is presented, proceed directly to the related questions. Determine what information is needed to answer the questions, and then return to the data table or graph and seek the information. Sometimes, although the data appear daunting, the questions are actually very easy.
- Make a light mark in your test booklet next to any questions you can't answer. Return to these questions after you reach the end of Section I. Sometimes questions that appear later in the test will refresh your memory on a particular topic, and you will be able to answer one or more of those earlier questions.
- Always read the entire question carefully, and underline key words or ideas. You might wish to double underline words such as NOT or EXCEPT in that type of multiple-choice question.
- Read each and every one of the answer choices carefully before you make your final selection.
- Use the process of elimination to help you arrive at the correct answer. Even if you are quite sure of an answer, cross out the letters of incorrect choices in your test booklet as you eliminate them. This cuts down on the incorrect choices and allows you to narrow the remaining choices even further.

- If you can eliminate even one answer choice, it is usually better to make an educated guess than to leave the answer blank.
- Become completely familiar with the instructions for the multiple-choice questions before you take the exam. By knowing the instructions cold, you'll save yourself the time of reading them carefully on exam day.

Section II: Strategies for Free-Response Questions

Below is a list of strategies that you can use to increase your chances of excelling on the free-response section of the exam.

- You will have a 10-minute period to review the essay questions before you receive a response book. During this time, you should organize your thoughts and outline your essays on the sheet provided. After the preparation time, you will be given the response book. Each essay question is repeated within the response book, and this is where you should record each answer. You have about 22 minutes to spend on each essay.
- Read the question; then read the question again. Be sure you answer the question that is asked and that you address each part of the question. As you read a question, underline any directive words (usually the first word in an essay) that indicate how you should answer and focus the material in your essay. Some of the most frequently used directives on the AP Biology Exam are listed below, along with descriptions of what you need to do in your writing to answer the question.

  - *Analyze* (show relationships between events; explain)
  - *Compare* (discuss similarities between two or more things)
  - *Contrast* (discuss points of difference or divergence between two or more things)
  - *Describe* (give a detailed account)
  - *Design* (create an experiment and convey its ideas)
  - *Explain* (clarify; tell the meaning)

- Make an outline of your response. Reread the question as many times as necessary to make sure that you will cover each aspect of the topic. Free-response questions frequently have several parts, so you will need to take this into account as you outline your ideas.
- Write an essay! As the exam states clearly in the directions to free-response questions, a diagram or graph by itself is never an acceptable way to answer a free-response question. However, you should think about whether you could use a labeled diagram or graph to develop your written answer in some useful way.
- The essay you craft for this exam is not the same type of essay you should write for an English course. Yes, it should be well-organized; however, introductory sentences and conclusions are absolutely not necessary. Readers are interested in what you know and how well you express your knowledge. Spend your time packing the essay with the biological information you have worked so hard to learn.

- If the question has several parts, answer the parts in the sequence given. Use a letter or some other indication for each part, so that the faculty consultant does not overlook a section of your response.
- If you are asked to perform a calculation, be sure to show the steps used to arrive at your answer. You have heard this before: show your work!
- If you cannot remember a specific term, describe the structure or process.
- Define any scientific term that you use that is directly related to your response. For example, if you discuss hydrogen bonding and how it relates to properties of water, be sure to explain what hydrogen bonds are, and then describe or define adhesion, cohesion, and so on.
- Your handwriting can affect your results. Although faculty consultants make every attempt to read each essay, sometimes it is impossible to decipher messy handwriting. When your handwriting is poor, the reader may lose concentration or patience and miss an important word or phrase.
- Don't leave any part of any essay blank. Every point made is worth twice as much as each multiple-choice point.
- If time allows, proofread your essays. Don't worry about crossing out material—readers understand that your responses are first drafts and that you are writing down ideas under the pressure of time.

The success of your four free-response essays will depend a great deal on how clearly and extensively you answer the questions posed. Of course, the structure of your essays will depend entirely on your knowledge of the subjects at hand. Take a look at an example of a free-response question below.

2. In cancer, the cell's reproductive machinery experiences a loss of control that makes cancer cells reproduce continually, and eventually form a tumor.
   (a) **Describe** three DNA-related cellular events that could lead to the loss of cell division control that contributes to cancer.
   (b) **Describe** why tumors are detrimental to the body.
   (c) **Discuss** several cell processes that you think should be studied more closely in finding a cure for cancer.

In order to answer this question, you should isolate exactly what it is that you must answer. You may want to underline the relevant information in the question to remind yourself of your focus:

2. In cancer, the <u>cell's reproductive machinery</u> experiences a loss of control that makes <u>cancer cells reproduce</u> continually, and eventually <u>form a tumor</u>.
   (a) **Describe** <u>three DNA-related cellular events</u> that could <u>lead to the loss of cell division control</u> that <u>contributes to cancer</u>.
   (b) **Describe** why <u>tumors</u> are <u>detrimental</u> to the body.
   (c) **Discuss** <u>several cell processes</u> that you think should be <u>studied more closely</u> in finding a <u>cure for cancer</u>.

In order to answer the first part of the question, you'll need to identify three appropriate DNA-related cellular events.

1. A mutation or change in the original DNA sequence
2. Errors in DNA replication that go undetected by the cell's proofreading devices
3. A translocation

Under each of the three events, you should list any and all details you remember about those events to use in your description. When you flesh out these details, you'll need to clearly connect them to the concept of the loss of cell division control leading to cancer.

To answer the second part of the question, you'll need to list as many reasons as you can think of as to why tumors are harmful to the body. These might include cancerous cells' ability to metastasize; tumors' ability to occur almost anywhere in the body; their tendency to block the flow of blood when they grow near blood vessels; disruption of the natural function of any organ in the body; and endangering of homeostasis. Of course, after you list reasons, you'll need to add details to each item in your list.

To answer the final part of the question, you'll need to consider first what causes cancer. Then you must think creatively in order to suggest possible approaches to dealing with each specific cause.

Part II of this book contains a review of everything that you learned in your textbook that could be on the AP Biology test. Many questions will be posed along the way so that you can get used to being tested on the concepts in the way that the College Board will test you. In Part III, all twelve labs have been reviewed, including sample test questions to check your understanding. In Part IV, there are two practice tests for you to try on your own, along with complete answers and explanations.

AP Topic Correlation to Campbell and Reece *BIOLOGY*, Eighth Edition

The following chart is intended to help you study for the AP Biology Exam. The left column includes a series of AP Biology topics with which you should be familiar before you take the AP Biology Exam. The right column includes a detailed breakdown of corresponding chapters and Key Concepts in your Campbell and Reece *BIOLOGY*, Eighth Edition, textbook. You may want to use this chart throughout the year to review what you've learned. It is also an excellent place to begin your pre-exam review of subjects.

AP BIOLOGY TOPICS	CORRELATION TO *BIOLOGY,* 8E AP EDITION
I. Molecules and Cells	**Units 1 and 2**
A. Chemistry of Life	**Chapters 3, 4, 5, 8**
1. Water	Concepts 3.1–3.3
2. Organic molecules in organisms	Concepts 4.1–4.3, 5.1–5.5
3. Free energy changes	Concepts 8.1–8.3
4. Enzymes	Concepts 8.4–8.5
B. Cells	**Chapters 6, 7, 11, 12**
1. Prokaryotic and eukaryotic cells	Concepts 6.1–6.7, 27.1–27.6
2. Membranes	Concepts 6.2, 6.4, 7.1–7.5, 11.1–11.4
3. Subcellular organization	Concepts 6.2–6.7
4. Cell cycle and its regulation	Concepts 12.1–12.3
C. Cellular Energetics	**Chapters 8, 9, 10**
1. Coupled reactions	Concepts 8.3, 9.1–9.4
2. Fermentation and cellular respiration	Concepts 9.1–9.6
3. Photosynthesis	Concepts 10.1–10.4
II. Heredity and Evolution	**Units 3, 4, and 5**
A. Heredity	**Chapters 13–15**
1. Meiosis and gametogenesis	Concepts 13.1–13.4
2. Eukaryotic chromosomes	Concepts 15.1–15.3, 16.3
3. Inheritance patterns	Concepts 14.1–14.4, 15.2–15.5
B. Molecular Genetics	**Chapters 16–20**
1. RNA and DNA structure and function	Concepts 16.1–16.2, 17.1–17.4
2. Gene regulation	Concepts 18.1–18.5
3. Mutation	Concepts 15.4, 17.5, 18.5, 21.5
4. Viral structure and replication	Concepts 19.1–19.3
5. Nucleic acid technology and applications	Concepts 20.1–20.4, 21.1–21.2
C. Evolutionary Biology	**Chapters 22–26**
1. Early evolution of life	Concepts 25.1–25.4
2. Evidence for evolution	Concepts 22.2–22.3, 24.2–24.4, 25.2–25.5, 26.4–26.5
3. Mechanisms of evolution	Concepts 22.1–22.3, 23.1–23.4, 24.1–24.4, 25.4–25.5, 26.4, 26.6

III. Organisms and Populations

Units 4, 5, 6, 7, and 8

A. Diversity of Organisms

    Chapters 25–34

  1. Evolutionary patterns

    Concepts 26.1, 26.3, 29.1–29.2, 32.1–32.4

  2. Survey of the diversity of life

    Concepts 27.1–27.4, 28.1–28.6, 29.2–29.3, 30.1–30.3, 31.1–31.4, 32.1, 32.4, 33.1–33.5, 34.1–34.8

  3. Phylogenetic classification

    Concepts 26.1–26.3, 26.6, 27.4, 28.1, 29.1–29.3, 30.2–30.3, 31.3–31.4, 32.4, 33.1–33.5, 34.1–34.8

  4. Evolutionary relationships

    Concepts 26.1, 26.6

B. Structure and Function of Plants and Animals

    Chapters 29, 30, 35–39, 40–51

  1. Reproduction, growth, and development (plants)

    Concepts 29.2–29.3, 30.1–30.3, 35.1–35.5, 38.1–38.2

  2. Reproduction, growth, and development (animals)

    Concepts 46.1–46.6, 47.1–47.3

  3. Structural, physiological, and behavioral adaptations (plants)

    Concepts 29.1–29.3, 30.1–30.3, 35.1–35.2, 36.1–36.6, 37.1–37.3, 38.1–38.2, 39.1–39.5

  4. Structural, physiological, and behavioral adaptations (animals)

    Concepts 40.1–40.4, 41.1–41.5, 42.1–42.7, 43.1–43.3, 44.1–44.6, 45.1–45.4, 46.1–46.6, 48.1–48.4, 49.1–49.3, 50.5–50.6, 51.2–51.5

  5. Response to the environment (plants)

    Concepts 39.1–39.5

  6. Response to the environment (animals)

    Concepts 40.3, 43.1–43.2, 44.1–44.2, 45.2–45.4, 48.1–48.4, 49.1–49.3, 50.1–50.4, 51.1–51.2

C. Ecology

    Chapters 52–56

  1. Population dynamics

    Concepts 53.1–53.6

  2. Communities and ecosystems

    Concepts 52.3–52.4, 54.1–54.5, 55.1–55.5

  3. Global issues

    Concepts 52.1–52.2, 53.6, 55.5, 56.1–56.5

A Review of Topics with Sample Questions, Answers, and Explanations

Part II is keyed to *Biology*, Eighth Edition, by Campbell and Reece. It gives an overview of important information and provides sample multiple-choice and free-response questions, along with answers and explanations. Use the topical review and the summary of key concepts sections at the end of each chapter in your textbook before attempting the practice questions. Be sure to review the answers thoroughly to prepare yourself for the range of questions you will encounter on the AP Biology Examination.

The Chemistry of Life

Chapter 2: The Chemical Context of Life

YOU MUST KNOW

- The three subatomic particles and their significance.
- The types of bonds, how they form, and their relative strengths.

Concept 2.1 Matter consists of chemical elements in pure form and in combinations called compounds

▮ **Matter** is anything that takes up space and has mass.

▮ An **element** is a substance that cannot be broken down to other substances by chemical reactions. *Examples:* gold, copper, carbon, and oxygen

▮ A **compound** is a substance consisting of two or more elements combined in a fixed ratio. *Examples:* water (H_2O) and table salt (NaCl).

▮ **C, H, O, N** make up 96% of living matter. About 25 of the 92 natural elements are known to be essential to life.

▮ **Trace elements** are those required by an organism in only minute quantities (e.g., iron and iodine).

Concept 2.2 An element's properties depend on the structure of its atoms

▮ **Atoms** are the smallest unit of an element that still retains the property of the element. Atoms are made up of neutrons, protons, and electrons.

▮ **Protons** are positively charged particles. They are found in the nucleus and determine the element.

▮ **Electrons** are negatively charged particles that are found in *electron shells* around the nucleus. They determine the chemical properties and reactivity of the element.

▮ **Neutrons** are particles with no charge. They are found in the nucleus. Their number can vary in the same element, resulting in *isotopes. Example:* ^{12}C and ^{14}C are isotopes of carbon. Both have 6 protons, but ^{12}C has 6 neutrons while ^{14}C has 8 neutrons.

▮ The **atomic number** is the number of protons an element possesses. This number is unique to every element. (See Figure 1.1)

▮ The **mass number** of an element is the sum of its protons and neutrons.

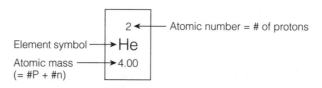

Figure 1.1

Concept 2.3 *The formation and function of molecules depend on chemical bonding between atoms*

▌ **Chemical bonds** are defined as interactions between the valence electrons of different atoms. Atoms are held together by chemical bonds to form molecules.

▌ A **covalent bond** occurs when valence electrons are shared by two atoms.

- **Nonpolar covalent bonds** occur when the electrons being shared are shared equally between the two atoms. *Examples:* O=O, H-H.
- Atoms vary in their *electronegativity*, a tendency to attract electrons of a covalent bond. Oxygen is strongly electronegative.
- In **polar covalent bonds**, one atom has greater electronegativity than the other, resulting in an unequal sharing of the electrons. *Example:* Refer to Figure 1.2 and note that within each molecule of H_2O the electrons are shared unequally, resulting in the region of the oxygen atom being slightly negative, while the regions about the hydrogen atoms are slightly positive.

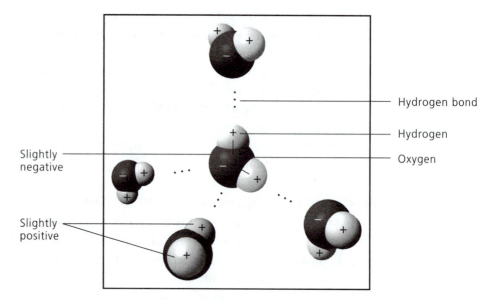

Figure 1.2

▌ **Ionic bonds** are ones in which two atoms attract valence electrons so unequally that the more electronegative atom steals the electron away from the less electronegative atom.

- An **ion** is the resulting charged atom or molecule.
- **Ionic bonds** occur because these ions will be either positively or negatively charged, and will be attracted to each other by these opposite charges.

- **Hydrogen bonds** are relatively weak bonds that form between the positively charged hydrogen atom of one molecule and the strongly electronegative oxygen or nitrogen of *another* molecule.
- **Van der Waals interactions** are very weak, transient connections that are the result of asymmetrical distribution of electrons within a molecule. These weak interactions contribute to the three-dimensional shape of large molecules.

Chapter 3: Water and the Fitness of the Environment

YOU MUST KNOW

- The importance of hydrogen bonding to the properties of water.
- Four unique properties of water, and how each contributes to life on Earth.
- How to interpret the pH scale.
- The importance of buffers in biological systems.

Concept 3.1 *The polarity of water molecules results in hydrogen bonding*

- The **structure of water** is the key to its special properties. Water is made up of one atom of oxygen and two atoms of hydrogen, bonded to form a molecule.
- Water molecules are **polar**. The end bearing the oxygen atom has a slightly negative charge, whereas the end bearing the hydrogen atoms has a slightly positive charge.
- **Hydrogen bonds** form between water molecules. The slightly negative oxygen atom from one water molecule is attracted to the slightly positive hydrogen end of *another* water molecule.
- Each water molecule can form a maximum of four hydrogen bonds at a time.

Concept 3.2 *Four emergent properties of water contribute to Earth's fitness for life*

- The key to each of these properties is hydrogen bonds. This is what makes water so unique.

  1. **Cohesion.** Cohesion is the linking of like molecules. Think "water molecule joined to water molecule" and visualize a water strider walking on top of a pond due to the surface tension that is the result of this property.
     - **Adhesion** is the clinging of one substance to another. Think "water molecule attached to some other molecule" such as water droplets adhering to a glass windshield.
     - **Transpiration** is the movement of water molecules up the very thin xylem tubes and their evaporation from the stomates in plants. The water molecules cling to each other by *cohesion*, and to the walls of the xylem tubes by *adhesion*.

2. Moderation of temperature is possible because of water's high specific heat.
    *   **Specific heat** is the amount of heat required to raise or lower the temperature of a substance by 1 degree Celsius. Relative to most other materials, the temperature of water changes less when a given amount of heat is lost or absorbed. This high specific heat makes the temperature of Earth's oceans relatively stable and able to support vast quantities of both plant and animal life.

3. Insulation of bodies of water by floating ice.
    *   Water is less dense as a solid than in its liquid state, whereas the opposite is true of most other substances. Because ice is less dense than liquid water, ice floats. This keeps large bodies of water from freezing solid and therefore moderates temperature.

4. Water is an important solvent. (The substance that something is dissolved in is called the solvent, while the substance being dissolved is called the *solute*. Together they are called the *solution*.)
    *   **Hydrophilic** substances are water-soluble. These include ionic compounds, polar molecules (e.g., sugars), and some proteins.
    *   **Hydrophobic** substances such as oils are nonpolar and do not dissolve in water.

Concept 3.3 *Acidic and basic conditions affect living organisms*

▌ The **pH** scale runs between 0 and 14 and measures the relative acidity and alkalinity of aqueous solutions. (See Figure 1.3.)

Figure 1.3

▌ **Acids** have an excess of H+ ions and a pH below 7.0. $[H+] > [OH-]$
▌ **Bases** have an excess of OH− ions, and pH above 7.0. $[H+] < [OH-]$
▌ Pure water is neutral, which means it has a pH of 7. $[H+] = [OH-]$
▌ **Buffers** are substances that minimize changes in pH. They accept H^+ from solution when they are in excess and donate H^+ when they are depleted.
▌ **Carbonic acid (H_2CO_3)** is an important buffer in living systems. It moderates pH changes in blood plasma and the ocean.

Chapter 4: Carbon and the Molecular Diversity of Life

YOU MUST KNOW

• The properties of carbon that make it so important.

Concept 4.2 Carbon atoms can form diverse molecules by bonding to four other atoms

▌ Carbon is unparalleled in its ability to form molecules that are large, complex and diverse. Why?

    1. It has 4 valence electrons.
    2. It can form up to 4 covalent bonds.
    3. These can be single, double, or triple covalent bonds.
    4. It can form large molecules.
    5. These molecules can be chains, ring-shaped, or branched.

▌ **Isomers** are molecules that have the same molecular formula but differ in their arrangement of these atoms. These differences can result in molecules that are very different in their biological activities.

Concept 4.3 Characteristic chemical groups help control how biological molecules function

▌ **Functional groups** attached to the carbon skeleton have diverse properties. The behavior of organic molecules is dependent on the identity of their functional groups.

▌ Some common functional groups are listed below:

Functional Group Name/Structure	Organic Molecules with the Functional Group and Items of Note about Functional Group
Hydroxyl, —OH	Alcohols such as ethanol, methanol; helps dissolve molecules such as sugars
Carboxyl, —COOH	Carboxylic acids such as fatty acids and sugars; acidic properties because it tends to ionize; source of H+ ions
Carbonyl, $\langle$CO	Ketones and aldehydes such as sugars
Amino, —NH_2	Amines such as amino acids
Phosphate, PO_3	Organic phosphates, including ATP, DNA, and phospholipids
Sulfhydryl, —SH	This group is found in some amino acids; forms disulfide bridges in proteins

Chapter 5: The Structure and Function of Macromolecules

YOU MUST KNOW

- The role of **dehydration synthesis** in the formation of organic compounds and **hydrolysis** in the digestion of organic compounds.
- How to recognize the four biologically important organic compounds (carbohydrates, lipids, proteins, and nucleic acids) by their structural formulas.
- The cellular functions of all four organic compounds.
- The four structural levels that proteins can go through to reach their final shape (**conformation**) and the **denaturing** impact that heat and pH can have on protein structure.

Concept 5.1 Most macromolecules are polymers, built from monomers

▌ **Polymers** are long chain molecules made of repeating subunits called **monomers**. *Examples:* Starch is a polymer composed of glucose monomers. Proteins are polymers composed of amino acid monomers. (See Figure 1.4.)

Figure 1.4

▌ **Condensation or dehydration reactions** create polymers from monomers. Two monomers are joined by removing one molecule of water. *Example:* $C_6H_{12}O_6 + C_6H_{12}O_6 \rightarrow C_{12}H_{22}O_{11} + H_2O)$

▌ **Hydrolysis** occurs when water is added to split large molecules. This occurs in the reverse of the above reaction.

Concept 5.2 Carbohydrates serve as fuel and building material

▌ **Carbohydrates** include both simple sugars (glucose, fructose, galactose, etc.) and polymers such as starch made from these and other subunits. All carbohydrates exist in a ratio of 1 carbon: 2 hydrogen: 1 oxygen or CH_2O

▌ **Monosaccharides** are the monomers of carbohydrates. Examples include glucose ($C_6H_{12}O_6$) and ribose ($C_5H_{10}O_5$). Notice the 1:2:1 ratio discussed above.

▌ **Polysaccharides** are polymers of monosaccharides. Examples are starch, cellulose, and glycogen.

▌ Two functions of polysaccharides are **energy storage** and **structural support.**

1. **Energy Storage Polysaccharides**
   • **Starch** is a storage polysaccharide found in plants (e.g., potatoes).
   • **Glycogen** is a storage polysaccharide found in animals, vertebrate muscle, and liver cells.

2. **Structural Support Polysaccharides**
   • **Cellulose** is a major component of plant cell walls.
   • **Chitin** is found in the exoskeleton of arthropods, such as lobsters and insects and the cell walls of fungi. It gives cockroaches their "crunch."

Concept 5.3 *Lipids are a diverse group of hydrophobic molecules*

▌ Lipids are all **hydrophobic.** They aren't polymers, as they are assembled from a variety of components. *Examples* include **waxes, oils, fats,** and **steroids.** (See Figure 1.5.)

▌ **Fats** (also called triglycerides) are made up of a **glycerol** molecule and three **fatty acid** molecules.

▌ **Fatty acids** include hydrocarbon chains of variable lengths. These chains are nonpolar and therefore hydrophobic.

1. **Saturated fatty acids**
   • have no double bonds between carbons
   • tend to pack solidly at room temperature
   • are linked to cardiovascular disease
   • are commonly produced by animals
   • *Examples* are butter and lard

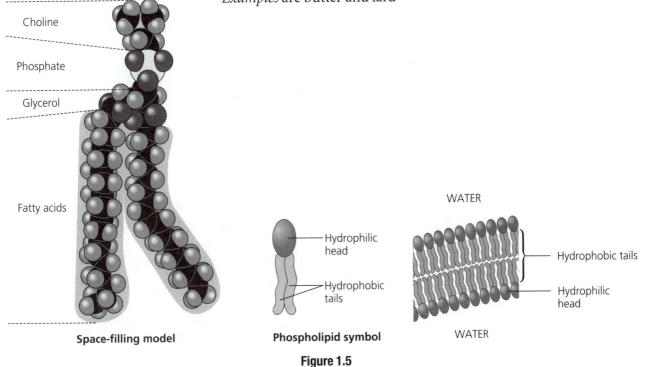

Choline

Phosphate

Glycerol

Fatty acids

Space-filling model

Hydrophilic head

Hydrophobic tails

Phospholipid symbol

WATER

Hydrophobic tails

Hydrophilic head

WATER

Figure 1.5

2. **Unsaturated fatty acids**
   - have some C=C (carbon double bonds); this results in kinks
   - tend to be liquid at room temperature
   - are commonly produced by plants
   - *Examples* are corn oil and olive oil

▌ **Functions:**

   ▪ *Energy storage.* Fats store twice as many calories/gram as carbohydrates!
   ▪ *Protection* of vital organs and *insulation.* In humans and other mammals, fat is stored in **adipose cells.**

▌ **Phospholipids** make up cell membranes. They

   ▪ have a glycerol backbone (head), which is hydrophilic
   ▪ have two fatty acid tails, which are hydrophobic
   ▪ are arranged in a bilayer in forming the cell membrane, with the hydrophilic heads pointing toward the watery cytosol or extra-cellular environment, and hydrophobic tails sandwiched in between

▌ **Steroids** are made up of four rings that are fused together.

   ▪ **Cholesterol** is a steroid. It is a common component of cell membranes.
   ▪ **Estrogen** and **testosterone** are steroid hormones.

Concept 5.4 Proteins have many structures, resulting in a wide range of functions

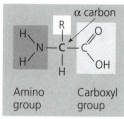

Figure 1.6

▌ **Proteins** are polymers made up of amino acid monomers.

▌ **Amino acids** contain a central carbon bonded to a carboxyl group, an amino group, a hydrogen atom, and an R group (variable group or side chain). (See Figure 1.6.)

▌ **Peptide bonds** link amino acids. They are formed by dehydration synthesis. The function of a protein depends on the order and number of amino acids.

▌ **There are four levels of protein structure** (See Figure 1.7):

   ▪ **Primary structure** is the unique sequence in which amino acids are joined.
   ▪ **Secondary structure** refers to one of two three-dimensional shapes that are the result of hydrogen bonding.
      ▪ **Alpha helix** is a coiled shape.
      ▪ **Beta pleated sheet** is an accordion shape.
   ▪ **Tertiary structure** results in a complex globular shape, due to interactions between R-groups, such as hydrophobic interactions, van der Waals interactions, hydrogen bonds, and disulfide bridges.
      ▪ Globular proteins such as enzymes are held in position by these R-group interactions.

- **Quaternary structure** refers to the association of two or more polypeptide chains into one large protein. Hemoglobin is a globular protein with quaternary structure, as it is composed of four chains.

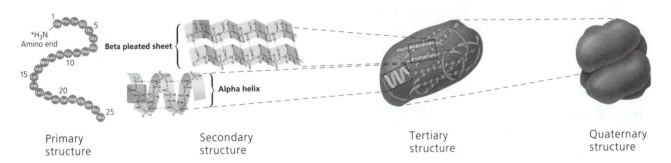

| Primary structure | Secondary structure | Tertiary structure | Quaternary structure |

Figure 1.7

- Protein shape is crucial to protein function. When a protein does not fold properly, its function is changed. This can be the result of a single amino acid substitution, such as that seen in the abnormal hemoglobin typical of sickle-cell disease.
- **Chaperonins** are protein molecules that assist in the proper folding of proteins within cells. They provide an isolating environment in which a polypeptide chain may attain final conformation.
- **Denaturation** occurs when a protein A protein is **denatured** when it loses its shape and ability to function due to **heat,** a **change in pH,** or some other disturbance.

Concept 5.5 Nucleic acids store and transmit hereditary information

- **DNA** (deoxyribonucleic acid) and **RNA** (ribonucleic acid) are the two nucleic acids. Their monomers are nucleotides.
- **Nucleotides** are made up of three parts (See Figure 1.8):
  - **Nitrogenous base** (adenine, thymine, cytosine, guanine, and uracil)
  - **Pentose** (5 carbon) sugar (deoxyribose in DNA or ribose in RNA)
  - **Phosphate group**

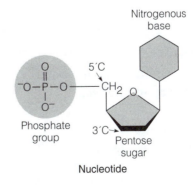

Figure 1.8

- **DNA** is the molecule of heredity.
  - It is double-stranded helix.
  - Its nucleotides are adenine, thymine, cytosine, and guanine.
  - Adenine nucleotides will hydrogen bond to thymine nucleotides, and cytosine to guanine.
- **RNA** is single-stranded. Its nucleotides are adenine, uracil, cytosine, and guanine. Note that it does not have thymine.

SUMMARY TABLE

Macromolecules/ Polymers	Monomers/Components	Examples	Functions
Carbohydrates	Monosaccharides	Sugars, starch, glycogen, and cellulose	Energy, energy storage, structural
Lipids	Fatty acids and glycerol	Fats, oils	Important energy source; insulation
Proteins	Amino acids	Hemoglobin, pepsin	Enzymes, movement
Nucleic Acids	Nucleotides (sugar, phosphate group, nitrogenous base)	DNA, RNA	Heredity; code for amino acid sequence

Multiple-Choice Questions

1. Which list of components characterizes RNA?
   - (A) a PO_3 group, deoxyribose, and uracil
   - (B) a PO_3 group, ribose, and uracil
   - (C) a PO_3 group, ribose, and thymine
   - (D) a PO_2 group, deoxyribose, and uracil
   - (E) a PO_2 group, deoxyribose, and thymine

2. Which of the following molecules would contain a polar covalent bond?
   - (A) Cl_2
   - (B) $NaCl$
   - (C) H_2O
   - (D) CH_4
   - (E) $C_6H_{12}O_6$

3. Which of the following is an example of a hydrogen bond?
   - (A) the bond between C and H in methane
   - (B) the attraction between the H of one water molecule and the O of another water molecule
   - (C) the bond between Na and Cl in salt
   - (D) the bond between the two hydrogen atoms
   - (E) the bond between Mg and Cl in $MgCl_2$

4. Three terms associated with the travel of water from the roots up through the vascular tissues of plants are
   - (A) adhesion, cohesion, and translocation.
   - (B) adhesion, cohesion, and transcription.
   - (C) cohesion, hybridization, and transpiration.
   - (D) cohesion, adhesion, and transpiration.
   - (E) transpiration, neutralization, and adhesion.

Directions: The group of questions below consists of five lettered choices followed by a list of numbered phrases or sentences. For each numbered phrase or sentence, select the one choice that is most closely related to it. Each choice may be used once, more than once, or not at all.

Questions 5–9
   - (A) Lipids
   - (B) Peptide bonds
   - (C) Alpha helix
   - (D) Unsaturated fatty acids
   - (E) Cellulose

5. Contain one or more double bonds which "kink" the carbon backbone

6. The major class of biological molecules that are not polymers

7. Linkages between the monomers of proteins

8. A secondary structure of proteins

9. A structural carbohydrate found in plants

10. The process by which protein conformation is lost or broken down is
    (A) dehydration synthesis.
    (B) translation.
    (C) denaturation.
    (D) hydrolysis.
    (E) protein synthesis.

11. An organic compound that is composed of carbon, hydrogen, and oxygen in a 1:2:1 ratio is known as a
    (A) lipid.
    (B) carbohydrate.
    (C) salt.
    (D) nucleic acid.
    (E) protein.

12. If three molecules of a fatty acid that has the formula $C_{16}H_{22}O_2$ are joined to a molecule of glycerol ($C_3H_8O_3$), then the resulting molecule would have the formula
    (A) $C_{48}H_{96}O_6$.
    (B) $C_{48}H_{98}O_8$.
    (C) $C_{51}H_{68}O_6$.
    (D) $C_{51}H_{106}O_8$.
    (E) $C_{51}H_{104}O_9$.

13. Which of the macromolecules below could be structural parts of the cell, enzymes, or involved in cell movement or communication?
    (A) nucleic acids
    (B) proteins
    (C) lipids
    (D) carbohydrates
    (E) minerals

14. Which macromolecule is the main component of all cell membranes?
    (A) DNA
    (B) phospholipids
    (C) carbohydrates
    (D) steroids
    (E) glucose

15. The partial negative charge at one end of a water molecule is attracted to a partial positive charge of another water molecule. What is this type of attraction called?
    (A) a polar covalent bond
    (B) an ionic bond
    (C) a hydration shell
    (D) a hydrogen bond
    (E) a hydrophobic bond

16. Polymers of carbohydrates and proteins are all synthesized from monomers by
    (A) the joining of monosaccharides.
    (B) hydrolysis.
    (C) dehydration reactions.
    (D) ionic bonding of monomers.
    (E) cohesion.

17. If the pH of a solution is decreased from 7 to 6, it means that the
    (A) concentration of H+ has decreased to ¹⁄₁₀ of what it was at pH7.
    (B) concentration of H+ has increased 10 times what it was at pH7.
    (C) concentration of OH− has increased 10 times what it was at pH7.
    (D) concentration of OH− has increased by ½ of what it was.
    (E) solution has become more basic.

18. Which of the following is NOT considered to be an emergent property of water?
    (A) cohesion
    (B) transpiration
    (C) moderation of temperature
    (D) insulation of bodies of water by floating ice
    (E) a versatile solvent

19. Which two functional groups are always found in amino acids?
    (A) amine and sulfhydryl
    (B) carbonyl and carboxyl
    (C) carboxyl and amine
    (D) alcohol and aldehyde
    (E) ketone and amine

20. Hydrolysis is involved in which of the following?
    (A) formation of starch
    (B) hydrogen bond formation between nucleic acids
    (C) peptide bond formation of proteins
    (D) the hydrophilic interactions of lipids
    (E) the digestion of maltose to glucose

Free-Response Question

1. *Phospholipids are a critical component of the cell membrane. The cell membrane is selectively permeable, allowing only certain substances in and out, and in certain amounts.*

   (a) Describe why phospholipids are important components of cell membranes, based on their structure and properties.
   (b) Explain why proteins are an important component of the cell membrane, based on their structure and properties.
   (*Note:* A strong response to this item requires an understanding of topics from Units 1 and 2 of the textbook.)

ANSWERS AND EXPLANATIONS

Multiple-Choice Questions

▌ **1. (B) is correct.** RNA is made up of a phosphate group, a ribose sugar, and one of the following four nitrogenous bases: cytosine, guanine, uracil, and adenine. The phosphate group of RNA contains a phosphate atom and three atoms of oxygen, not two. DNA is similar to RNA in many ways but different in two important ones: it contains deoxyribose instead of ribose as its sugar and it contains the base thymine instead of uracil.

▌ **2. (C) is correct.** The answer is water, H_2O. Polar covalent bonds are those in which valence electrons are shared between atoms, but unequally. (The more electronegative atom will attract the electrons more strongly, and that end of the molecule will have a slightly negative charge, whereas the less electronegative atom will attract the electron less strongly and be slightly positive.) The two atoms involved in the bond must differ in electronegativity in order to form a polar covalent bond.

▌ **3. (B) is correct.** Remember the definition of a hydrogen bond—it is between the hydrogen on one molecule and a strongly electronegative oxygen or nitrogen of another molecule.

▌ **4. (D) is correct.** The three terms you should keep in mind as you think of water traveling up through the xylem of a plant are transpiration (in which water evaporates from the plant's leaves); cohesion, in which the water molecules stick together due to the hydrogen bonds; and adhesion, whereby the water molecules stick to plant cell walls and resist the downward pull of gravity.

▌ **5. (D) is correct.** Unsaturated fatty acids contain one or more carbon-carbon double bonds, whereas saturated fatty acids contain no double bonds.

■ 6. (A) is correct. Lipids are the only one of the four major classes of biological molecules that are not polymers. They are grouped together because they are hydrophobic. Nucleic acids are polymers of nucleotide monomers, proteins are polymers of amino acid monomers, and carbohydrates are polymers of monosaccharide monomers.

■ 7. (B) is correct. The linkages between the amino acids of proteins are peptide bonds. Peptide bonds are covalent bonds formed in dehydration reactions. The carboxyl group of one amino acid is joined to the amino group of an adjacent amino acid, resulting in the loss of one molecule of water.

■ 8. (C) is correct. One common secondary structure of proteins is the α helix; another is the β pleated sheet. The secondary structure of a protein refers to a section of the polypeptide chain that is repeatedly folded or coiled in a regular pattern. The patterns are the result of regular hydrogen bonding between segments of the polypeptide backbone.

■ 9. (E) is correct. Cellulose is the polysaccharide that forms the strong cell walls of plant cells. It is a polymer of glucose.

■ 10. (C) is correct. Denaturation is the process by which proteins lose their overall structure, or conformation, as a result of changes in pH, temperature, or salt concentration. Denatured proteins are biologically inactive.

■ 11. (B) is correct. The ratio of carbon, hydrogen, and oxygen atoms in monosaccharides is 1:2:1.

■ 12. (C) is correct. To get this correct, you should first add up all the C, H, and O in three fatty acid chains plus one glycerol. This would be 51C, 74H and 9O. Then, recall that to join to molecules by dehydration synthesis, one molecule of water must be removed. Since there are four molecules involved, three molecules of water must be removed. Subtract 6H and 3O to arrive at the answer.

■ 13. (B) is correct. Proteins have many functions, which encompass most of a cell's metabolic activity.

■ 14. (B) is correct. Phospholipids are unique macromolecules. Their hydrophilic heads and hydrophobic tails contribute to the semipermeability of cell membranes.

■ 15. (D) is correct. The negative charge comes from the electronegative oxygen of one water molecule attracted to the partial positive of hydrogen of another water molecule.

■ 16. (C) is correct. The monomers in macromolecules are joined when a molecule of water is removed during dehydration, or condensation reactions.

■ 17. (B) is correct. Since the pH scale is logarithmic, each unit change is by a factor of 10. A drop of pH means the solution is more acidic and has 10 times more H+ ions.

■ 18. (B) is correct. Transpiration refers to the evaporation of water from pores in leaves. Transpiration is possible because of cohesion and adhesion, but it is not an emergent property of water.

19. (C) is correct. An amino acid is composed of a central carbon, bonded to a hydrogen, with a variable (R) group, and with a carboxyl (the acid part) at one end, an amino group at the other end (the amino part). Aldehydes and ketones are found in sugars; the sulfydryl group is found in only some amino acids, not all.

20. (E) is correct. Recall that hydrolysis means to use water to split a molecule, so look for a large molecule reduced to its monomers. Maltose is a disaccharide; glucose is a monosaccharide.

Free-Response Question

(a) A phospholipid molecule contains a negatively charged hydrophilic "head" (containing a glycerol molecule and a phosphate group) and two hydrophobic fatty acid tails. In cell membrane surfaces, phospholipids are arranged in a bilayer, in which the hydrophilic heads are in contact with the cell's watery interior and exterior, while the tails are pointed away from water and toward each other in the interior of the membrane. The fatty acid chains of phospholipids can contain double bonds, which makes them unsaturated. Because of the kinks in the tails, phospholipids aren't packed together tightly, which contributes to the fluidity of the membrane. The fluidity of the cell membrane is very important in its function; the less fluid the membrane is, the more impermeable it is. There is an optimum permeability for the cell membrane, at which all the substances necessary for metabolism can pass into and out of the cell.

   The fluidity of cell membranes enables hydrophobic molecules such as hydrocarbons, carbon dioxide, and oxygen to dissolve in the bilayer and easily cross the membrane. However, ions and polar molecules (including water, glucose, and other sugars) cannot pass through because of the hydrophobic interior. Protein channels and transport proteins allow these required substances to cross membranes.

(b) Proteins function as cell membrane transporters because they act as channels; substances that bind to them can help alter their conformation to permit the passage of molecules through them, and into the cell interior.

   There are many different ways by which proteins can permit the passage of ionic and polar molecules through the lipid bilayer. Proteins associated with the membrane are either integral proteins, which actually penetrate the lipid bilayer (ones that completely go through the bilayer are called transmembrane proteins), or they are "peripheral proteins" that are associated with the outside of the membrane. Transmembrane proteins can form hydrophilic channels that permit the passage of certain hydrophilic substances that otherwise would not be able to get across. Other functions that membrane proteins serve are to attach the cell to the extracellular matrix, to stabilize it, and to function in cell-cell recognition. Membrane proteins are also important in cell-cell signaling; some have enzyme function and carry out important metabolic reactions, and they aid in joining adjacent cells.

This response shows thorough knowledge of the processes of the structure of phospholipids, cell membrane structure and components, and movement across membranes. A strong response to this item requires an understanding of topics from Units 1 and 2 of the textbook. Note that the response includes the following key terms in context, showing the writer's knowledge of their meanings and relatedness:

phospholipids
hydrophilic head
glycerol
phosphate
hydrophobic tails
lipid bilayer
double bonds
unsaturated
permeability

metabolism
protein channels
transporters
conformation
integral proteins
transmembrane proteins
extracellular matrix
cell-cell signaling

The Cell

Chapter 6: A Tour of the Cell

YOU MUST KNOW

- The differences between prokaryotic and eukaryotic cells.
- The structure and function of organelles common to plant and animal cells.
- The structure and function of organelles found only in plant cells or only in animal cells.

Concept 6.2 *Eukaryotic cells have internal membranes that compartmentalize their functions*

▌ The table below organizes the major characteristics of prokaryotic and eukaryotic cells.

Characteristics	Prokaryotic Cells	Eukaryotic Cells
Plasma membrane	yes	yes
Cytosol with organelles	yes	yes
Ribosomes	yes	yes
Nucleus	no	yes
Size	1 µm–10 µm	10 µm–100 µm
Internal membranes	no	yes

▌ Prokaryotic cells include the domains Bacteria and Archaea. Eukaryotic cells belong to the domain Eukarya and include animals, fungi, plants, and protists.

▌ Three key details to remember about prokaryotes include

1. Chromosomes are grouped together in a region called the nucleoid, but there is no nuclear membrane and therefore no true nucleus.
2. No membrane-bounded organelles are found in the cytosol. (Ribosomes are found, but they are not membrane-bound.)
3. From the table above, notice how much smaller prokaryotes are than eukaryotes.

■ Three corresponding details about eukaryotic cells:

1. A membrane-enclosed nucleus contains the cell's chromosomes.
2. Many membrane-bounded organelles are found in the cytoplasm.
3. On average, eukaryotes are much larger than prokaryotes.

■ Use Figures 2.1 and 2.2 to locate each component of a plant or animal cell as they are reviewed.

■ The **plasma membrane** forms the boundary for a cell; selectively permits the passage of materials into and out of the cell; and is made up of *phospholipids*, *proteins*, and associated *carbohydrates*.

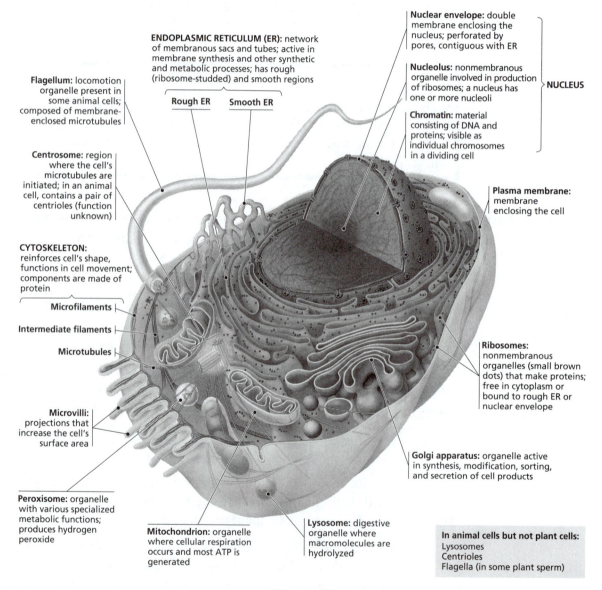

Figure 2.1

■ The **nucleus** has the following key characteristics:

1. The nucleus contains most of the cell's DNA. It is in the nucleus where DNA is used as the template to make messenger RNA (mRNA), which contains the code to produce a protein. Because the nucleus contains the genetic information, it is referred to as the control center of the cell.
2. The nucleus is the most noticeable organelle in the cell because of its large relative size. The nucleus is surrounded by a double membrane, the **nuclear envelope**. Note that the nuclear envelope is continuous with the rough endoplasmic reticulum. The nuclear envelope contains **nuclear pores** that control what may enter or leave the nucleus.

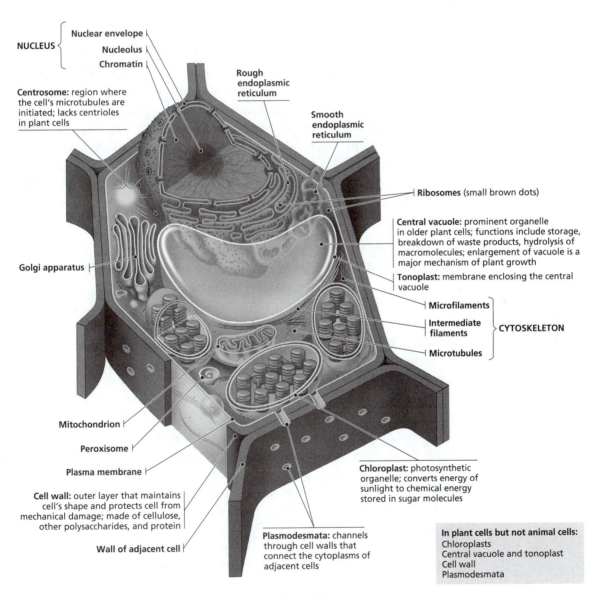

Figure 2.2

3. **Chromatin** is the complex of DNA and protein housed in the nucleus. As a cell gets ready for cell division, chromatin condenses into chromosomes.

4. The **nucleolus** is a region of the nucleus where ribosomal RNA (rRNA) complexes with proteins to form ribosomal subunits.

▌ **Ribosomes** are sites of protein synthesis in the cell. Ribosomes consist of a large and small subunit and may be found floating free in the cytosol (when making proteins for use within the cell) or bound to rough endoplasmic reticulum (when making proteins for export or use in the cell membranes).

▌ **Endoplasmic reticulum** (ER) makes up more than half the total membrane structure in many cells. The ER is a network of membranes and sacs whose internal area is called the cisternal space. There are two types of ER:

1. **Smooth ER** has three primary functions: synthesis of lipids, metabolism of carbohydrates, and detoxification of drugs and poisons.

2. **Rough ER** is so called because its associated ribosomes make the structure appear rough under the microscope. Ribosomes associated with ER synthesize proteins that are generally secreted by the cell. As the proteins are produced on the ER-bound ribosomes, the polypeptide chains travel across the ER membrane and into the cisternal space. Within the cisternal space the proteins can be concentrated before they are moved by *transport vesicles* to the Golgi apparatus for additional modification.

▌ The **Golgi apparatus** operates something like the postal system—proteins from the transport vesicles are modified, stored, and shipped. As Figures 2.1 and 2.2 show, the Golgi apparatus consists of flattened sacs of membranes, again called cisternae, arranged in stacks. Golgi stacks have polarity—the *cis* face receives vesicles, whereas the *trans* face ships vesicles.

▌ **Mitochondria** are organelles in which cellular respiration takes place. In cellular respiration ATP is created, so mitochondria are often referred to as the powerhouses of the cell. Mitochondria are enclosed by a double membrane; the inner membrane has infolds called cristae. The details of cellular respiration are covered in Chapter 9.

▌ **Peroxisomes** are single-membrane-bound compartments in the cell responsible for various metabolic functions that involve the transfer of hydrogen from compounds to oxygen, producing hydrogen peroxide (H_2O_2). Peroxisomes break down fatty acids to be sent to the mitochondria for fuel and detoxify alcohol by transferring hydrogen from the poison to oxygen.

▌ The **cytoskeleton** is a network of protein fibers that run throughout the cytoplasm where it is responsible for support, motility, and regulating some biochemical activities. Three types of fibers make up the cytoskeleton:

1. **Microtubules,** made of the protein tubulin, are the largest of the cytoskeleton fibers. Microtubules shape and support the cell and also serve as tracks along which organelles equipped with *motor molecules* can move. They also separate chromosomes during mitosis and meio-

sis (forming the spindle) and are the structural components of cilia and flagella (found primarily in animal cells).

2. **Microfilaments** are composed of the protein actin. Much smaller than microtubules, microfilaments function in smaller scale support. When coupled with the motor molecule *myosin*, microfilaments can be involved with movement. Examples include ameboid movement, cytoplasmic streaming, and contraction of muscle cells.

3. **Intermediate filaments** are slightly larger than microfilaments and smaller than microtubules. Intermediate fibers are more permanent fixtures in the cell, where they are important in maintaining the shape of the cell and fixing the position of certain organelles.

▮ **Centrosomes** are a region located near the nucleus, from which microtubules grow (the area is also called the microtubule organizing center). Centrosomes contain centrioles in animal cells.

These are the cell structures associated with animal cells only

▮ **Lysosomes** are membrane-bound sacs of hydrolytic enzymes that can digest large molecules, including proteins, polysaccharides, fats, and nucleic acids. They have digestive enzymes that break down macromolecules to organic monomers that are released into the cytosol and thus recycled by the cell. The digestive or hydrolytic enzymes work best in the acidic environment found in lysosomes. If a lysosome breaks open or leaks, the enzymes are not very active in the neutral pH of the cell. This is a good example of the importance of cell compartmentalization.

▮ **Centrioles** are located within the centrosome of animal cells, where they replicate before cell division.

▮ A specialized arrangement of microtubules is responsible for the beating of flagella and cilia.

1. **Flagella** are usually long and few in number. Many unicellular eukaryotic organisms are propelled through the water by flagella, as are the sperm of animals, algae, and some plants.

2. **Cilia** are usually much shorter and more numerous than flagella. Cilia can also be used in locomotion or, when held in place as part of a tissue layer, they can move fluid over the surface of the tissue. For example, the lining of the trachea moves mucus-trapped debris out of the lungs in this manner.

▮ Though different in length, number per cell, and beating pattern, cilia and flagella share a common ultrastructure. Nearly all eukaryotic cilia and flagella have nine pairs of microtubules surrounding a central core of two microtubules. This arrangement is referred to as the "9 + 2 pattern."

▮ **Extracellular matrix** (ECM) of animal cells is situated just external to plasma membrane; it is composed of glycoproteins secreted by the cell (most prominent of which is collagen). The ECM greatly strengthens tissues and serves as a conduit for transmitting external stimuli into the cell, which can turn genes on and modify biochemical activity.

◾ Animal cells have three types of intercellular junctions:

1. **Tight junctions** are sections of animal cell membrane where two neighboring cells are fused, making the membranes water-tight.
2. **Desmosomes** fasten adjacent animal cells together, functioning like rivets to fasten cells into strong sheets.
3. **Gap junctions** provide channels between adjacent animal cells through which ions, sugars, and other small molecules can pass.

These are the cell structures associated with plant cells only

◾ **Central vacuoles** are membrane-bounded organelles whose functions include storage and breakdown of some waste products. Comparing a plant cell to an animal cell, the large central vacuole is one of the striking differences between the two types of cells. In plants, a vacuole can make up as much as 80% of the cell.

◾ **Chloroplasts** are found in both plant and algae cells, where they are the sites of photosynthesis. Chapter 10 covers the details of photosynthesis.

◾ The **cell wall** of a plant protects the plant and helps maintain its shape. The primary component of cell walls is the carbohydrate *cellulose*.

◾ **Plasmodesmata** are channels that perforate adjacent plant cell walls and allow the passage of some molecules from cell to cell.

> **STUDY TIP:** Know the structure and function of each organelle and whether it is found in a plant cell, animal cell, or both. As an example, be prepared to discuss structures found in plant cells, but not in animal cells. (Plant cells have a large central vacuole, chloroplasts, and a cell wall.)

Chapter 7: Membrane Structure and Function

> **YOU MUST KNOW**
> • Why membranes are selectively permeable.
> • The role of phospholipids, proteins, and carbohydrates in membranes.
> • How water will move if a cell is placed in an isotonic, hypertonic, or hypotonic solution.
> • How electrochemical gradients are formed.

Concept 7.1 Cellular membranes are fluid mosaics of lipids and proteins

◾ The cell or **plasma membrane** is **selectively permeable**; that is, it allows some substances to cross it more easily than others.

◾ Membranes are predominantly made of phospholipids and proteins held together by weak interactions that cause the membrane to be fluid. The fluid mosaic model of the cell membrane describes the membrane as fluid, with proteins embedded in or associated with the phospholipid bilayer. Figure 2.3 shows the current model of an animal plasma membrane. Find each part of the membrane as the three primary organic molecules of the membrane are described:

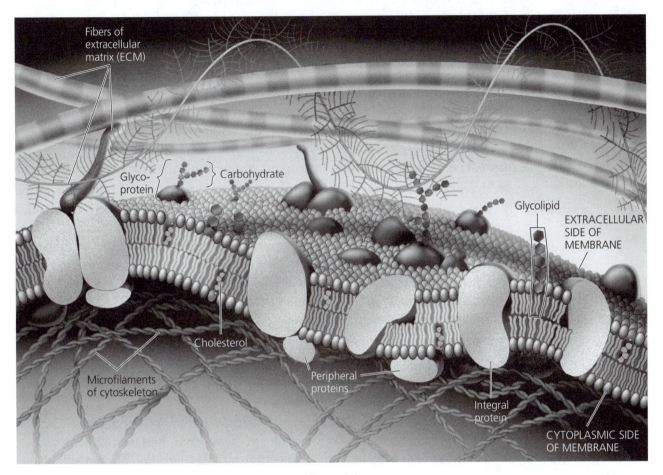

Figure 2.3

- The **phospholipids** in the membrane provide a hydrophobic barrier that separates the cell from its liquid environment. Hydrophilic molecules cannot easily enter the cell, but hydrophobic molecules can enter much more easily, hence, the selectively permeable nature of the membrane.
- There are both integral proteins and peripheral proteins in the cell membrane. **Integral proteins** are those that are completely embedded in the membrane, some of which are transmembrane proteins that span the membrane completely. **Peripheral proteins** are loosely bound to the membrane's surface.
- **Carbohydrates** on the membrane are crucial in cell-cell recognition (which is necessary for proper immune function) and in developing organisms (for tissue differentiation). Cell surface carbohydrates vary from species to species and are the reason that blood transfusions must be type-specific.

Concept 7.2 *Membrane structure results in selective permeability*

▌ Nonpolar molecules—such as hydrocarbons, carbon dioxide, and oxygen—are hydrophobic and can dissolve in the phospholipid bilayer and cross the membrane easily.

▌ The hydrophobic core of the membrane impedes the passage of ions and polar molecules, which are hydrophilic. However, hydrophilic substances can avoid the lipid bilayer by passing through **transport proteins** that span the membrane (see Figure 2.3).

▌ Perhaps the most important molecule to move across the membrane is water. Water moves through special transport proteins termed **aquaporins**. Aquaporins greatly accelerate the speed (3 billion water molecules per aquaporin per second!) at which water can cross membranes.

Concept 7.3 *Passive transport is diffusion of a substance across a membrane with no energy investment*

▌ Hydrocarbons, carbon dioxide, and oxygen are hydrophobic substances that can pass easily across the cell membrane by passive diffusion. In **passive diffusion,** a substance travels from where it is more concentrated to where it is less concentrated, diffusing down its **concentration gradient**. This type of diffusion requires that no work be done, and it relies only on the thermal motion energy intrinsic to the molecule in question. It is called "passive" because the cell expends no energy in moving the substances.

▌ The diffusion of water across a selectively permeable membrane is **osmosis**. A cell has one of three water relationships with the environment around it.

  ▪ In an **isotonic solution** there will be no net movement of water across the plasma membrane. Water crosses the membrane, but at the same rate in both directions.

  ▪ In a **hypertonic solution** the cell will lose water to its surroundings. The *hyper-* prefix refers to more solutes in the water around the cell, hence, the movement of water to the higher (hyper-) concentration of solutes. In this case the cell loses water to the environment, will shrivel, and may die.

  ▪ In a **hypotonic solution** water will enter the cell faster than it leaves. The *hypo-* prefix refers to fewer solutes in the water around the cell, hence, the movement of water into the cell where solutes are more heavily concentrated. In this case the cell will swell and may burst.

> **STUDY TIP:** AP Lab 1 deals with osmosis and diffusion. Work with these ideas until you can predict the direction of water movement based on the concentration of solutes inside and outside the cell.

▌ **Ions** and **polar molecules** cannot pass easily across the membrane. The process by which ions and hydrophilic substances diffuse across the cell membrane with the help of transport proteins is called **facilitated diffusion**. Transport proteins are specific (like enzymes) for the substances they transport. They work in one of two ways:

1. They provide a hydrophilic channel through which the molecules in question can pass.
2. They bind loosely to the molecules in question and carry them through the membrane.

Concept 7.4 Active transport uses energy to move solutes against their gradients

▌ In **active transport**, substances are moved against their concentration gradient—that is, from the side where they are *less* concentrated to the side where they are *more* concentrated. This type of transport requires energy, usually in the form of ATP.

▌ A common example of active transport is the **sodium-potassium pump**. This transmembrane protein pumps sodium out of the cell and potassium into the cell. The sodium-potassium pump is necessary for proper nerve transmission and is a major energy consumer in your body as you read this.

▌ The inside of the cell is negatively charged compared with outside the cell. The difference in electric charge across a membrane is expressed in voltage and termed the **membrane potential**. Because the inside of the cell is negatively charged, a positively charged ion on the outside, like sodium, is attracted to the negative charges inside the cell. Thus, two forces drive the diffusion of ions across a membrane:

1. *A chemical force*, which is the ion's concentration gradient.
2. And *a voltage gradient* across the membrane, which attracts positively charged ions and repels negatively charged ions.

This combination of forces acting on an ion forms an **electrochemical gradient**.

▌ A transport protein that generates voltage across the membrane is called an **electrogenic pump**. The sodium-potassium pump and the proton pump are examples of electrogenic pumps.

> *STUDY TIP:* Both photosynthesis and cellular respiration, the topics of two upcoming chapters, utilize electrochemical gradients as potential energy sources to generate ATP. By carefully studying electrochemical gradients now, you will be in a good position to understand more complex processes later.

▌ In **cotransport**, an ATP pump that transports a specific solute indirectly drives the active transport of other substances. In this process, the substance that was initially pumped across the membrane—a H$^+$ pumped by a proton pump, for example—can do work as it moves back across the membrane by diffusion and brings with it a second compound, like sucrose, against its gradient. This process is analogous to water that has been pumped uphill and performs work as it flows back down. Figure 2.4 will help you visualize the process. Note that this process has generated an electrochemical gradient, a source of potential energy that performs cell work.

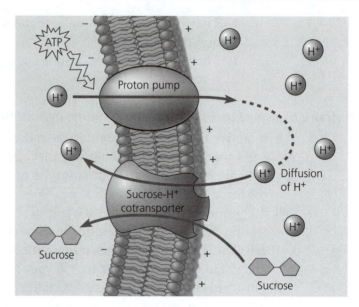

Figure 2.4

Concept 7.5 *Bulk transport across the plasma membrane occurs by exocytosis and endocytosis*

▌ Large molecules are moved across the cell membrane through exocytosis and endocytosis.

- ▪ In **exocytosis**, vesicles from the cell's interior fuse with the cell membrane, expelling their contents.
- ▪ In **endocytosis**, the cell forms new vesicles from the plasma membrane; this is basically the reverse of exocytosis, and this process allows the cell to take in macromolecules. There are three types of endocytosis:
  1. **Phagocytosis** ("cellular eating") occurs when the cell wraps pseudopodia around a solid particle and brings it into the cell.
  2. In **pinocytosis** ("cellular drinking"), the cell takes in small droplets of extracellular fluid within small vesicles. Pinocytosis is not specific, because any and all included solutes are taken into the cells.
  3. **Receptor-mediated endocytosis** is a very specific process. Certain substances (generally referred to as ligands) bind to specific receptors on the cell's surface (these receptors are usually clustered in coated pits), and this causes a vesicle to form around the substance and then to pinch off into the cytoplasm.

Chapter 8: An Introduction to Metabolism

YOU MUST KNOW

- The key role of ATP in energy coupling.
- That enzymes work by lowering the energy of activation.
- The catalytic cycle of an enzyme that results in the production of a final product.
- The factors that influence the efficiency of enzymes.

Concept 8.1 *An organism's metabolism transforms matter and energy, subject to the laws of thermodynamics*

▌ **Metabolism** is the totality of an organism's chemical reactions. Metabolism as a whole manages the material and energy resources of the cell.

- ▪ A **catabolic pathway** leads to the release of energy by the breakdown of complex molecules to simpler compounds. *Example:* Catabolic pathways occur when your digestive enzymes break down food and release energy.
- ▪ **Anabolic pathways** consume energy to build complicated molecules from simpler ones. *Example:* Anabolic pathways occur when your body links together amino acids to form muscle protein in response to physical exercise.

▌ **Energy** is defined as the capacity to do work. Anything that is moving is said to possess **kinetic energy**. An object at rest can possess **potential energy** if it has stored energy as a result of its position or structure. **Chemical energy**, a form of potential energy, is stored in molecules, and the amount of chemical energy a molecule possesses depends on its chemical bonds.

▌ **Thermodynamics** is the study of energy transformations that occur in matter.

- ▪ The **first law of thermodynamics** states that the energy of the universe is constant and that energy *can* be transferred and transformed, but it *cannot* be created or destroyed.
- ▪ The **second law of thermodynamics** states that every energy transfer or transformation increases the **entropy**, or the amount of disorder or randomness, in the universe.

Concept 8.2 The free-energy change of a reaction tells us whether the reaction occurs spontaneously

▌ **Free energy** is defined as the part of a system's energy that is able to perform work when the temperature of a system is uniform

  ■ An **exergonic reaction** is one in which energy is released. Exergonic reactions occur spontaneously (that does not necessarily mean quickly) and release free energy to the system.

  ■ An **endergonic reaction** is one that requires energy in order to proceed. Endergonic reactions absorb free energy; that is, they require free energy from the system.

Concept 8.3 ATP powers cellular work by coupling exergonic reactions to endergonic reactions

▌ A key feature in the way cells manage their energy resources to do cell work is **energy coupling**, the use of an exergonic process to drive an endergonic one.

▌ The primary source of energy for cells in energy coupling is **ATP (adenosine triphosphate)**. ATP is made up of the nitrogenous base adenine, bonded to ribose and a chain of three phosphate groups. When a phosphate group is hydrolyzed, energy is released in an exergonic reaction.

▌ Work in the cell is done by the release of a phosphate group from ATP. The exergonic release of the phosphate group is used to do the endergonic work of the cell. When ATP transfers one phosphate group through hydrolysis, it becomes **ADP (adenosine diphosphate)**.

Concept 8.4 Enzymes speed up metabolic reactions by lowering energy barriers

▌ **Catalysts** are substances that can change the rate of a reaction without being altered in the process. **Enzymes** are macromolecules that are biological catalysts. In this chapter all enzymes considered are proteins; however, in Chapters 17 and 26, RNA enzymes—termed *ribozymes*—are discussed.

▌ The **activation energy** of a reaction is the amount of energy it takes to start a reaction—the amount of energy it takes to break the bonds of the reactant molecules. Enzymes speed up reactions *by lowering the activation energy* of the reaction—but without changing the free energy change of the reaction. The reactant that the enzyme acts on is called a **substrate**. Figure 2.5 graphically depicts how enzymes function.

▌ A certain region in the enzyme, known as the **active site**, is the part of the enzyme that binds to the substrate. The enzyme and substrate form a complex called an **enzyme-substrate complex** that is generally held together by weak interactions. The substrate is then converted into **products**, and the products are released from the enzyme. Use Figure 2.6 to locate each step in the catalytic cycle of an enzyme.

> *STUDY TIP:* Over the last twenty years, questions from the enzyme section of the AP curriculum have appeared more consistently than questions from any other section. Laboratory 2 focuses on enzymes, so be sure to review the important steps in the experimental design of this lab. Using the figures as an aid, review thoroughly this section on enzyme function.

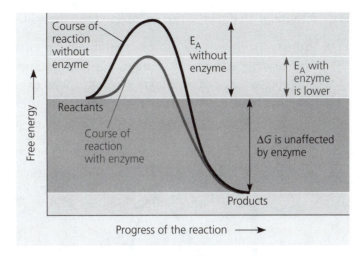

Figure 2.5

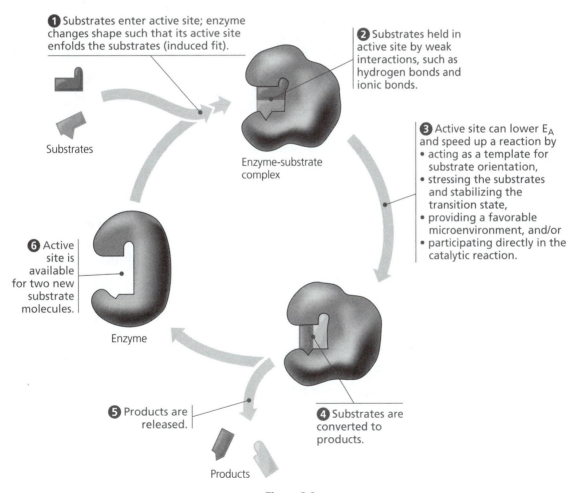

❶ Substrates enter active site; enzyme changes shape such that its active site enfolds the substrates (induced fit).

Substrates

❷ Substrates held in active site by weak interactions, such as hydrogen bonds and ionic bonds.

Enzyme-substrate complex

❸ Active site can lower E_A and speed up a reaction by
• acting as a template for substrate orientation,
• stressing the substrates and stabilizing the transition state,
• providing a favorable microenvironment, and/or
• participating directly in the catalytic reaction.

❻ Active site is available for two new substrate molecules.

Enzyme

❺ Products are released.

❹ Substrates are converted to products.

Products

Figure 2.6

■ The activity of an enzyme can be affected by several factors:

■ Protein enzymes have complicated three-dimensional shapes that are dramatically affected by changes in **pH** and **temperature**. Changes in the precise shape of an enzyme usually mean the enzyme will not be as effective. Note how the rate of the reaction is altered in the graphs in Figure 2.7 when temperature and pH are not optimal.

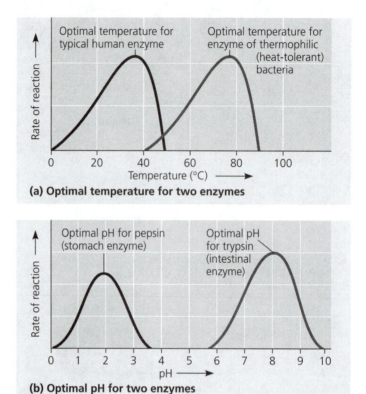

Figure 2.7

■ Many enzymes require nonprotein helpers, termed **cofactors**, to function properly. Cofactors include metal ions like zinc, iron, and copper and function in some crucial way to allow catalysis to occur. If the cofactor is organic, it is more properly referred to as a coenzyme. **Coenzymes** are organic cofactors; vitamins are examples of coenzymes.

■ **Competitive inhibitors** are reversible inhibitors that *compete with the substrate for the active site* on the enzyme. Competitive inhibitors are often chemically very similar to the normal substrate molecule and reduce the efficiency of the enzyme as it competes for the active site.

■ **Noncompetitive inhibitors** do not directly compete with the substrate molecule; instead, they impede enzyme activity by binding to another part of the enzyme. This causes the enzyme to change its shape, rendering the active site nonfunctional.

Concept 8.5 Regulation of enzyme activity helps control metabolism

▌ Many enzyme regulators bind to an **allosteric** site on the enzyme, which is a specific binding site, but not the active site. Once bound, the shape of the enzyme is changed, and this can either stimulate or inhibit enzyme activity.

▌ The end product on an enzymatic pathway can switch off its pathway by binding to the allosteric site of an enzyme in the pathway. This type of allosteric inhibition is termed **feedback inhibition**. Feedback inhibition increases the efficiency of the pathway by turning it off when the end product accumulates in the cell.

Chapter 11: Cell Communication

YOU MUST KNOW

• The three stages of cell communication: reception, transduction, and response.
• How G-protein-coupled receptors receive cell signals and start transduction.
• How receptor tyrosine kinase receive cell signals and start transduction.
• How a phosphorylation cascade amplifies a cell signal during transduction.
• How a cell response in the nucleus turns on genes while in the cytoplasm it activates enzymes.
• What apoptosis means and why it is important to normal functioning of multicellular organisms.

Concept 11.1 External signals are converted into responses within the cell

▌ In signaling, animal cells communicate by direct contact or by secreting local regulators, such as growth factors or neurotransmitters. There are three stages of cell signaling:

▪ **Reception**—The target cell's detection of a signal molecule coming from outside the cell.
▪ **Transduction**—The conversion of the signal to a form that can bring about a specific cellular response.
▪ **Response**—The specific cellular response to the signal molecule.

Concept 11.2 Reception: A signal molecule binds to a receptor protein, causing it to change shape

▌ The binding between a signal molecule (**ligand**) and a **receptor** is highly specific. A conformational change in a receptor is often the initial transduction of the signal. Receptors are found in two places.

▪ *Intracellular receptors* are found inside the plasma membrane in the cytoplasm or nucleus. The signal molecule must cross the plasma membrane and therefore must be hydrophobic, like the steroid testosterone, or very small, like nitric oxide (NO).
▪ *Plasma membrane receptors* bind to water-soluble ligands.

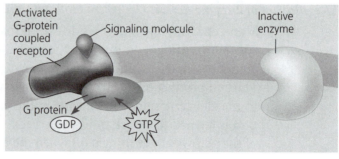

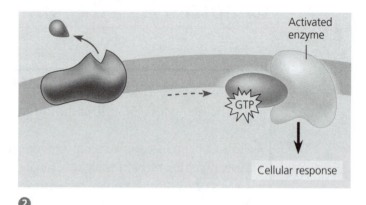

Figure 2.8

▌ A **G-protein-coupled receptor** is a membrane receptor that works with the help of a **G protein**. Follow Figure 2.8 to review how these receptors work.

  ▪ Note in step 1 of the figure that the ligand or signaling molecule has bound to the G-protein-coupled receptor. This causes a conformational change in the receptor so that it may now bind to an inactive G protein, causing a GTP to displace the GDP. This activates the G protein.
  ▪ In step 2 the G protein binds to a specific enzyme and activates it. When the enzyme is activated, it can trigger the next step in a pathway leading to a cellular response. All the molecular shape changes are temporary. To continue the cellular response, new signal molecules are required.

▌ A second type of membrane protein is the **receptor tyrosine kinase**. Follow Figure 2.9 to review how they function.

  ▪ Step 1 shows the binding of signal molecules to the receptors and the subsequent formation of a dimer. In the dimer configuration each tyrosine kinase adds a phosphate from an ATP molecule.
  ▪ Step 2 shows the fully activated receptor protein as it initiates a unique cellular response for each phosphorylated tyrosine. The ability of a single ligand to activate multiple cellular responses is a key difference between G-protein-coupled receptors and receptor tyrosine kinases.

▌ Specific signal molecules cause **ligand-gated ion channels** in a membrane to open or close, regulating the flow of specific ions.

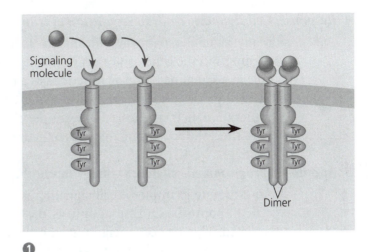

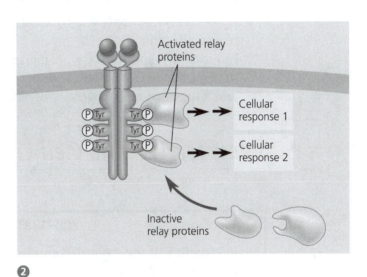

Figure 2.9

Concept 11.3 Transduction: Cascades of molecular interactions relay signals from receptors to target molecules in the cell

▌ Signal transduction pathways often involve a **phosphorylation cascade**. Because the pathway is usually a multistep one, the possibility of greatly amplifying the signal exists. At each step enzymes called **protein kinases** phosphorylate and thereby activate many proteins at the next level. This cascade of phosphorylation greatly enhances the signal, allowing for a large cellular response.

▌ Not all components of signal transduction pathways are proteins. Many signaling pathways involve small, nonprotein water-soluble molecules or ions called **second messengers**. Calcium ions and cyclic AMP are two common second messengers. The second messengers, once activated, can initiate a phosphorylation cascade resulting in a cellular response.

Concept 11.4 Response: Cell signaling leads to regulation of transcription or cytoplasmic activities

▌ Many signaling pathways ultimately regulate protein synthesis, usually by turning specific genes on or off in the nucleus. Often, the final activated molecule in a signaling pathway functions as a transcription factor.

▌ In the cytoplasm, signaling pathways often regulate the activity of proteins rather than their synthesis. For example, the final step in the signaling pathway may affect the activity of enzymes or cause cytoskeleton rearrangement.

Concept 11.5 Apoptosis (programmed cell death) integrates multiple cell signaling pathways

▌ An elaborate example of cell signaling is a program of controlled cell suicide called **apoptosis**. During apoptosis the cell is systematically dismantled and digested. This protects neighboring cells from damage that would occur if a dying cell merely leaked out its digestive and other enzymes.

■ Apoptosis is triggered by signals that activate a cascade of "suicide" proteins in the cells.

■ In vertebrates apoptosis is a normal part of development and is essential for a normal nervous system, for the operation of the immune system, and for normal morphogenesis of hands and feet in humans.

Chapter 12: The Cell Cycle

YOU MUST KNOW

- The structure of the replicated chromosome.
- The stages of mitosis.
- The role of kinases and cyclin in the regulation of the cell cycle.

Concept 12.1 Cell division results in genetically identical daughter cells

▌ The **cell cycle** is the life of a cell from the time it is first formed from a dividing parent cell until its own division into two cells.

▌ A cell's endowment of DNA, its genetic information, is called its **genome**. Before the cell can divide, the cell's genome must be copied.

■ All eukaryotic organisms have a characteristic number of chromosomes in their cell nuclei. As an example, human **somatic cells** (all body cells except gametes) have 46 chromosomes, which is the diploid chromosome number. *Mitosis* is the process by which somatic cells divide, forming daughter cells that contain the same chromosome number as the parent cell.

■ Human **gametes**—sperm and egg cells—are haploid and have half the number of chromosomes as a diploid cell. Human gametes have 23 chromosomes. A special type of cell division called meiosis (the topic of Chapter 13) results in gametes.

■ When the chromosomes are replicated, each duplicated chromosome consists of two **sister chromatids** attached by a **centromere**. Figure 2.10 will help you visualize this arrangement.

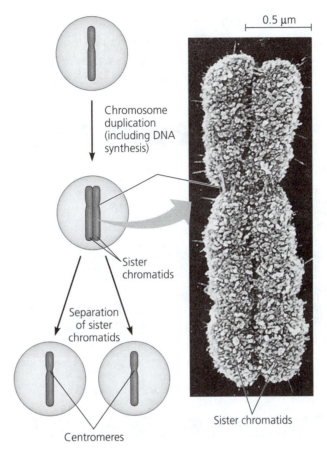

Figure 2.10

- ■ The two sister chromatids have identical DNA sequences.
- ■ Later, in the process of cell division, the two sister chromatids will separate and move into two new cells. Once the sister chromatids separate, they are considered individual chromosomes.

■ **Mitosis** is the division of the cell's nucleus. It may be followed by **cytokinesis**, which is the division of the cell's cytoplasm. Where there was one cell, there are now two, each the genetic equivalent of the parent cell.

Concept 12.2 *The mitotic phase alternates with interphase in the cell cycle*

■ The primary events of **interphase**, which is 90% of the cell cycle, follow:

- ■ In G_1 **phase** the cell grows while carrying out cell functions unique to its cell type.
- ■ In **S phase** the cell continues to carry out its unique functions but does one other important process—it duplicates its chromosomes. This

means it faithfully makes a copy of the DNA that makes up the cell's chromosomes.

- The **G₂ phase** is the gap after the chromosomes have been duplicated and just before mitosis.

▌ Mitosis can be broken down into five phases, not including cytokinesis. At each stage, find the specific references in Figure 2.11. You may be asked to identify stages by diagrams on the AP Biology Exam. To simplify your studying, key features of each phase are given.

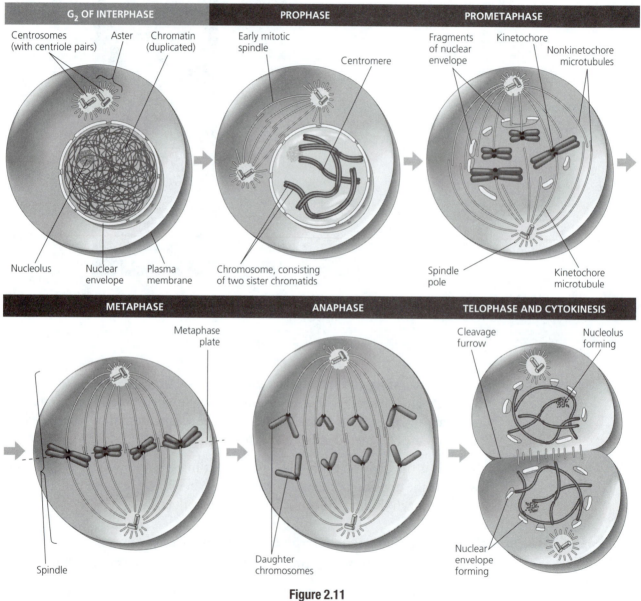

Figure 2.11

■ **Prophase:**

1. The chromatin becomes more tightly coiled into discrete chromosomes.
2. The nucleoli disappear.
3. The mitotic spindle (consisting of microtubules extending from the two centrosomes) begins to form in the cytoplasm.

■ **Prometaphase:**

1. The nuclear envelope begins to fragment, allowing the microtubules to attach to the chromosomes.
2. The two chromatids of each chromosome are held together by protein kinetochores in the centromere region.
3. The microtubules will attach to the kinetochores.

■ **Metaphase:**

1. The microtubules move the chromosomes to the metaphase plate at the equator of the cell. The microtubule complex is referred to as the spindle.
2. The centrioles have migrated to opposite poles in the cell, riding along on the developing spindle.

■ **Anaphase:**

1. Sister chromatids begin to separate, pulled apart by motor molecules interacting with kinetochore microtubules.
2. The cell elongates, as the nonkinetochore microtubules ratchet apart, again with the help of motor molecules.
3. By the end of anaphase, the opposite ends of the cell both contain complete and equal sets of chromosomes.

■ **Telophase:**

1. The nuclear envelopes re-form around the sets of chromosomes located at opposite ends of the cell.
2. The chromatin fiber of the chromosomes becomes less condensed.
3. **Cytokinesis** begins, during which the cytoplasm of the cell is divided. In animal cells, a **cleavage furrow** forms that eventually divides the cytoplasm; in plant cells, a **cell plate** forms that divides the cytoplasm.

Concept 12.3 The cell cycle is regulated by a molecular control system

▌ The steps of the cycle are controlled by a **cell cycle control system**. This control system moves the cell through its stages by a series of **checkpoints**, during which signals tell the cell either to continue dividing or to stop.

▌ The major cell cycle checkpoints include the G_1 **phase checkpoint**, G_2 **phase checkpoint**, and **M phase checkpoint**.

- The **G₁ phase checkpoint** seems to be most important. If the cell gets the go-ahead signal at this checkpoint, it usually completes the whole cell cycle and divides. If it does not receive the go-ahead signal, it enters a nondividing phase called G_0 *phase.*

  - **Kinases** are the protein enzymes that control the cell cycle. They exist in the cells at all times but are active only when they are connected to **cyclin** proteins. Thus, they are called **cyclin-dependent kinases (Cdk).** Specific kinases give the go-ahead signals at the G_1 and G_2 checkpoints.
  - As a specific example, cyclin molecules combine with Cdk molecules producing enough molecules of **MPF** to pass the G_2 checkpoint and initiate the events of mitosis.
  - How does the cell stop cell division? During anaphase, MPF switches itself off by starting a process that leads to the destruction of cyclin molecules. Without cyclin molecules Cdk molecules become inactive, bringing mitosis to a close.

- Normal cell division has two key characteristics:

  - **Density-dependent inhibition**—The process in which crowded cells stop dividing.
  - **Anchorage dependency**—Normal cells must be attached to a substratum, like the extracellular matrix of a tissue, to divide.

- Cancer cells exhibit neither density-dependent inhibition nor anchorage dependency. Cancer is covered in more depth in Concept 18.5, but several important points are made in this chapter.

  - The process that changes a normal cell to a cancer cell is **transformation**.
  - A **tumor** is a mass of abnormal cells within otherwise normal tissue. If the abnormal cells remain at the original site, the lump is called a **benign tumor.** A **malignant tumor** becomes invasive enough to impair the functions of one or more organs. An individual with a malignant tumor is said to have cancer. Malignant tumors may have cells that separate from the original tumor and enter blood vessels or lymph vessels. This spread of cancer cells is called **metastasis**.

For Additional Review

Compare the process of meiosis with the process of mitosis. In your comparison, include a study of the change in chromosomal number through the cell, the purposes of each process within an organism, and the starting material and product for each. *Note:* The details of meiosis are covered in Chapter 13, Unit 3 of the text.

Multiple-Choice Questions

1. Which structure could you observe with a light microscope?
   (A) a ribosome
   (B) a Golgi apparatus
   (C) a nucleus
   (D) an endoplasmic reticulum
   (E) a peroxisome

2. Prokaryotic and eukaryotic cells have all of the following structures in common EXCEPT
   (A) a plasma membrane.
   (B) DNA.
   (C) a nucleoid region.
   (D) ribosomes.
   (E) cytoplasm.

Directions: Questions 3–7 below consist of five lettered choices followed by a list of numbered phrases or sentences. For each numbered phrase or sentence, select the one choice that is most closely related to it. Each choice may be used once, more than once, or not at all in each group.

Questions 3–7
   (A) Peroxisomes
   (B) Golgi apparatus
   (C) Lysosomes
   (D) Endoplasmic reticulum
   (E) Mitochondria

3. An organelle that is characterized by extensive, folded membranes and is often associated with ribosomes

4. An organelle with a *cis* and *trans* face, which act as the packaging and secreting center of the cell

5. The sites of cellular respiration

6. Single-membrane structures in the cell that perform many metabolic functions and produce hydrogen peroxide in the process

7. Large membrane-bound structures that contain hydrolytic enzymes and that are found predominantly in animal cells

8. Which of the following molecules is a typical component of an animal cell membrane?
   (A) starch
   (B) glucose
   (C) nucleic acids
   (D) carbohydrates
   (E) vitamin K

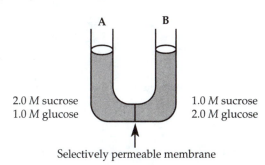

2.0 *M* sucrose 1.0 *M* sucrose
1.0 *M* glucose 2.0 *M* glucose

Selectively permeable membrane

U-Tube Setup

9. The drawing above shows two solutions of glucose and sucrose in a U-tube containing a semipermeable membrane (which allows the passage of sugars). Which of the following accurately describes what will take place next?
   (A) Glucose will diffuse from side A to side B.
   (B) Sucrose will diffuse from side B to side A.
   (C) No net movement of molecules will occur.
   (D) Glucose will diffuse from side B to side A.
   (E) There will be a net movement of water from side B to side A.

10. Which of the following is an example of passive transport across the cell membrane?
    (A) the stimulation of a muscle cell
    (B) the uptake of glucose by the microvilli of cells lining the stomach
    (C) the movement of insulin across the cell membrane
    (D) the movement of carbon dioxide across the cell membrane
    (E) the selective uptake of hormones across the cell membrane

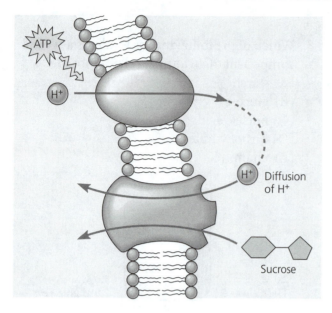

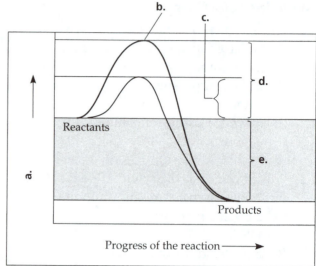

11. The figure above illustrates the process of
    (A) cotransport.
    (B) passive diffusion.
    (C) receptor-mediated endocytosis.
    (D) phagocytosis.
    (E) pinocytosis.

12. Large molecules are moved out of the cell by which of the following processes?
    (A) pinocytosis
    (B) phagocytosis
    (C) receptor-mediated endocytosis
    (D) cytokinesis
    (E) exocytosis

13. The above graph most accurately depicts the energy changes that take place in which of the following types of reactions?
    (A) hypothermic
    (B) hyperthermic
    (C) endergonic
    (D) exergonic
    (E) free range

14. Which of the following theories or laws states that every energy transfer increases the amount of entropy in the universe?
    (A) the free energy law
    (B) the first law of thermodynamics
    (C) the second law of thermodynamics
    (D) evolutionary theory
    (E) the law of increased chaos

15. Catalysts speed up chemical reactions by
    (A) decreasing the free energy change of the reaction.
    (B) increasing the free energy change of the reaction.
    (C) degrading the competitive inhibitors in a reaction.
    (D) lowering the activation energy of the reaction.
    (E) raising the activation energy of the reaction.

Directions: The group of questions below consists of five lettered choices followed by a list of numbered phrases or sentences. For each numbered phrase or sentence, select the one choice that is most closely related to it. Each choice may be used once, more than once, or not at all.

Questions 16–20
    (A) Allosteric interactions
    (B) Feedback inhibition
    (C) Competitive inhibitor
    (D) Noncompetitive inhibitor
    (E) Cooperativity

16. Describes interactions by an enzyme that is capable of either activating or inhibiting a metabolic pathway

17. A reversible inhibitor that looks similar to the normal substrate and competes for the active site of the enzyme

18. The process by which the binding of the substrate to the enzyme triggers a favorable conformation change, which causes a similar change in all of the proteins' subunits

19. The process by which a metabolic pathway is shut off by the product it produces

20. Binds to the enzyme at a site other than the active site, causing the enzyme to change shape and be unable to bind substrate.

21. $A + B \rightarrow AB +$ Energy
    Which of the following best characterizes the reaction represented above?
    (A) metabolism
    (B) anabolism
    (C) catabolism
    (D) endergonic reaction
    (E) exergonic reaction

22. The purpose of cellular respiration in a eukaryotic cell is to
    (A) synthesize carbohydrates from CO_2.
    (B) synthesize fats and proteins from CO_2.
    (C) break down carbohydrates to provide energy for the cell in the form of ATP.
    (D) break down carbohydrates to provide energy for the cell in the form of ADP.
    (E) provide oxygen to the cell.

23. In cell signaling, how is the flow of specific ions regulated?
    (A) opening and closing of ligand-gated ion channels
    (B) transduction
    (C) cytoskeleton rearrangement
    (D) endocytosis
    (E) phosphorylation cascades

24. What is a G protein?
    (A) a specific type of membrane-receptor protein
    (B) a protein on the cytoplasmic side of a membrane that becomes activated by a receptor protein
    (C) a membrane-bound enzyme that converts ATP to cAMP
    (D) a tyrosine kinase relay protein
    (E) a guanine nucleotide that converts between GDP and GTP to activate and inactivate relay proteins

25. Which of the following can activate a protein by transferring a phosphate group to it?
    (A) cAMP
    (B) G protein
    (C) phosphodiesterase
    (D) protein kinase
    (E) protein phosphatase

26. Many signal transduction pathways use second messengers to
    (A) transport a signal through the lipid bilayer portion of the plasma membrane.
    (B) relay a signal from the outside to the inside of the cell.
    (C) relay the message from the inside of the membrane throughout the cytoplasm.
    (D) amplify the message by phosphorylating proteins.
    (E) dampen the message once the signal molecule has left the receptor.

27. Which of the following signal molecules pass through the plasma membrane and bind to intracellular receptors that move into the nucleus and function as transcription factors to regulate gene expression?
    (A) epinephrine
    (B) growth factors
    (C) yeast mating factors α and a
    (D) testosterone, a steroid hormone
    (E) neurotransmitter released into synapse between nerve cells

Directions: The group of questions below consists of five lettered choices followed by a list of numbered phrases or sentences. For each numbered phrase or sentence, select the one choice that is most closely related to it. Each choice may be used once, more than once, or not at all in each group.

Questions 28–32
    (A) Telophase
    (B) Interphase
    (C) Cytokinesis
    (D) Prometaphase
    (E) Anaphase

28. Cytokinesis begins during this final stage of mitosis.

29. Division of the cytoplasm of the cell

30. Sister chromatids begin to separate.

31. The genetic material of the cell replicates to prepare for cell division.

32. Microtubules begin to attach to the centromeres of the sister chromatids.

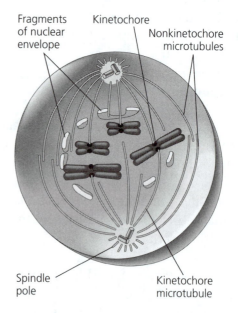

33. What stage of mitosis is represented in this figure?
    (A) prophase
    (B) prometaphase
    (C) metaphase
    (D) anaphase
    (E) telophase

34. After which of the following checkpoints in the cell cycle is the cell most likely fated to divide?
    (A) G_2 phase checkpoint
    (B) M phase checkpoint
    (C) interphase checkpoint
    (D) G_1 phase checkpoint
    (E) MPF checkpoint

Free-Response Question

1. *Prokaryotic and eukaryotic cells are physiologically different in many ways, but both represent functional collections of living matter.*

(a) It has been theorized that the organelles of eukaryotic cells evolved from prokaryotes living symbiotically within a larger cell. Compare and contrast the structure of the prokaryotic cell with eukaryotic cell organelles, and make an argument for or against this theory.

(b) Trace the path of a protein in a eukaryotic cell from its formation to its excretion from the cell.

ANSWERS AND EXPLANATIONS

Multiple-Choice Questions

▌ **1. (C) is correct.** Light microscopes are good for viewing objects that are 0.2 μm or larger. With a light microscope, you can observe animal and plant cells, some bacterial cells, and some larger organelles such as nuclei and mitochondria. To see the other organelles in the list of choices, you would need an electron microscope.

▌ **2. (C) is correct.** The nucleoid region is the only cell structure on this list not found in both prokaryotes and eukaryotes. Eukaryotic cells have a true nucleus, which is surrounded by a membrane called a nuclear envelope. The genetic material of prokaryotes is localized in a clump in one particular region of the cell that is not enclosed by a membrane.

▌ **3. (D) is correct.** The endoplasmic reticulum (ER) is an organelle characterized by extensive, folded membranes, and it is often associated with ribosomes.

▌ **4. (B) is correct.** The Golgi apparatus is the organelle that has a *cis* and *trans* face, and it acts as the packaging and secreting center of the cell. It consists of a series of flattened sacs of membranes called cisternae.

▌ **5. (E) is correct.** Mitochondria are the powerhouses of the cell; cellular respiration takes place in the mitochondria, forming ATP, the cell's energy currency. Mitochondria are bound by double membranes, and the proteins involved in ATP production are embedded in the inner membranes of the mitochondria.

▌ **6. (A) is correct.** Peroxisomes perform many metabolic functions in the cell, including the production of hydrogen peroxide in the process. The hydrogen peroxide, a poison, is immediately broken down in the peroxisome by the enzyme catalase. You may recall that catalase was the enzyme used in AP Lab 2.

▌ **7. (C) is correct.** Lysosomes are characteristic of animal cells but not most plant cells. They are large membrane-bound structures that contain hydrolytic enzymes, and they are responsible for the breakdown of proteins, polysaccharides, fats, and nucleic acids. They function best at a low pH (around 5), so they pump hydrogen ions from the cytosol into their lumen to maintain this acidic pH.

▌ **8. (D) is correct.** The only answer choice listed that names a molecule typically found in the plasma membranes of animal cells is *D*, carbohydrates. The major components of animal cell membranes are phospholipids, integral and peripheral proteins, and carbohydrates. One of the main functions of carbohydrates in the cell membrane is cell-cell recognition, which means carbohydrates are an important component of the immune system. Cell surface carbohydrates are unique to each organism.

▌ **9. (D) is correct.** Substances will move down their concentration gradient until their concentration is equal on either side of a membrane. For this reason, because the concentration of glucose on side B of the tube is 2.0 *M*, while the concentration of glucose on the A side of the tube is 1.0 *M*, glucose will move to side A.

▌ **10. (D) is correct.** The only substance listed that can passively diffuse through the cell membrane is carbon dioxide. Remember that passive diffusion occurs without the cell doing any work, and that the substances that can passively diffuse across the membrane are small nonpolar molecules, such as carbon dioxide and oxygen. Other substances need the processes of facilitated diffusion and the help of transport proteins to cross the membrane. This is true of all of the answer choices listed except for *D*.

▌ **11. (A) is correct.** This figure illustrates the process of cotransport. In cotransport, a pump that is powered by ATP transports a specific solute, protons in this case, out of the cell. The protons then travel down their concentration gradient back into the cell, passing through another transport protein and indirectly providing energy for the movement of another substance (sucrose in this case) against its concentration gradient.

▌ **12. (E) is correct.** Large molecules are moved out of the cell by exocytosis. In exocytosis, vesicles that are to be exported from the cell (often coming from the Golgi apparatus) fuse with the plasma membrane, and their contents are expelled into the extracellular matrix. Pinocytosis, phagocytosis, and receptor-mediated endocytosis are types of endocytosis; cytokinesis is cell division.

▌ **13. (D) is correct.** The shape of the curve in the art shown most closely depicts an exergonic reaction. The potential energy of the products is lower than that of the reactants—meaning that in the course of the reaction, energy is given off. This is characteristic of exergonic reactions. Conversely, in an endergonic reaction, energy is taken in during the course of the reaction.

▌ **14. (C) is correct.** The second law of thermodynamics states that every energy transfer that occurs increases the amount of entropy in the universe. The first law of thermodynamics states that the amount of energy in the universe is constant, and therefore energy can be neither created nor destroyed. Evolutionary theory refers to the myriad changes that have taken place to transform living organisms from the beginning of life on Earth until today.

▌ **15. (D) is correct.** Catalysts speed up chemical reactions by providing an alternate reaction pathway that lowers the activation energy of the reaction. Less energy is required to start the reaction, so it runs more quickly.

▌ **16. (A) is correct.** In allosteric regulation, the enzyme is usually composed of more than one polypeptide chain with more than one allosteric site (remote from the active site), and the enzyme usually oscillates between an inactive conformation and an active one. When an allosteric activator binds to the allosteric site, the protein assumes a stable conformation with a functional active site, and the reaction can proceed. When an allosteric inhibitor binds, this stabilizes the inactive conformation of the protein.

17. (C) is correct. Competitive inhibitors compete for the active site of the enzyme. They are able to bind because they closely resemble the normal substrate. One way to overcome the effects of competitive inhibitors is to increase the amount of substrate so that chances are greater that a substrate molecule (rather than the competitive inhibitor) will bind.

18. (E) is correct. In cooperativity, the enzyme in question has more than one subunit with more than one active site, and it is able to bind more than one substrate—so multiple reactions can be taking place at once in the enzyme. The binding of one substrate molecule to the enzyme causes a conformation change that makes the binding of other substrate molecules, at the other active sites, more favorable.

19. (B) is correct. In feedback inhibition, the product of a metabolic pathway switches off the pathway by binding to and inhibiting an enzyme involved somewhere along the pathway.

20. (D) is correct. In noncompetitive inhibition, the inhibitor binds to a site other than the active site of the enzyme, and this causes the enzyme to change shape. The change in conformation makes the substrate unable to bind to the active site of the enzyme, and this prevents the reaction from taking place.

21. (E) is correct. The reaction shown here is an exergonic reaction. An exergonic reaction is a spontaneous chemical reaction in which there is a net release of free energy. Energy is given off in the course of the reaction shown.

22. (C) is correct. The purpose of cellular respiration in eukaryotes is to produce energy for cellular work in the form of ATP. Respiration is an aerobic process, meaning that it requires oxygen. Answer choices *A* and *B* are incorrect because respiration involves the breakdown (not the synthesis) of carbohydrates, fats, and proteins. Choice *D* is wrong because ADP is the product of the dephosphorylation of ATP—it is left over after the energy from ATP has been released. Choice *E* is wrong because oxygen is required for cellular respiration. Cellular respiration can also break down lipids and proteins for cellular energy.

23. (A) is correct. When a signal molecule binds to the receptor protein, the gate of the ion channel opens or closes, allowing or blocking the flow of specific ions.

24. (B) is correct. Answer *A* is a reference to a G-protein-coupled receptor, but is a quick pick if the question is not read carefully. The G-protein is activated by the G-protein-coupled receptor, which is a protein and eliminates *E* as a possible answer.

25. (D) is correct. Kinase enzymes are involved with ATP. Protein kinase enzymes are used to amplify the signal during the transduction phase of cell signaling by activating cell proteins with a phosphate from ATP.

26. (C) is correct. Many signaling pathways involve small, nonprotein water-soluble molecules or ions called second messengers. Calcium ions and cyclic AMP are two common second messengers. The second messengers, once activated (and always found on the inside of the membrane), can initiate a phosphorylation cascade resulting in a cellular response.

27. (D) is correct. Intracellular receptors work with signal molecules that are hydrophobic compounds and therefore able to cross the plasma membrane.

Testosterone, as indicated in the question, is a steroid hormone and thus hydrophobic. Intracellular receptors often act as transcription factors.

■ **28. (A) is correct.** In telophase, nuclear envelopes begin to form around the sets of chromosomes, which are now located at opposite ends of the cell. The chromatin becomes less condensed, and cytokinesis begins—the cytoplasm of the cell is divided.

■ **29. (C) is correct.** During cytokinesis, the cytoplasm of the cell is divided approximately equally as the cell membrane pinches off (in animal cells), forming two daughter cells; a cell plate forms in plant cells.

■ **30. (E) is correct.** In anaphase, the sister chromatids, which were lined up along the equator of the cell, begin to separate, pulled apart by the retracting microtubules. By the end of anaphase, the opposite ends of the cell contain complete and equal sets of chromosomes.

■ **31. (B) is correct.** Interphase is not a part of mitosis; rather it is the part of the cell cycle when the cell gets ready to divide by replicating its DNA. There are three stages in interphase: G_1 phase, S phase, and G_2 phase. The genetic material is replicated in S phase.

■ **32. (D) is correct.** Prometaphase is the phase of mitosis in which the nuclear envelope begins to fragment so that the microtubules can begin to attach to the kinetochores of the chromatids, which by this time are very condensed.

■ **33. (B) is correct.** The depicted cell is in prometaphase. As you can see, the nuclear envelope is fragmenting, and the microtubules have already attached to some of the kinetochores at the centromeres of the chromosomes. The chromosomes are condensed and beginning to line up along the cell's equator.

■ **34. (D) is correct.** The most crucial checkpoint of the cell cycle is the G_1 checkpoint. In the cell cycle, a checkpoint is a point at which there can be a signal to stop or to go ahead with division. If a cell receives the signal to go ahead at the G_1 checkpoint, it will usually complete the cycle and divide. If it does not receive the go-ahead signal, it will enter the (nondividing) G_0 phase for an indeterminate period of time.

Free-Response Question

(a) Some eukaryotic cell organelles might have evolved from free-living prokaryotic organisms. First of all, prokaryotic cells are much smaller than eukaryotic cells—they range from 100 nm to 10 μm, compared to the average size of eukaryotic cells: 10 μm–100 μm. However, mitochondria (organelles unique to eukaryotic cells, and functioning in the creation of ATP in cellular respiration) and eukaryotic cell nuclei are comparable in size to prokaryotic cells, ranging from about 1 μm–10 μm.

Another interesting characteristic of organelles that may tie them to prokaryotes is their structure and cell contents. To illustrate this, let's consider the structure of mitochondria. With few exceptions, mitochondria are found in all animal cells, plant cells, fungi, and protists. They can exist in great numbers in these cells, or cells can contain just one mitochondrion (depending on the metabolic activity of the cell). It has been observed that mitochondria can move around, alter their shape, and even divide in two—all of which are characteristic of living cells. Their structure consists of a double membrane exterior (the membrane is a typical combination of phospholipids and proteins, like the membrane of the cell itself); the outer membrane is relatively smooth, but the interior membrane has infoldings called cristae. This creates two different compartments in mitochondria: the inner compartment is the mitochondrial matrix, and the compartment in between the two membranes is called the intermembrane space. Mitochondria also contain mitochondrial DNA. Not very much DNA is contained in mitochondria, but the presence of DNA could be evidence that they were independent organisms at some time. Also similar to prokaryotes, mitochondria do not contain many interior structures other than their genetic material (which is not enclosed in a nucleus) and their cell membranes. All of the above indicates a close evolutionary relationship between prokaryotic cells and mitochondria.

(b) In order to trace the path of proteins in the cell from their creation to their expulsion, we must start in the nucleus. In the nucleus, mRNA is transcribed from DNA, and mRNA travels out of the nucleus to the cytoplasm, ending up at ribosomes, some of which are associated with the endoplasmic reticulum (called rough endoplasmic reticulum because of this association). Here, they are translated into proteins, which then undergo folding to assume their final shape, or conformation. Proteins then either carry out their metabolic function in the cell, whether they act as structural components, enzymes, etc., or they are packaged for secretion from the cell.

Secretory proteins travel from the endoplasmic reticulum to the series of flattened membranous sacs known as the Golgi apparatus. They enter at the *cis* face and eventually bud from the *trans* face, after undergoing a series of modifications to prepare them for secretion. The vesicles may then fuse with the cell membrane, and the contents are released from the cell in a process called exocytosis.

This response shows that the writer used the following key terms in context, showing the writer's knowledge of their meanings and relatedness:

organelles	*mRNA*
prokaryote	*ribosomes*
eukaryote	*endoplasmic reticulum*
mitochondria	*conformation*
phospholipids	*enzymes*
proteins	*secretory proteins*
cristae	*Golgi apparatus*
mitochondrial matrix	cis/trans *face*
DNA	*vesicle*

The response also contains an explanation of the following subjects and processes:

—*size comparison of prokaryotes and eukaryotes (with data)*
—*structure, content, and behavior of prokaryotes versus mitochondria*
—*location of transcription and translation*
—*pathway of proteins*

Respiration and Photosynthesis

Chapter 9: Cellular Respiration: Harvesting Chemical Energy

YOU MUST KNOW

- The difference between fermentation and cellular respiration.
- The role of glycolysis in oxidizing glucose to two molecules of pyruvate.
- The process that brings pyruvate from the cytosol into the mitochondria and introduces it into the citric acid cycle.
- How the process of chemiosmosis utilizes the electrons from NADH and $FADH_2$ to produce ATP.

Oxidation-reduction reactions, fermentation, cellular respiration, and photosynthesis are covered in one of the most technically challenging sections of your textbook. Here, we will focus on the major steps of each of the processes, as well as the results. Questions on the AP* Biology Exam are likely to focus on the net results of photosynthesis and respiration—not on the exact reactions that create the products. As you work through these chapters, compare and contrast the two fundamental cell processes.

Concept 9.1 Catabolic pathways release energy by oxidizing organic fuels

▌ **Catabolic pathways** occur when molecules are broken down and their energy is released. Two types of catabolism are

1. Fermentation—The partial degradation of sugars that occurs without the use of oxygen.
2. Cellular respiration—The most prevalent and efficient catabolic pathway, in which oxygen is consumed as a reactant along with the organic fuel. This is also termed **aerobic respiration**, as oxygen is required.

- Carbohydrates, fats, and proteins can all be broken down to release energy in cellular respiration. However, glucose is the primary nutrient molecule that is used in cellular respiration. The standard way of representing the process of cellular respiration shows glucose being broken down in the following reaction:

$$C_6H_{12}O_6 + 6\,O_6 \rightarrow 6\,CO_2 + 6\,H_2O + \text{Energy (686 kcal/mole of glucose)}$$

- The exergonic release of energy from glucose is used to phosphorylate ADP to ATP. Life processes constantly consume ATP; cellular respiration burns fuels and uses the energy to regenerate ATP.
- The reactions of cellular respiration are of a type termed **oxidation-reduction (redox)** reactions. In redox reactions electrons are transferred from one reactant to another.

  - The loss of one or more electrons from a reactant is called **oxidation**. When a reactant is *oxidized*, it loses electrons and, consequently, energy.
  - The gain of one or more electrons is **reduction**. When a reactant is *reduced*, it gains electrons and, therefore, energy.

- At key steps in cellular respiration, electrons are stripped from glucose. Each electron travels with a proton, thereby forming a hydrogen atom. The hydrogen atoms are not transferred directly to oxygen, as the formula might suggest, but instead are usually passed to an electron carrier, the coenzyme **NAD⁺** (a derivative of the B vitamin niacin). Within the cell **NAD⁺** accepts two electrons, plus the stabilizing hydrogen ion, to form NADH. Note that NADH has been reduced and therefore has gained energy.
- Figure 3.1 shows the three stages of cellular respiration. Each stage is separately featured in the next three concepts. Use this figure to begin to develop an overall concept of the process of cellular respiration.

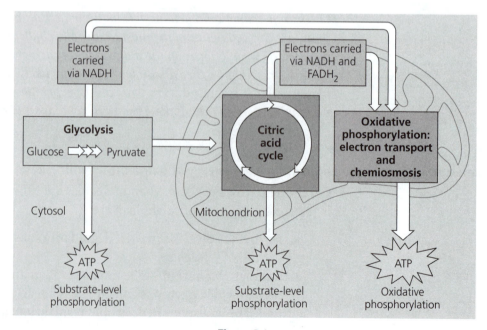

Figure 3.1

Concept 9.2 Glycolysis harvests chemical energy by oxidizing glucose to pyruvate

▌ In **glycolysis** (which occurs in the cytosol), the degradation of glucose begins as it is broken down into two pyruvate molecules. The six-carbon glucose molecule is split into two three-carbon sugars through a long series of steps.

> ***STUDY TIP:*** Use Figure 3.2 as a guide to the important features of glycolysis. It is not necessary to understand each chemical step in glycolysis.

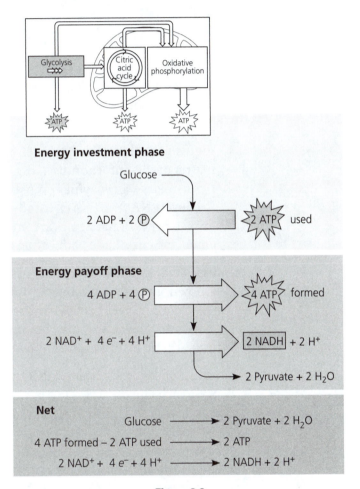

Figure 3.2

▌ In the course of glycolysis, there is an ATP-consuming phase and an ATP-producing phase. In the ATP-consuming phase, two ATP molecules are consumed, which helps destabilize glucose and make it more reactive. Later in glycolysis 4 ATP molecules are produced; thus, glycolysis results in a net gain of 2 ATPs. Two NADHs are also produced, which will be utilized in the electron transport system (see Figure 3.1) to produce ATP.

▌ Notice the *net* energy gain in glycolysis as indicated in Figure 3.2—2 ATP molecules and 2 NADH molecules. Most of the potential energy of the glucose molecule still resides in the two remaining pyruvates, which will now feed into the citric acid cycle, as discussed in the next concept.

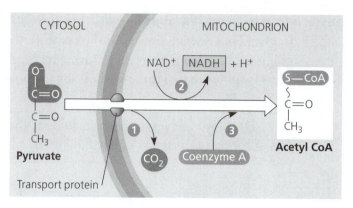

Figure 3.3

Concept 9.3 *The citric acid cycle completes the energy-yielding oxidation of organic molecules*

▌ The junction between glycolysis and the citric acid cycle is shown in Figure 3.3. Note the following:

1. Pyruvate, in the cytosol, uses a transport protein to move into the matrix of the mitochondria.
2. In the matrix, an enzyme complex removes a CO_2, strips away electrons to convert NAD^+ to NADH, and adds coenzyme A to form acetyl CoA.
3. Two acetyl CoA molecules are produced per glucose. Acetyl CoA now enters the enzymatic pathway termed the citric acid cycle.

▌ In the **citric acid cycle** (which occurs in the mitochondrial matrix), the job of breaking down glucose is completed with CO_2 released as a waste product. Each turn of the citric acid cycle requires the input of one acetyl CoA. The citric acid cycle must make two turns before the glucose is completely oxidized.

▌ The citric acid cycle results in the following:

■ One turn of the cycle results in **2CO_2, 3NADH, 1FADH$_2$,** and **1ATP**.
■ Because each glucose yields *two* pyruvates, the total products of the citric acid cycle are usually listed as the result of two cycles:

4CO_2, 6NADH, 2FADH$_2$, and **2ATP**

▌ At the end of the citric acid cycle note that the 6 original carbons in glucose have been released as CO_2. (You are exhaling this gas as you study.) Only 2 ATP molecules, however, have been produced. Where is all the energy? The energy is held in the electrons in the electron carriers, NADH and FADH$_2$. These electrons will by utilized by the electron transport system, explained in the next concept.

Concept 9.4 *During oxidative phosphorylation, chemiosmosis couples electron transport to ATP synthesis*

▌ Use Figure 3.4 as a map to understand the process of electron transport.

1. The electron transport chain is embedded in the inner membrane of the mitochondria. Notice that it is composed of three transmembrane proteins that work as hydrogen pumps and two carrier molecules that

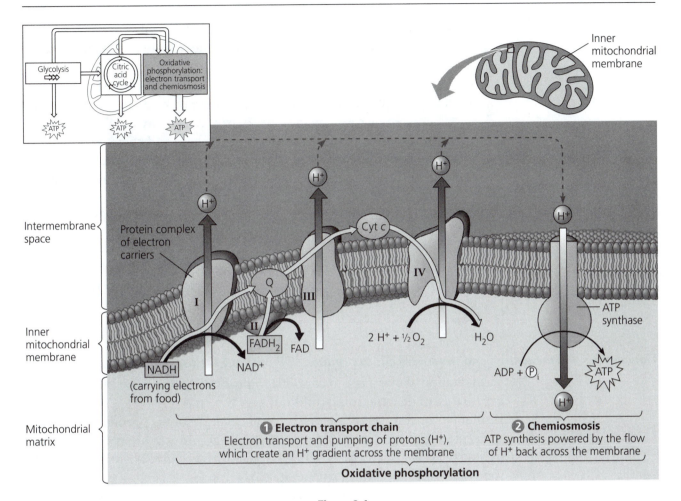

Figure 3.4

transport electrons between hydrogen pumps. There are thousands of such electron transport chains in the inner mitochondrial membrane.

2. The electron transport chain is powered by electrons from the electron carrier molecules NADH and $FADH_2$ ($FADH_2$ is also a B vitamin coenzyme that functions as an electron acceptor in the citric acid cycle). As the electrons flow through the electron chain, the loss of energy by the electrons is used to power the pumping of protons across the inner membrane. At the end of the electron chain, the electrons combine with two hydrogen ions and oxygen to form water. Notice that O_2 is the final electron acceptor, and when it is not available, the transport of electrons comes to a screeching halt! No hydrogen ions are pumped and no ATP is produced.

3. The hydrogen ions flow back down their gradient through a channel in the transmembrane protein known as ATP synthase. **ATP synthase** harnesses the **proton motive force**—the gradient of hydrogen ions—to phosphorylate ADP, forming ATP. The proton motive force is in place because the inner membrane of the mitochondria is impermeable to

hydrogen ions. Like water behind a dam with its only exit being a spill-way, electrons are held behind the inner membrane with their only exit ATP synthase.

4. This process is referred to as chemiosmosis. **Chemiosmosis** is an energy-coupling mechanism that uses energy stored in the form of an H^+ gradient across a membrane to drive cellular work (ATP synthesis in our example). The electron transport chain and chemiosmosis compose **oxidative phosphorylation**. This specific term is used because ADP is phosphorylated and oxygen is necessary to keep the electrons flowing.

5. The ATP yield per molecule of glucose is between 36 and 38 ATPs. Oxidative phosphorylation produces 32 to 34 of the total.

> **STUDY TIP:** Sketch this process and explain it verbally. This is a fundamental biological process that you should understand.

Concept 9.5 *Fermentation enables some cells to produce ATP without the use of oxygen*

▌ Fermentation allows a cell to continue to produce ATP without the use of oxygen, that is, under **anaerobic** conditions.

▌ Fermentation consists of glycolysis (recall that glycolysis produces 2 net ATP molecules) and reactions that regenerate NAD^+. In glycolysis oxygen is not needed to accept electrons; NAD^+ is the electron acceptor. Therefore, the pathways of fermentation must regenerate NAD^+.

▌ The two common types of fermentation are alcohol fermentation and lactic acid fermentation:

 ▪ In **alcohol fermentation,** pyruvate is converted to ethanol, releasing CO_2 and oxidizing NADH in the process to create more NAD^+.
 ▪ In **lactic acid fermentation,** pyruvate is reduced by NADH (NAD^+ is formed in the process), and lactate is formed as a waste product.

▌ **Facultative anaerobes** are organisms that can make ATP by aerobic respiration if oxygen is present, but that can switch to fermentation under anaerobic conditions.

Concept 9.6 *Glycolysis and the citric acid cycle connect to many other metabolic pathways*

▌ In addition to glucose and other sugars, proteins and fats are often used to generate ATP through cellular respiration. Organic molecules are also used in **biosynthesis,** the building of macromolecules through anabolic pathways. For example, amino acids from the hydrolysis of proteins in food can be incorporated into the consumer's own proteins. Compounds formed as intermediates of glycolysis and the citric acid cycle can be diverted into anabolic pathways to help provide the building blocks of necessary macromolecules.

Chapter 10: Photosynthesis

> **YOU MUST KNOW**
>
> - How photosystems convert solar energy to chemical energy.
> - How linear electron flow in the light reactions results in the formation of ATP, NADPH, and O_2.
> - How chemiosmosis generates ATP in the light reactions.
> - How the Calvin cycle uses the energy molecules of the light reactions to produce G3P.
> - The metabolic adaptations of C_4 and CAM plants to arid, dry regions.

Concept 10.1 Photosynthesis converts light energy to the chemical energy of food

▌ Before you look at the molecular details of photosynthesis, it is important to think of photosynthesis in an ecological context.

  ■ Life on Earth is solar powered by autotrophs. **Autotrophs** are "self-feeders"; they sustain themselves without eating anything derived from other organisms. Autotrophs are the ultimate source of organic compounds and are therefore known as *producers.*

  ■ **Heterotrophs** live on compounds produced by other organisms and are thus known as *consumers.* Animals immediately come to mind as heterotrophs, but also remember that decomposers like fungi and many prokaryotes are heterotrophs. Heterotrophs are dependent on the process of photosynthesis for both food and oxygen.

▌ **Chloroplasts** are the specific sites of photosynthesis in plant cells.

  ■ Chloroplasts are organelles that are mostly located in the cells that make up the **mesophyll** tissue found in the interior of the leaf.

  ■ Use Figure 3.5 to become familiar with the structure of chloroplasts. An envelope of two membranes encloses the **stroma,** which is a dense fluid-filled area. Within the stroma is a vast network of interconnected membranous sacs called **thylakoids.** The thylakoids segregate the stroma from another compartment, the **thylakoid space.**

  ■ **Chlorophyll** is located in the thylakoid membranes and is the light-absorbing pigment that drives photosynthesis and gives plants their green color.

▌ The exterior of the lower epidermis of a leaf contains many tiny pores called **stomata,** through which carbon dioxide enters and oxygen and water vapor exit the leaf. Notice carbon dioxide on the reactant side of the following equation and oxygen on the products side.

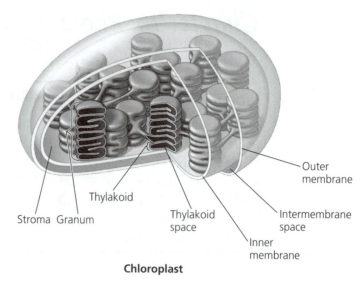

Stroma Granum Thylakoid Thylakoid space Inner membrane Intermembrane space Outer membrane

Chloroplast

Figure 3.5

▎The overall reaction of photosynthesis looks like this:

$$6\,CO_2 + 6\,H_2O + \text{Light energy} \rightarrow C_6H_{12}O_6 + 6O_2$$

- Notice that the overall chemical change during photosynthesis is the reverse of the one that occurs during cellular respiration.
- All the oxygen you breathe was formed in the process of photosynthesis when a water molecule was split! Water is split for its electrons, which are transferred along with hydrogen ions from water to carbon dioxide, reducing it to sugar. This process requires energy (an endergonic process), which is provided by the sun.

▎Photosynthesis is a chemical process that requires two stages to complete.

- The **light reactions** occur in the thylakoid membranes where solar energy is converted to chemical energy. Light is absorbed by chlorophyll and drives the transfer of electrons from water to $NADP^+$, forming NADPH. Water is split during these reactions, and O_2 is released. The light reactions also generate ATP, using chemiosmosis to power the addition of a phosphate group to ADP, a process called **photophosphorylation.** The net products of the light reactions are **NADPH** (which stores electrons), **ATP,** and **oxygen.**
- The Calvin cycle occurs in the stroma, where CO_2 from the air is incorporated into organic molecules in **carbon fixation.** The Calvin cycle uses the fixed carbon plus NADPH and ATP from the light reactions in the formation of new sugars.

> *STUDY TIP:* Use Figure 3.6 to help in understanding where reactions occur and the overall purpose of the two stages of photosynthesis. If you understand the big picture, the details will be easier to comprehend.

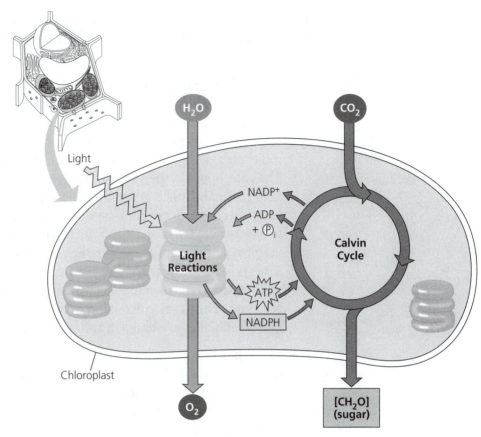

Figure 3.6

Concept 10.2 *The light reactions convert solar energy to the chemical energy of ATP and NADPH*

▌ Not surprisingly, light is an important concept in photosynthesis.

- Light is electromagnetic energy, and it behaves as though it is made up of discrete particles, called **photons**—each of which has a fixed quantity of energy.
- Substances that absorb light are called **pigments,** and different pigments absorb light of different wavelengths. Chlorophyll is a pigment that absorbs violet-blue and red light while transmitting and reflecting green light. This is why we see summer leaves as green.
- A graph plotting a pigment's light absorption versus wavelength is called an **absorption spectrum.** The absorption spectrum of chlorophyll provides clues to the effectiveness of different wavelengths for driving photosynthesis. This is confirmed by an action spectrum. An **action spectrum** for photosynthesis graphs the effectiveness of different wavelengths of light in driving the process of photosynthesis. Note examples of both of these graphs in your text, Figure 10.9.

▌ Photons of light are absorbed by certain groups of pigment molecules in the thylakoid membrane of chloroplasts. These groups are called **photosystems** and consist of two parts:

1. A **light-harvesting complex** made up of many chlorophyll and carotenoid molecules (accessory pigments in the thylakoid membrane); this allows the complex to gather light effectively. When chlorophyll absorbs light energy in the form of photons, one of the molecule's electrons is raised to an orbital of higher potential energy. The chlorophyll is then said to be in an "excited" state.

2. Like a human "wave" at a sports arena, the energy is transferred to the **reaction center** of the photosystem. The reaction center consists of two chlorophyll *a* molecules, which donate the electrons to the second member of the reaction center, the **primary electron acceptor.** The solar-powered transfer of an electron from the reaction center chlorophyll *a* pair to the primary electron acceptor is the first step of the light reactions. This is the conversion of light energy to chemical energy, and what makes photoautotrophs the producers of the natural world.

▌ Thylakoid membranes contain two photosystems that are important to photosynthesis—**photosystem I** (PS I) and **photosystem II** (PS II). PS I is sometimes designated P700 because the chlorophyll *a* in the reaction center of this photosystem absorbs red light of this wavelength the best; PS II is sometimes referred to as P680 for the same reason. Don't let switches in designation be confusing.

▌ Following are the major steps of the light reactions of photosynthesis. The key to the light reactions is a flow of electrons through the photosystems in the thylakoid membrane, a process called **linear (noncyclic) electron flow.** Find each step in Figure 3.7 as you read the following summary:

1. Photosystem II absorbs light energy, allowing the P680 reaction center of two chlorophyll *a* molecules to donate an electron to the primary electron acceptor. The reaction center chlorophyll is oxidized and now requires an electron.

2. An enzyme splits a water molecule into two hydrogen ions, two electrons, and an oxygen atom. The electrons are supplied to the P680 chlorophyll *a* molecules. The oxygen combines with another oxygen molecule, forming the O_2 that will be released into the atmosphere.

3. The original excited electron passes from the primary electron acceptor of photosystem II to photosystem I through an electron transport chain (similar to the electron chain in cellular respiration).

4. The energy from the transfer of electrons down the electron transport chain is used to pump protons, creating a gradient that is used in chemiosmosis to phosphorylate ADP to ATP. ATP will be used as energy in the formation of carbohydrates in the Calvin cycle.

5. Meanwhile, light energy has also activated PS I, resulting in the donation of an electron to its primary electron acceptor. The electrons just donated by PS I are replaced by the electrons from PS II. (Keep in mind that the ultimate source of electrons is water.)

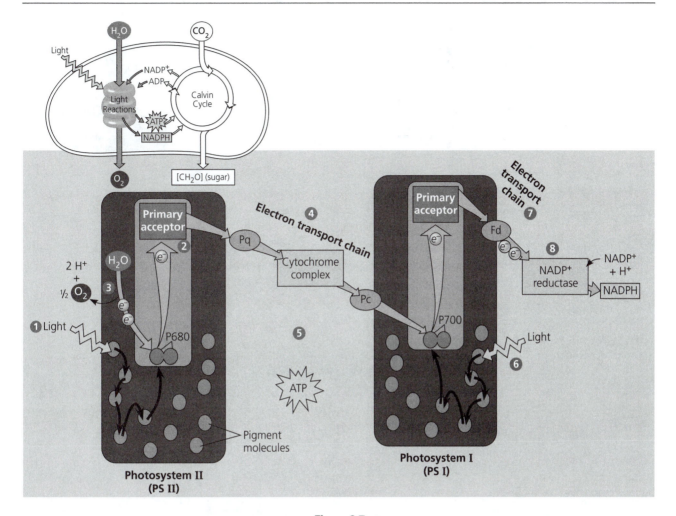

Figure 3.7

6. The primary electron acceptor of photosystem I passes the excited electrons along to another electron transport chain, which transmits them to $NADP^+$, which is reduced to NADPH—the second of the two important light-reaction products. The high energy electrons of NADPH are now available for use in the Calvin cycle.

▌ An alternative to linear electron flow is **cyclic electron flow.** Cyclic electron flow uses PS I, but not PS II. Cyclic electron flow uses a short circuit of linear electron flow by cycling the excited electrons back to their original starting point in PS I. Cyclic electron flow produces ATP by chemiosmosis, but no NADPH is produced and no oxygen is released.

▌ Chloroplasts and mitochondria generate ATP by the same basic mechanism: chemiosmosis. Examining Figure 3.8 will quickly demonstrate the same basic chemiosmotic plan as cellular respiration. Use Figure 3.8 to illustrate the following:

   ▪ An electron transport chain uses the flow of electrons to pump protons across the thylakoid membrane.

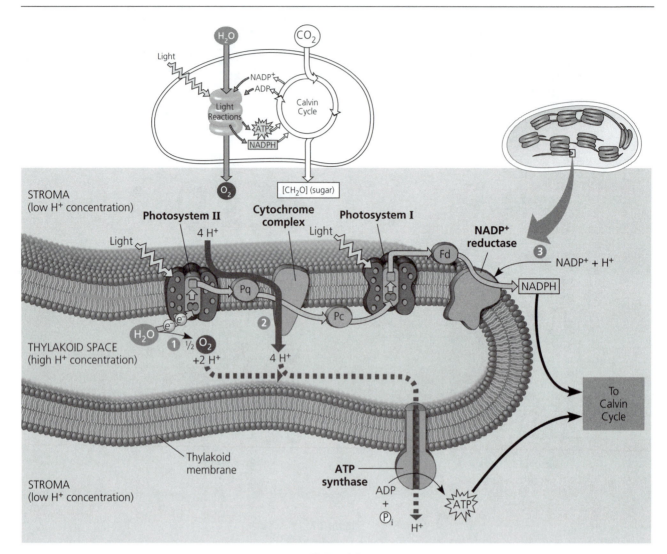

STROMA
(low H⁺ concentration)

THYLAKOID SPACE
(high H⁺ concentration)

STROMA
(low H⁺ concentration)

Figure 3.8

- A proton motive force is created within the thylakoid space that can be utilized by ATP synthase to phosphorylate ADP to ATP. Notice that the proton motive force is generated in three places: (1) hydrogen ions from water; (2) hydrogen ions pumped across the membrane by the cytochrome complex; (3) the removal of a hydrogen ion from the stroma when $NADP^+$ is reduced to NADPH.
- Although similar, chemiosmosis in cellular respiration and photosynthesis are not identical. In addition to some spatial differences, the key conceptual difference is that mitochondria use chemiosmosis to transfer chemical energy from food molecules to ATP, whereas chloroplasts transform light energy into chemical energy in ATP. This is the essence of the difference between a consumer and a producer.

Concept 10.3 The Calvin cycle uses ATP and NADPH to convert CO_2 to sugar

▮ In the course of the **Calvin cycle,** carbon enters in the form of CO_2 and leaves in the form of a sugar. The cycle spends ATP as an energy source and consumes NADPH as reducing power for adding high energy electrons to make the sugar. Use Figure 3.9 to chart each step summarized in the outline below. You must note that in order to net one molecule of G3P, the cycle must go through three rotations and fix three molecules of CO_2.

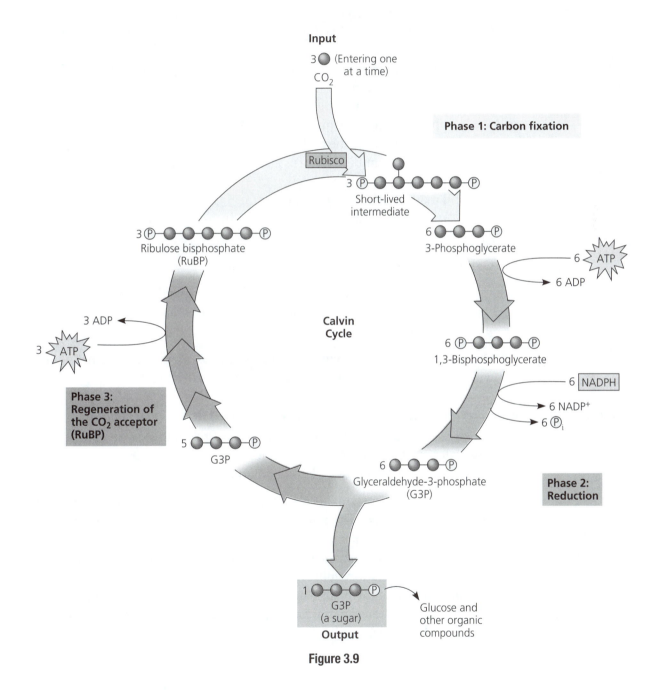

Figure 3.9

■ These are the major steps of the Calvin cycle. It is not important that you memorize the intermediate organic molecules, but it is important to understand the conceptual scheme of reducing CO_2 to a sugar.

1. Three CO_2 molecules are attached to three molecules of the 5-carbon sugar **ribulose bisphosphate (RuBP).** These reactions are catalyzed by the enzyme **rubisco** (probably the most common protein in the biosphere) and produce an unstable product that immediately splits into two three-carbon compounds called 3-phosphoglycerate. At this point carbon has been fixed—the incorporation of CO_2 into an organic compound.
2. The 3-phosphoglycerate molecules are phosphorylated to become 1, 3-bisphosphoglycerate.
3. Next, 6 NADPH reduce the six 1, 3-bisphosphoglycerates to six **glyceraldehyde-3-phosphate (G3P).**
4. One G3P leaves the cycle to be used by the plant cell. Two G3P molecules can combine to form glucose, which is generally listed as the final product of photosynthesis.
5. Finally, RuBP is regenerated as the 5 G3Ps are reworked into 3 of the starting molecules, with the expenditure of 3 ATP molecules.

■ The Calvin cycle, as explained here and resulting in the formation of one net G3P, requires the following energy molecules:

■ Nine molecules of ATP are consumed (to be replenished by the light reactions) along with six molecules of NADPH (also to be replenished by the light reactions).

■ One of the six G3P molecules produced in the Calvin cycle is a net gain, and will be used for biosynthesis or the energy needs of the cell.

Concept 10.4 Alternative mechanisms of carbon fixation have evolved in hot, arid climates

■ The overall problem is that CO_2 enters the leaf through stomata, the same pore that water exits the leaf in transpiration.
■ The specific problem for C_3 plants is as follows:

■ On hot dry days C_3 plants produce less sugar because the declining levels of CO_2 in the leaf starves the Calvin cycle. This occurs because the plant must keep its stomata closed to conserve water, thus no CO_2 uptake.
■ Additionally, the enzyme rubisco can bind O_2 in place of CO_2. This causes the oxidation or breakdown of RuBP, resulting in a loss of energy and carbon for the plant—a metabolic process called **photorespiration.** Photorespiration can drain away as much as 50% of the carbon fixed by the Calvin cycle!

- How can hot, arid regions have any plants? They have metabolic adaptations (as well as structural ones covered in Topic 8) that reduce photorespiration. The two most important of these adaptations are C_4 and CAM plants.
- C_4 plants have two kinds of photosynthetic cells: bundle-sheath cells and mesophyll cells. The **bundle-sheath cells** are grouped around the leaf's veins, and the mesophyll cells are dispersed elsewhere around the leaf.
- Use Figure 10.19 in your text as a guide for the following steps to C_4 photosynthesis:

    1. CO_2 is added to **phosphoenolpyruvate** (PEP) to form the four-carbon compound **oxaloacetate** (hence the designation C_4 plant). This reaction is catalyzed by **PEP carboxylase,** which does not combine with O_2 and does not participate in photorespiration. At this point carbon is fixed.
    2. The mesophyll cells export the oxaloacetate to the bundle-sheath cells, which break down the oxaloacetate, releasing CO_2. The C_4 pathway acts as a CO_2 pump.
    3. The CO_2 in the bundle-sheath cells is then converted into carbohydrates through the normal Calvin cycle.

        - In C_4 plants, the mesophyll cells pump CO_2 into the bundle-sheath cells, keeping the CO_2 concentration high enough so that rubisco will be more likely to bind to CO_2 rather than O_2. This reduces photorespiration. In C_4 plants the two stages of photosynthesis are separated structurally, providing a process that minimizes photorespiration and enhances sugar production.

- **CAM photosynthesis** is another adaptation to hot, dry climates.

    - These plants keep their stomata closed during the day to prevent excessive water loss. Of course, this also prevents gas exchange. At night, the stomata open and CO_2 is fixed in organic acids and stored in vacuoles. In the morning when the stomata close, the plant cells release the stored CO_2 from the acids and proceed with photosynthesis.
    - In CAM plants the two stages of photosynthesis are separated temporally.

- In both C_4 and CAM photosynthesis, CO_2 is first transformed into an organic intermediate before it enters the Calvin cycle. All of the processes—C_3, C_4, and CAM photosynthesis—use the Calvin cycle; they just have different methods for getting there.

For Additional Review

Compare and contrast the process of chemiosmosis in both the mitochondrion and the chloroplast. Note how the H+ gradient is established and the orientation of the ATP synthase molecules.

Multiple-Choice Questions

1. The purpose of cellular respiration in a eukaryotic cell is to
   (A) synthesize carbohydrates from CO_2.
   (B) synthesize fats and proteins from CO_2.
   (C) break down carbohydrates to provide energy for the cell in the form of ATP.
   (D) break down carbohydrates to provide energy for the cell in the form of ADP.
   (E) provide oxygen to the cell.

$$2 K + Br_2 \rightarrow 2 K^+ + 2 Br^-$$

2. In the course of the above reaction, potassium is
   (A) neutralized.
   (B) oxidized.
   (C) reduced.
   (D) sublimated.
   (E) recycled.

3. The net result of glycolysis is
   (A) 4 ATP and 4 NADH.
   (B) 4 ATP and 2 NADH.
   (C) 2 ATP and 4 NADH.
   (D) 2 ATP and 2 NADH.
   (E) 4 ATP and 8 NADH.

4. One glucose molecule provides enough carbons for two trips through the citric acid cycle. How many molecules of ATP are directly produced in this process?
   (A) 1
   (B) 2
   (C) 3
   (D) 4
   (E) 5

5. The process that produces the greatest amount of ATP during respiration is
   (A) glycolysis.
   (B) fermentation.
   (C) the citric acid cycle.
   (D) oxidative phosphorylation.
   (E) lactic acid formation.

Directions: The group of questions below consists of five lettered choices followed by a list of numbered phrases or sentences. For each numbered phrase or sentence, select the one choice that is most closely related to it. Each choice may be used once, more than once, or not at all in each group.

Questions 6–10
   (A) Chemiosmosis
   (B) Electron transport chain
   (C) The citric acid cycle
   (D) Glycolysis
   (E) Fermentation

6. The process by which glucose is split into pyruvate

7. The process by which a hydrogen ion gradient is used to produce ATP

8. A process that makes a small amount of ATP and can produce lactic acid as a by-product

9. A series of membrane-embedded electron carriers that ultimately create the hydrogen ion gradient to drive the synthesis of ATP

10. The process by which the chemical breakdown of glucose is completed and CO_2 is produced

11. Muscle fatigue is caused when the process of fermentation in oxygen-depleted cells produces which of the following?
    (A) ADP
    (B) Ethanol
    (C) Lactic acid
    (D) Uric acid
    (E) Pyruvate

12. Groups of photosynthetic pigment molecules situated in the thylakoid membrane are called
    (A) photosystems.
    (B) carotenoids.
    (C) chlorophyll.
    (D) grana.
    (E) CAM plants.

13. The main products of the light reactions of photosynthesis are
    (A) NADPH and $FADH_2$.
    (B) NADPH and ATP.
    (C) ATP and $FADH_2$.
    (D) ATP and CO_2.
    (E) ATP and H_2O.

14. The process in photosynthesis that bears the most resemblance to chemiosmosis and oxidative phosphorylation in cell respiration is called
    (A) cyclic phosphorylation.
    (B) linear photophosphorylation.
    (C) ATP synthase coupling.
    (D) preemptive photophosphorylation.
    (E) dark reaction phosphorylation.

15. The major product of the Calvin cycle is
    (A) rubisco.
    (B) oxaloacetate.
    (C) ribulose bisphosphate.
    (D) pyruvate.
    (E) glyceraldehyde-3-phosphate.

16. All of the following statements are false EXCEPT
    (A) C_3 plants grow better in hot, arid conditions than do C_4 plants.
    (B) C_4 plants grow better in cold, moist conditions than do C_3 plants.
    (C) C_3 plants grow better in hot, arid conditions than do CAM plants.
    (D) CAM plants grow better in cold, moist conditions than do C_3 plants.
    (E) CAM plants and C_4 plants both grow better in hot, arid conditions than do C_3 plants.

17. All of the following statements about photosynthesis are true EXCEPT
    (A) the light reactions convert solar energy to chemical energy in the form of ATP and NADPH.
    (B) the Calvin cycle uses ATP and NADPH to convert CO_2 to sugar.
    (C) photosystem I contains P700 chlorophyll *a* molecules at the reaction center; photosystem II contains P680 molecules.
    (D) in chemiosmosis, electron transport chains pump protons (H^+) across a membrane from a region of high H^+ concentration to a region of low H^+ concentration.
    (E) the steps of the Calvin cycle are sometimes referred to as the dark reactions because they do not directly require light in order to take place.

18. Which of the following is mismatched with its location?
    (A) light reactions—grana.
    (B) electron transport chain—thylakoid membrane.
    (C) Calvin cycle—stroma.
    (D) ATP synthesis—double membrane surrounding chloroplast.
    (E) splitting of water—thylakoid space.

19. The chlorophyll known as P680 has its electron "holes" filled by electrons from
    (A) photosystem I.
    (B) photosystem II.
    (C) water.
    (D) NADPH

20. CAM plants avoid photorespiration by
    (A) fixing CO_2 into organic acids during the night; these acids then release CO_2 during the day.
    (B) Performing the Calvin cycle at night.
    (C) Fixing CO_2 into four-carbon compounds in the mesophyll, which release CO_2 in the bundle-sheath cells.
    (D) Using PEP carboxylate to fix CO_2 to ribulose bisphosphate (RuBP).
    (E) Keeping their stomata open during the day.

21. How many "turns" of the Calvin cycle are required to produce one molecule of glucose?
    (A) 1
    (B) 2
    (C) 3
    (D) 6
    (E) 12

22. What are the final electron acceptors for the electron transport chains in the light reactions of photosynthesis and in cellular respiration?
    (A) O_2 in both
    (B) CO_2 in both
    (C) H_2O in the light reactions and O_2 in respiration
    (D) P700 and NADP+ in the light reactions and NAD+ or FAD in repiration
    (E) NADP+ in the light reactions and O_2 in respiration

Free-Response Question

1. *Cellular respiration and photosynthesis are basic cellular processes. Below are key events in cellular respiration and/or photosynthesis.*

   (a) Explain how a photosystem converts light energy to chemical energy.
   (b) Explain the specific role of glycolysis in cellular respiration.
   (c) Describe the function of water in cellular respiration and photosynthesis.

ANSWERS AND EXPLANATIONS

Multiple-Choice Questions

▌ **1. (C) is correct.** The purpose of cellular respiration in eukaryotes is to produce energy for cellular work in the form of ATP. Respiration is an aerobic process, meaning that it requires oxygen. Answer choices *A* and *B* are incorrect because respiration involves the breakdown (not the synthesis) of carbohydrates, fats, and proteins. Choice *D* is wrong because ADP is the product of the dephosphorylation of ATP—it is left over after the energy from ATP has been released. Choice *E* is wrong because oxygen is required for cellular respiration.

▌ **2. (B) is correct.** In the course of the reaction shown, potassium (K) is oxidized. Oxidation involves the loss of an electron while reduction is the gain of an electron by an atom or molecule. In this reaction, potassium is oxidized and bromine is reduced. Both cellular respiration and photosynthesis involve numerous oxidation-reduction (redox) reactions.

■ **3. (D) is correct.** The net energy result of glycolysis is the production of two molecules of ATP and two molecules of NADH. Glycolysis is the first of the three stages of respiration—the second being the citric acid cycle and the third being oxidative phosphorylation. During glycolysis, glucose is broken down and oxidized to form two molecules of pyruvate. Glycolysis occurs in the cytosol, and the pyruvate it produces travels to the mitochondria, where it is used in the citric acid cycle.

■ **4. (B) is correct.** In the citric acid cycle, 2 ATPs are produced. The citric acid cycle takes in a molecule called acetyl CoA (pyruvate is converted into acetyl CoA before it enters the citric acid cycle), and this is joined to a four-carbon molecule of oxaloacetate to form a six-carbon compound citrate that is then broken down again to produce oxaloacetate; the oxaloacetate reenters the cycle. In the course of the citric acid cycle, the following are produced: $4 \, CO_2$, 2 ATP, 6 NADH, and $2 \, FADH_2$.

■ **5. (D) is correct.** The process that produces the most ATP during cellular respiration is oxidative phosphorylation. $FADH_2$ and NADH donate electrons to the electron transport chain, which is coupled to ATP synthesis by chemiosmosis; the movement of electrons down the electron transport chain creates an H^+ gradient across the mitochondrial membrane, which drives the synthesis of ATP from ADP. About 34 ATP are produced per glucose molecule.

■ **6. (D) is correct.** In glycolysis, glucose is oxidized to two molecules of pyruvate. This is the first step in cellular respiration, and it also produces 2 ATP and 2 NADH.

■ **7. (A) is correct.** In chemiosmosis, the hydrogen ion gradient created by the transfer of electrons in the electron transport chain provides the power to synthesize ATP from ADP.

■ **8. (E) is correct.** Fermentation is an anaerobic alternative to cellular respiration. It consists of glycolysis and several reactions that serve to regenerate NAD^+. Electrons are transferred from NADH to pyruvate or its derivatives, and the NAD^+ oxidizes sugar in glycolysis. There are two main types of fermentation: alcohol fermentation (which creates ethanol as a product) and lactic acid fermentation (which creates lactate).

■ **9. (B) is correct.** The electron transport chain is a series of inner mitochondrial matrix membrane-embedded molecules that are capable of being oxidized and reduced as they pass along electrons. The energy produced from the passage of these electrons down the chain is used to create an H^+ gradient across the membrane, and the flow of H^+ down the gradient and back across the membrane powers the phosphorylation reaction of ADP to form ATP.

■ **10. (C) is correct.** The citric acid cycle includes the final reactions for the breakdown of glucose that began in glycolysis. The pyruvate from glycolysis is converted into acetyl CoA, which enters the cycle and is joined to oxaloacetate to create citrate, which is then converted to oxaloacetate again and reused. This cycle gives off CO_2 and forms 1 ATP, 3 NADH, and $1 \, FADH_2$. The cycle goes through one rotation to break down each of the molecules of pyruvate produced in glycolysis (which of course is first converted to acetyl CoA), so the net result of the breakdown of one glucose molecule is 2 ATP, 6 NADH, and $2 \, FADH_2$.

11. (C) is correct. When muscle cells in the body are depleted of oxygen, they switch from cellular respiration to lactic acid fermentation. In lactic acid fermentation, NADH reduces pyruvate directly, and lactate is formed as a waste product. This lactic acid fermentation occurs when all of the cell's metabolic machinery is using available oxygen to break down sugars in the cell.

12. (A) is correct. Groups of photosynthetic pigment molecules in the thylakoid membrane are called photosystems. The two photosystems involved in photosynthesis are photosystem I and photosystem II. Both contain chlorophyll molecules and many proteins and other organic molecules, and both have a light-harvesting complex that harnesses incoming light. Each of these photosystems contains a reaction center, where chlorophyll a and the primary electron acceptor are located.

13. (B) is correct. The main products of the light reactions of photosynthesis are NADPH and ATP. NADPH and ATP are used to convert CO_2 to sugar in the Calvin cycle. The enzyme rubisco combines CO_2 with ribulose bisphosphate (RuBP), and electrons from NADPH and energy from ATP to synthesize a three-carbon molecule called glyceraldehyde-3-phosphate.

14. (B) is correct. The process in photosynthesis that bears the closest resemblance to chemiosmosis and oxidative phosphorylation in cellular respiration is linear photophosphorylation. In this process, energy from the transfer of electrons down the electron transport chain is used to create a hydrogen ion gradient used in the making of ATP. Later, the energy stored in this ATP is used during the formation of carbohydrates in the Calvin cycle.

15. (E) is correct. The organic product of the Calvin cycle, which may be used later to build large carbohydrates in the cell, is glyceraldehyde-3-phosphate, or G3P. This molecule is created as a result of the fixation of three molecules of CO_2, which costs the cell ATP and NADPH that were created in the light reactions of photosynthesis.

16. (E) is correct. C_4 and CAM plants both grow better than do C_3 plants under conditions of increased median air temperature and decreased relative humidity. Both C_4 and CAM plants use an alternative method of carbon fixation that enables them to fix carbon into an acid intermediate for later deposit into the Calvin cycle.

17. (D) is correct. The electron transport chains pump protons across membranes from regions of low H^+ concentrations to regions of high H^+ concentrations. This proton pumping occurs in both mitochondria and chloroplasts, and the protons then diffuse (with the concentration gradient) back across the membrane through ATP synthases. This drives the synthesis of ATP.

18. (D) is correct. ATP synthase is located in the thylakoid membrane. Notice the direction of flow of protons through ATP synthase in photosynthesis: the protons flow from the thylakoid space to the stroma. ATP is produced in the stroma, where it will be used by the Calvin cycle.

19. (C) is correct. When water is split, three products are formed: two protons, an oxygen atom that immediately bonds with another oxygen to form O_2, and two electrons. The electrons immediately feed the P680 chlorophyll a in the reaction center of Photosystem II. The ultimate electron donor in photosynthesis is water.

■ **20. (A) is correct.** CAM plants separate the two stages of photosynthesis temporally to reduce photorespiration. This is accomplished by fixing CO_2 at night using PEP carboxylase and storing the carbon in organic acids. During the day when CAM plants have their stomata closed to conserve water, the carbon from the organic acids is chemically released and used in the Calvin cycle.

■ **21. (D) is correct.** Each turn of the Calvin cycle involves the enzyme rubisco fixing one atom of carbon. It follows that it would take six turns to produce the six carbon sugar glucose. Students sometimes miss this question by confusing the Calvin cycle with the Krebs or citric acid cycle of cellular respiration. Read these questions carefully, being disciplined enough to carefully identify what the question is asking.

■ **22. (E) is correct.** The light reactions of photosynthesis move electrons from their low energy state in water to a higher energy level when the electrons are donated to $NADP^+$ to make NADPH. In cellular respiration electrons pass down the electron transport chain from high to low potential energy, ultimately combining with O_2 and hydrogen ions to form water.

Free-Response Question

(a) Although plants have two photosystems, they both work in the same way. Both photosystems have two components: the light harvesting complexes and the reaction center. The light-harvesting complex is made up of many chlorophyll and accessory pigment molecules. When one of the pigment molecules absorbs light energy in the form of photons, one of the molecule's electrons is raised to an orbital of higher potential energy. The pigment molecule is then said to be in an "excited" state. The increase in potential energy is transferred to the reaction center of the photosystem. The reaction center consists of two chlorophyll *a* molecules, which use the increased potential energy passed to them by the photosynthetic pigments to donate electrons to the primary electron acceptor. The solar-powered transfer of an electron from the reaction center chlorophyll *a* pair to the primary electron acceptor is the first step of the light reactions. This is the conversion of light energy to chemical energy.

(b) Glycolysis is the first stage of cellular respiration and occurs in the cytoplasm. Glycolysis involves the breakdown of glucose to two pyruvate molecules. To accomplish this, 2 ATP molecules are invested, which helps to destabilize glucose, making it more reactive and allowing glucose to break into two three-carbon molecules. By the time the pathway has produced pyruvate, 4 ATP molecules have been produced along with 2 NADH molecules. This gives a net energy gain of 2 ATPs and 2 NADHs. Thus, one important role of glucose is to produce energy molecules for the cell to use in its life processes. The second role is to produce pyruvate, which can feed into the citric acid cycle in the mitochondria and ultimately into the electron transport chain where most of the ATP in cellular respiration is produced.

(c) In cellular respiration water is a product of the reaction, whereas in photosynthesis water is a reactant. In cellular respiration water is formed when the electrons at the end of the electron transport chain in the cristae membrane of the mitochondria combine with hydrogen ions and an atom of oxygen to form water. Water is the ultimate electron acceptor in cellular respiration. In photosynthesis an enzyme splits a water molecule into two electrons, two hydrogen ions, and an oxygen atom. The electrons are supplied as needed directly to the chlorophyll molecules in the reaction center of photosystem II. In photosynthesis water is the ultimate electron donor.

This response shows that the writer used the following key terms in context, showing the writer's knowledge of their meanings and relatedness:

photosystems	*accessory pigments*
light harvesting complexes	*primary electron acceptor*
reaction center	*ATP*
chlorophyll	*NADPH*

The response also contains an explanation of the following subjects and processes:

—*role of photosystems in photosynthesis*
—*how light energy becomes transformed to chemical energy*
—*function of glycolysis in overall scheme of cellular respiration*
—*ATP and NADPH as important energy molecules in the cell*
—*role of water as an electron sink or an electron donor*

Mendelian Genetics

Chapter 13: Meiosis and Sexual Life Cycles

YOU MUST KNOW

- The differences between asexual and sexual reproduction.
- The role of meiosis and fertilization in sexually reproducing organisms.
- The importance of homologous chromosomes to meiosis.
- How the chromosome number is reduced from diploid to haploid through the stages of meiosis.
- Three important differences between mitosis and meiosis.
- The importance of crossing over, independent assortment, and random fertilization to increasing genetic variability.

Concept 13.1 *Offspring acquire genes from parents by inheriting chromosomes*

▌ **Genes** are segments of DNA that code for the basic units of heredity and are transmitted from one generation to the next. In animals and plants, reproductive cells that transmit genes from one generation to the next are called **gametes**.

▌ A **locus** (plural, loci) is the location of a gene on a chromosome. See Figure 4.1.

    ▪ In **asexual reproduction** a single parent is the sole parent and passes copies of all its genes to its offspring. In asexual reproduction the new offspring arise by mitosis and have virtually exact copies of the parent's genome. An individual that reproduces asexually gives rise to a **clone,** a group of genetically identical individuals.

    ▪ In **sexual reproduction,** two individuals (parents) contribute genes to the offspring. This form of reproduction results in greater genetic variation in the offspring than asexual reproduction.

Concept 13.2 *Fertilization and meiosis alternate in sexual life cycles*

▌ A **life cycle** is the generation-to-generation sequence of stages in the reproductive history of an organism, from conception to production of its own offspring.

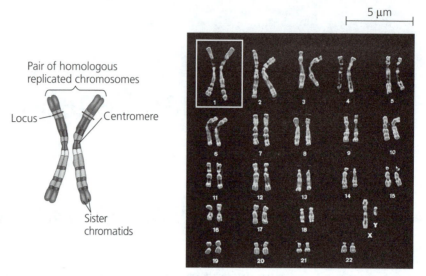

Figure 4.1

■ **Somatic cells** are any cells in the body that are not gametes. Each somatic cell in humans has 46 chromosomes. Liver cells and neurons are somatic cells.

  ▪ The **karyotype** of an organism refers to a picture of its complete set of chromosomes, arranged in pairs of homologous chromosomes from the largest pair to the smallest pair. Figure 4.1 is a karyotype made from a human somatic cell. Notice that the 46 chromosomes are paired into 23 homologous chromosomes.

  ▪ In **homologous chromosomes** both chromosomes of each pair carry genes that control the same inherited characteristics. If a gene for eye color is found at a specific locus on one chromosome, its homologues will have the same gene at the same locus.

    ▪ Homologous chromosomes are similar in length and centromere position, and they have the same staining pattern.

    ▪ One homologous chromosome from each pair is inherited from each parent; in other words, half of the set of 46 chromosomes in your somatic cells was inherited from your mother, and the other half was inherited from your father.

  ▪ Exceptions to the rule that all chromosomes are part of a homologous pair may be found with the sex chromosomes—in humans, it is the **X** and **Y**. Human females have a homologous pair of chromosomes, XX, but males have one X chromosome and one Y chromosome. Nonsex chromosomes are called **autosomes**.

  ▪ *What sex did the somatic cell come from that was used to make the karyotype in Figure 4.1?* (Check your answer at the end of this topic section.)

  ▪ **Gametes**—meaning sperm and ova (eggs)—are haploid cells. Haploid cells contain half the number of chromosomes of somatic cells. In humans, gametes contain 22 autosomes plus a single sex chromosome (either X or Y), giving them a haploid number of 23. The haploid number of chromosomes is symbolized by *n*.

- **Meiosis** and **fertilization** are the key events in sexually reproducing life cycles. The human life cycle in Figure 4.2 is typical of a sexually reproducing animal.

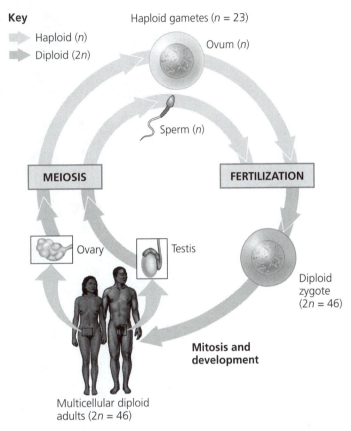

Key

➤ Haploid (*n*)

➤ Diploid (2*n*)

Haploid gametes (*n* = 23)

Ovum (*n*)

Sperm (*n*)

MEIOSIS

FERTILIZATION

Ovary

Testis

Diploid zygote (2*n* = 46)

Mitosis and development

Multicellular diploid adults (2*n* = 46)

Figure 4.2

- During **fertilization** (the combination of a sperm cell and an egg cell), one haploid gamete from each parent fuse. The result is a fertilized egg called a **zygote**. It is **diploid** (has two sets of chromosomes) and may be symbolized by 2*n*.
- **Meiosis** is the type of cell division that reduces the numbers of sets of chromosomes from two to one. Fertilization restores the diploid number as the gametes are combined. Fertilization and meiosis alternate in the life cycles of sexually reproducing organisms.

Concept 13.3 *Meiosis reduces the number of chromosome sets from diploid to haploid*

- Meiosis and mitosis look similar—both are preceded by the replication of the cell's DNA, for instance, but in meiosis this replication is followed by two stages of cell division, meiosis I and meiosis II.
- The final result of meiosis is four **daughter cells,** each of which has half as many chromosomes as the parent cell.

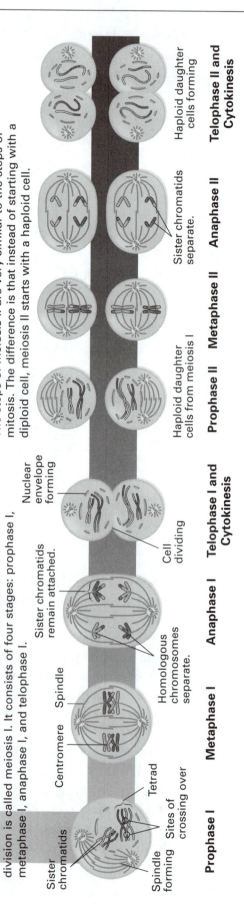

Meiosis I:

In contrast to mitosis, meiosis involves two divisions. The first division is called meiosis I. It consists of four stages: prophase I, metaphase I, anaphase I, and telophase I.

Interphase

Just as in mitosis, the cell duplicates its DNA. Each chromosome then consists of two identical sister chromatids that can be seen more clearly in prophase.

Centrosomes
Centrioles
Chromatin
Nuclear envelope

Sister chromatids
Spindle forming
Sites of crossing over
Tetrad
Centromere
Spindle
Sister chromatids remain attached.
Homologous chromosomes separate.
Nuclear envelope forming
Cell dividing

Prophase I **Metaphase I** **Anaphase I** **Telophase I and Cytokinesis**

Meiosis II:

The steps of meiosis II are very similar to the steps of mitosis. The difference is that instead of starting with a diploid cell, meiosis II starts with a haploid cell.

Sister chromatids separate.
Haploid daughter cells from meiosis I
Haploid daughter cells forming

Prophase II **Metaphase II** **Anaphase II** **Telophase II and Cytokinesis**

Figure 4.3

Carefully follow the stages in Figure 4.3 as they are explained:

▌ **Interphase:** Each of the chromosomes makes a copy of itself; that is, each chromosome replicates its DNA, roughly doubling the amount of DNA in the cell. The centrosome also divides during this phase.

STUDY TIP: Understanding prophase I is critical to understanding meiosis. Study the unique events of prophase I carefully.

▌ **Meiosis I:** The first cellular division in meiosis is referred to as meiosis I.

  ▪ **Prophase I:** The chromosomes condense, resulting in two sister chromatids attached at their centromeres.

    ▪ **Synapsis** occurs—that is, the joining of homologous chromosomes along their length. This newly formed structure is called a *tetrad* and precisely aligns the homologous chromosomes gene by gene. This perfect alignment is necessary for the next step—crossing over.

    ▪ In **crossing over** the DNA from one homologue is cut and exchanged with an exact portion of DNA from the other homologue. Essentially, a small part of the DNA from one parent is exchanged with the DNA from the other parent. *The result of crossing over is to increase genetic variation.* Where crossing over has occurred (two to three times per homologous pair), criss-crossed regions termed **chiasmata** form, which hold the homologues together until anaphase I.

    ▪ After crossing over, the centrioles move away from each other, the nuclear envelope disintegrates, and spindle microtubules attach to the kinetochores forming on the chromosomes that begin to move to the metaphase plate of the cell.

ORGANIZE YOUR THOUGHTS

In Prophase I:

1. Synapsis occurs, forming tetrads.
2. Tetrads undergo crossing over.
3. Crossing over increases genetic variation.
4. Areas of crossing over form chiasmata.
5. The nuclear envelope disintegrates, allowing the spindle to attach to the homologues.

  ▪ **Metaphase I:** The homologous pairs of chromosomes are lined up at the metaphase plate, and microtubules from each pole attach to each member of the homologous pairs in preparation for pulling them to opposite ends of the cell. How many homologous pairs are found in the cell in Figure 4.3?

- **Anaphase I:** The spindle apparatus helps to move the chromosomes toward opposite ends of the cell; sister chromatids stay connected and move together toward the poles.
- **Telophase I** and **cytokinesis:** The homologous chromosomes move until they reach the opposite poles. Each pole, then, contains a haploid set of chromosomes, with each chromosome still consisting of two sister chromatids.
  - Cytokinesis is the division of the cytoplasm and occurs during telophase. A **cleavage furrow** occurs in animal cells, and **cell plates** (the forming new cell wall) occur in plant cells. Both result in the formation of two haploid cells.

- **Meiosis II:** The second cellular division in meiosis is referred to as meiosis II.

  - **Prophase II:** A spindle apparatus forms, and sister chromatids move toward the metaphase plate.
  - **Metaphase II:** The chromosomes are lined up on the metaphase plate, and the kinetochores of each sister chromatid prepare to move to the opposite poles.
  - **Anaphase II:** The centromeres of the sister chromatids separate, and individual chromosomes move to opposite ends of the cell.
  - **Telophase II** and **cytokinesis:** The chromatids have moved all the way to opposite ends of the cell, nuclei reappear, and cytokinesis occurs. Each of the four daughter cells has the haploid number of chromosomes and is genetically different from the other daughter cells and from the parent cell.

STUDY TIP: Be prepared to cite the following three examples when asked to explain differences between mitosis and meiosis.

- Three events occur during meiosis I that do not occur during mitosis.

  1. Synapsis and crossing over normally do not occur during mitosis.
  2. At metaphase I, paired homologous chromosomes (tetrads) are positioned on the metaphase plate, rather than individual replicated chromosomes, as in mitosis.
  3. At anaphase I, duplicated chromosomes of each homologous pair move toward opposite poles, but the sister chromatids of each duplicated chromosome stay attached. In mitosis, the chromatids separate.

Concept 13.4 Genetic variation produced in sexual life cycles contributes to evolution

> **STUDY TIP:** There are three important processes that contribute to variation. Be able to list and explain them.

▌ **Crossing over:** During prophase I the exchange of genetic material on homologous chromosomes between nonsister chromatids occurs. Use Figure 4.3 to help make this unique feature of meiosis clear. Notice that all four chromatids that make up the tetrad are different due to crossing over. In metaphase II when sister chromatids separate, each chromatid is unique, thus increasing variation.

▌ **Independent assortment of chromosomes:** In metaphase I, when the homologous chromosomes are lined up on the metaphase plate, they can pair up in any combination, with any of the homologous pairs facing either pole. This means that there is a 50-50 chance that a particular daughter cell will get a maternal chromosome or a paternal chromosome from the homologous pair.

▌ **Random fertilization:** Because each egg and sperm is different, as a result of independent assortment and crossing over, each combination of egg and sperm is unique.

Chapter 14: Mendel and the Gene Idea

> ## YOU MUST KNOW
>
> - Terms associated with genetics problems: P, F_1, F_2, dominant, recessive, homozygous, heterozygous, phenotypic, and genotypic.
> - How to derive the proper gametes when working a genetics problem.
> - The difference between an allele and a gene.
> - How to read a pedigree.

Concept 14.1 Mendel used the scientific approach to identify two laws of inheritance

▌ True-breeding parents in a genetic cross are called the **P (parental) generation;** their offspring are called the F_1 **(first filial) generation**. If the F_1 population is crossed, their offspring are called the F_2 **(second filial) generation**.

▌ The following are four related concepts that make up Mendel's model explaining the 3:1 inheritance pattern that he observed among F_2 offspring.

1. **Alternative versions of genes cause variations in inherited characteristics among offspring.** For example, consider flower color in peas.

The gene for flower color in pea plants comes in two versions: white and purple. These alternative versions of the gene, called **alleles,** are the result of slightly different DNA sequences.

2. **For each character, every organism inherits one allele from each parent.**
3. **If the two alleles are different, then the *dominant allele* will be fully expressed in the offspring, whereas *recessive allele* will have no noticeable effect on the offspring.**
4. **The two alleles for each character separate during gamete production.** If the parent has two of the same alleles, then the offspring will all get that version of the gene, but if the parent has two different alleles for a gene, each offspring has a 50% chance of getting one of the two alleles. This is Mendel's **law of segregation.**

> *STUDY TIP:* Use Figure 4.4 to find each of the four basic concepts of Mendel's model.

▌ **Law of independent assortment** was Mendel's second law. It states that each pair of alleles will segregate (separate) independently during gamete formation.

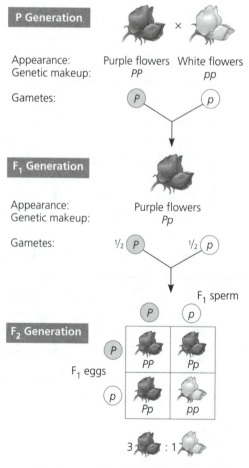

Figure 4.4   Mendel's Law of Segregation

- **Homozygous** organisms have two of the same alleles for a particular trait. If the dominant allele for a trait is designated as *R* (dominant traits are generally capitalized), and the recessive allele is designated *r* (recessive traits are generally not capitalized), then an individual could be homozygous for the dominant trait (*RR*) or homozygous for the recessive trait (*rr*).

- A **heterozygous** organism has two different alleles for a trait (*Rr*).

- **Phenotype** refers to an organism's expressed physical traits, and **genotype** refers to an organism's genetic makeup. For example, the phenotype of a seed might be round, and its genotype could be *RR* or *Rr*.

- A **testcross** is done to determine if an individual showing a dominant trait is homozygous or heterozygous. A homozygous dominant parent will yield all dominant phenotypes in the offspring (*RR* X *rr*), whereas a heterozygous parent will give a ratio of one dominant character to one recessive character.

- A **monohybrid cross** is a cross involving the study of only one character (e.g., flower color), while a **dihybrid cross** is a cross intended to study two characters (e.g., flower color and seed shape).

- The diagram below shows the results of a dihybrid cross. In this case, in the parental generation two homozygous plants are crossed: one homozygous dominant for light gray and round seeds (*YYRR*) and one homozygous recessive for dark gray and wrinkled seeds (*ppyy*). The only gamete type the first parent can produce is *YR*, and the only gamete the second parent can produce is *yr*. The F$_1$ generation, therefore, is composed of individuals with genotype *YyRr*. Crossing *YyRr* with a second *YyRr* gives an F$_2$ generation that completes the cross and looks like Figure 4.5.

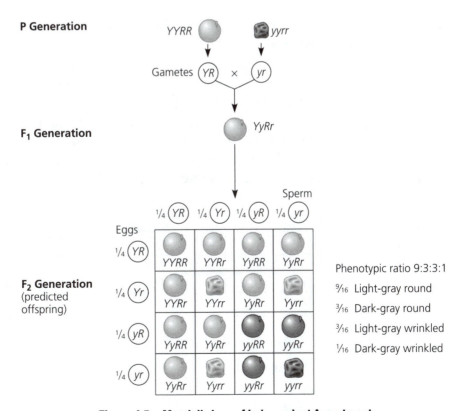

Figure 4.5   Mendel's Law of Independent Assortment

> **STUDY TIP:** When working genetic problems, be sure your have written the genotypes of the parents correctly and derive the gametes correctly. Study Figure 4.5, paying special attention to how the gametes were obtained for the F_2 generation.

Concept 14.2 The laws of probability govern Mendelian inheritance

▌ Understanding how to predict genetic crosses involves familiarity with the basic laws of probability. There are two laws that you will use directly in solving genetics problems:

- ■ **The rule of multiplication:** When calculating the probability that two or more independent events will occur together in a specific combination, multiply the probabilities of each of the two events. Thus, the probability of a coin landing face up two times in two flips is $\frac{1}{2} \times \frac{1}{2} = \frac{1}{4}$. If you cross two organisms with the genotypes *AABbCc* and *AaBbCc*, the probability of an offspring having the genotype *AaBbcc* is $\frac{1}{2} \times \frac{1}{2} \times \frac{1}{4} = \frac{1}{16}$.
- ■ **The rule of addition:** When calculating the probability that any of two or more mutually exclusive events will occur, you need to add together their individual probabilities. For example, if you are tossing a die, what is the probability that it will land on either the side with 4 spots or the side with 5 spots?

$$\frac{1}{6} + \frac{1}{6} = \frac{1}{3}.$$

Concept 14.3 Inheritance patterns are often more complex than predicted by simple Mendelian genetics

- ▌ **Complete dominance** is dominance in which the heterozygote and the homozygote for the dominant allele are indistinguishable. A *Yy* yellow seed is just as yellow as a *YY* yellow seed.
- ▌ **Codominance** occurs when two alleles are dominant and affect the phenotype in two different but equal ways. The traditional example for this type of dominance is human blood types. Notice in the chart below that in blood type AB both alleles are completely expressed.

Phenotype	Genotype	Antibodies Expressed
A	$I^A I^A$ or $I^A i$	Anti-B
B	$I^B I^B$ or $I^B i$	Anti-A
AB	$I^A I^B$	None
0	ii	Anti-A, Anti-B

- **Incomplete dominance** is a type of dominance in which the F_1 hybrids have an appearance that is in between that of the two parents. For example, if two plants, one with white flowers and one with red flowers, were crossed and all of the offspring had pink flowers, you could conclude that the trait for flower color exhibits incomplete dominance. Breeding two of the hybrids with incomplete dominance gives a flower ratio of 1 red: 2 pink: 1 white.

- **Multiple alleles** occur when a gene has more than two alleles. Again, a good example of this is seen in human blood types. The chart shows the three alleles in human blood types: I^a, I^b, and i. Notice how the three alleles combine to form different blood types.

- **Pleiotropy** is the property of a gene that causes it to have multiple phenotypic effects. For example, sickle-cell disease has multiple symptoms all due to a single defective gene.

- In **epistasis,** a gene at one locus alters the effects of a gene at another locus. For example, an individual may have genes for heavy skin pigmentation, but if a separate gene that produces the pigment is defective, the genes for pigment deposition will not be expressed. This would lead to a condition known as albinism.

- In **polygenic inheritance,** two or more genes have an additive effect on a single character in the phenotype (such as height or skin color in humans). When several genes are involved, the phenotype usually is described by a bell-shaped curve, with fewer individuals at each extreme and most individuals clustered in the middle.

Concept 14.4 Many human traits follow Mendelian patterns of inheritance

- A **pedigree** is a diagram that shows the relationship between parents and off-spring across two or more generations. (See Figure 4.6.) In a typical pedigree, circles represent women, and squares represent men. White open circles or squares indicate that the individual did not or does not express a particular trait, whereas black indicates that the individual expresses or expressed that trait. Through the patterns they reveal, pedigrees can help determine the genome of individuals that comprise them; pedigrees can also help predict the genome of future offspring.

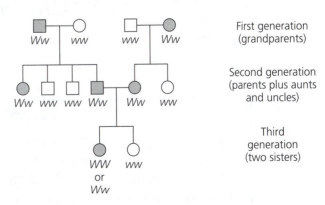

Figure 4.6   Human Pedigree Analysis

- **Recessively inherited disorders** require two copies of the defective gene for the disorder to be expressed. Examples include the following:
  - **Cystic fibrosis** is caused by a mutation in an allele that codes for a cell membrane protein that functions in the transport of chloride ions into and out of cells. The resulting high extracellular levels of chloride cause mucus to be thicker and stickier, leading to organ malfunction and recurrent bacterial infections.
  - **Tay-Sachs** disease is caused by an allele that codes for a dysfunctional enzyme, which is unable to break down certain lipids in the brain. As these lipids accumulate in the brain cells, the child suffers from blindness, seizures, and degeneration of brain function, leading to death.
  - **Sickle-cell disease** is caused by an allele that codes for a mutant hemoglobin molecule that forms long rods when the oxygen levels in the blood are low. These long rods cause the red blood cell to sickle, clogging small blood vessels and leading to pain, organ damage, and even paralysis.

- **Lethal dominant alleles** require only one copy of the allele in order for the disorder to be expressed. Usually, only late-acting lethal alleles are passed on.
- **Huntington's disease** is caused by a lethal dominant allele. It is a degenerative disease of the nervous system, which usually doesn't affect the individual until he or she is over 40 years old.
- Genetic testing may be used on a fetus to detect certain genetic disorders. Two common tests are amniocentesis and chorionic villus sampling (CVS).

  1. **Amniocentesis** occurs when the physician removes amniotic fluid from around the fetus. The amniotic fluid can be utilized to detect some genetic disorders, and the cells in the fluid can be cultured for a karyotype.
  2. **Chorionic villus sampling** involves using a narrow tube inserted through the cervix to suction out a tiny sample of the placenta that contains only fetal cells. A karyotype can immediately be developed from these cells.

Chapter 15: The Chromosomal Basis of Inheritance

YOU MUST KNOW

- How the chromosome theory of inheritance connects the physical movement of chromosomes in meiosis to Mendel's laws of inheritance.
- The unique pattern of inheritance in sex-linked genes.
- How alteration of chromosome number or structurally altered chromosomes (deletions, duplications, etc.) can cause genetic disorders.

Concept 15.1 Mendelian inheritance has its physical basis in the behavior of chromosomes

▌ The **chromosome theory of inheritance** states that genes have specific locations (called *loci*) on chromosomes and that it is chromosomes that segregate and assort independently. It is important to connect this physical movement of chromosomes in meiosis to Mendel's laws of inheritance.

    ▪ A **sex-linked gene** is one located on a sex chromosome (X or Y in humans). After the chromosome theory of inheritance was formed, *Thomas Hunt Morgan* discovered the existence of sex-linked genes.

Concept 15.2 Sex-linked genes exhibit unique patterns of inheritance

▌ In humans, there are two types of sex chromosomes, X and Y. Normal females have two X chromosomes, whereas normal males have one X and one Y chromosome.

▌ Figure 4.7 shows the unique pattern of inheritance in sex-linked genes. Notice that the sex-linked gene is found on the X chromosome, but not on the Y chromosome. In addition to tracking the gene from one generation to the next, it is also necessary to track the sex of the offspring. Figure 4.7 is based on work with *Drosophila* (fruit flies) performed by Thomas Hunt Morgan, but the pattern is the same in humans.

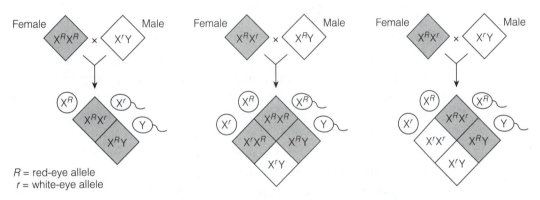

R = red-eye allele
r = white-eye allele

Figure 4.7　Patterns of Inheritance with Sex-Linked Traits

▌ Each egg or ovum contains an X chromosome; there are two types of sperm: those with an X chromosome and those with a Y chromosome. In fertilization, there is a 50-50 chance that a sperm carrying an X or Y will reach and penetrate the egg first. Thus, gender is determined by chance and by the male sperm cell in humans.

▌ Fathers pass sex-linked genes on to their daughters but not to their sons; fathers pass the Y chromosome to their sons.

▌ Females will express a sex-linked trait exactly like any other trait, but males, with only one X chromosome, will express the allele on the X chromosome they inherited from their mother. The terms *homozygous* and *heterozygous* do not apply to a male pattern of sex-linked genes.

▌ The vast majority of genes on the X chromosome are not related to sex.

- Several sex-linked disorders have medical significance:
  - **Duchenne muscular dystrophy** is a sex-linked disorder characterized by a progressive weakening of the muscles and loss of coordination. Affected individuals rarely live past their early 20s.
  - **Hemophilia** is a sex-linked disorder characterized by having blood with an inability to clot normally, caused by the absence of proteins required for blood clotting.

- **X-inactivation** regulates gene dosage in females. Although female mammals inherit two X chromosomes, one of the X chromosomes (randomly chosen) in each cell of the body becomes inactivated during embryonic development by **methylation**. As a result, males and females have the same effective dose of genes with loci on the X chromosome.

- The inactive chromosome condenses into a **Barr body,** which lies along the inside of the nuclear envelope. Still, females are not affected as heterozygote carriers of problematic alleles, because half of their sex chromosomes are normal and produce the necessary proteins.

Concept 15.3 Linked genes tend to be inherited together because they are located near each other on the same chromosome

- **Linked genes** are located on the same chromosome and therefore tend to be inherited together during cell division.

- **Genetic recombination** is the production of offspring with a new combination of genes inherited from the parents. Many genetic crosses yield some offspring with the same phenotype as one of the parents (these offspring are referred to as **parental types**) and some offspring with phenotypes different from either parent (these offspring are referred to as **recombinants**).

- **Crossing over** can explain why some linked genes get separated during meiosis. During meiosis, unlinked genes follow independent assortment because they are located on different chromosomes. Linked genes are located on the same chromosome and would not be predicted to follow independent assortment. However, sometimes genetic crosses give results that seem to indicate some independent assortment has occurred, even when genes are on the same chromosome. These results are not due to independent assortment but can be explained by crossing over. Research further indicates that the farther apart two genes are on a chromosome, the higher the probability that crossing over will occur between them. The likelihood of crossing over between different genes on the same chromosome is expressed as a percent.

- A **linkage map** is a genetic map that is based on the percentage of cross-over events.

- A **map unit** is equal to a 1% recombination frequency. Map units are used to express relative distances along the chromosome.

Concept 15.4 Alterations of chromosome number or structure cause some genetic disorders

▍ **Nondisjunction** occurs when the members of a pair of homologous chromosomes do not separate properly during meiosis I, or sister chromatids don't separate properly during meiosis II.

▍ As a result of nondisjunction, one gamete receives two copies of the chromosome, while the other gamete receives none. In the next step, if the faulty gametes engage in fertilization, the offspring will have an incorrect chromosome number. This is known as **aneuploidy**.

▍ Fertilized eggs that have received three copies of the chromosome in question are said to be **trisomic;** those that have received just one copy of a chromosome are said to be **monosomic** for the chromosome.

▍ **Polyploidy** is the condition of having more than two complete sets of chromosomes, forming a 3n or 4n individual. Rare in animals, this condition is fairly frequent in plants.

▍ Portions of a chromosome may also be lost or rearranged, resulting in the following mutations:

  ▪ A **deletion** occurs when a chromosomal fragment is lost, resulting in a chromosome with missing genes.

  ▪ A **duplication** occurs when the chromosome fragment that broke off (causing the deletion above) becomes attached to its sister chromatid. In this case, the zygote will get a double dose of the genes located on that chromosome.

  ▪ An **inversion** occurs when a chromosome fragment breaks off and reattaches to its original position—but backward, so that the part of the fragment that was originally at the attachment point is now at the end of the chromosome.

  ▪ A **translocation** occurs when the deleted chromosome fragment joins a *nonhomologous* chromosome.

▍ Human disorders caused by chromosome alterations include the following:

  ▪ **Down syndrome:** An aneuploid condition that is the result of having an extra chromosome 21 (Trisomy 21). Down syndrome includes characteristic facial features, short stature, heart defects, and mental retardation.

  ▪ **Klinefelter syndrome:** An aneuploid condition in which a male possesses the sex chromosomes XXY (an extra X). Klinefelter males have male sex organs but are sterile.

  ▪ **Turner syndrome:** A monosomic condition in which the female has just one sex chromosome, an X. Turner syndrome females are sterile because the reproductive organs do not mature. Turner syndrome is the only known viable monosomy in humans.

Multiple-Choice Questions

1. A couple has six children, all daughters. If the woman has a seventh child, what is the probability that the seventh child will be a daughter?
   (A) ⁶⁄₇
   (B) ½
   (C) ½₆
   (D) ¼₉
   (E) ½

2. If alleles *R* and *S* are linked, and the probability of gamete *R* segregating into a gamete is ¼, while the probability of allele *S* segregating into a gamete is ½, what is the probability that both will segregate into the same gamete?
   (A) ¼ × ½
   (B) ¼ ÷ ½
   (C) ¼ + ½
   (D) ¼ + ¼
   (E) ½

3. In llamas, coat color is controlled by a gene that exists in two allelic forms. If a homozygous yellow llama is crossed with a homozygous brown llama, the offspring have gray coats. If two of the gray-coated offspring were crossed, what percentage of their offspring would have brown coats?
   (A) 100%
   (B) 75%
   (C) 50%
   (D) 25%
   (E) 0%

4. Which of the following is NOT true of meiosis?
   (A) During metaphase, spindle microtubules first come into contact with chromosomes.
   (B) The chromosome number in the newly formed cells is half that of the parent cell.
   (C) The homologous chromosomes line up along the metaphase plate, or equator of the cell.
   (D) The cytoplasm of the cell and all its organelles are divided approximately in half.
   (E) In anaphase II, the sister chromatids travel to opposite ends of the cell.

5. In rabbits, the trait for short hair (*S*) is dominant, and the trait for long hair (*s*) is recessive. The trait for green eyes (*G*) is dominant, and the trait for blue eyes (*g*) is recessive. A cross between two rabbits produces a litter of six short-haired rabbits with green eyes, and two short-haired rabbits with blue eyes. What is the most likely genotype of the parent rabbits in this cross?
   (A) *ssgg* × *ssgg*
   (B) *SSGG* × *SSGG*
   (C) *SsGg* × *SsGg*
   (D) *SsGg* × *SSGg*
   (E) *ssGG* × *ssGG*

6. In humans, hemophilia is a sex-linked recessive trait. If a man and a woman have a son who is affected with hemophilia, which of the following is definitely true?
   (A) The mother carries an allele for hemophilia.
   (B) The father carries an allele for hemophilia.
   (C) The father is afflicted with hemophilia.
   (D) Both parents carry an allele for hemophilia.
   (E) The boy's paternal grandfather has hemophilia.

7. Which of the following explains a significantly low rate of crossing over between two genes?
   (A) They are located far apart on the same chromosome.
   (B) They are located on separate but homologous chromosomes.
   (C) The genes code for proteins that have similar functions.
   (D) The genes code for proteins that have very different functions.
   (E) The genes are located very close together on the same chromosome.

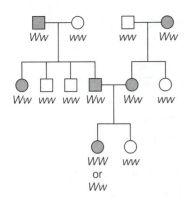

First generation (grandparents)

Second generation (parents plus aunts and uncles)

Third generation (two sisters)

8. In the pedigree above, circles represent females and squares represent males; those who express a particular trait are shaded, whereas those who do not are not shaded. Which pattern of inheritance best describes the pedigree for this trait?
   (A) Sex-linked recessive
   (B) Sex-linked dominant
   (C) Autosomal recessive
   (D) Autosomal dominant
   (E) Codominant

Questions 9–10 refer to an individual with blood type O, whose mother has blood type A.

9. The father must have which of the following blood types?
   (A) A, B, or O
   (B) AB or A
   (C) AB or B
   (D) AB only
   (E) O only

10. If the type O individual were to mate with a person with type AB blood, which of the following is the best calculation of the ratio of the offspring?
    (A) $3\ I^A i{:}1\ I^B i$
    (B) $2\ I^A i{:}1\ I^B i$
    (C) $I^A i{:}I^B i$
    (D) $1\ I^A i{:}2\ I^A I^B{:}1\ I^B i$
    (E) $9\ I^A I^B{:}3\ I^A i{:}3\ I^B i{:}1\ O$

11. Two yellow mice with the genotype Yy are mated. After many offspring, ⅔ are yellow and ⅓ are not yellow (a 2:1 ratio). Mendelian genetics dictates that this cross should produce offspring that were ¼ YY (yellow), ½ Yy (yellow), and ¼ yy (not yellow). What is the most likely conclusion from this experiment?
    (A) The mice did not bear enough offspring for the ratio calculation to be specific.
    (B) Y is lethal in the homozygous form and caused death early in development.
    (C) Nondisjunction occurred.
    (D) A mutation masked the effects of the Y allele.
    (E) A mutation masked the effects of the y allele.

12. All of the following contribute to genetic recombination EXCEPT
    (A) random fertilization.
    (B) independent assortment.
    (C) crossing over.
    (D) gene linkage.
    (E) random gene mutation.

13. In cucumbers, warty (W) is dominant over dull (w), and green (G) is dominant over orange (g). A cucumber plant that is homozygous for warty and green is crossed with one that is homozygous for dull and orange. The F_1 generation is then crossed. If a total of 144 offspring is produced in the F_2 generation, which of the following is the closest to the number of dull green cucumbers expected?
    (A) 3
    (B) 10
    (C) 28
    (D) 80
    (E) 111

14. The restoration of the diploid chromosome number after halving in meiosis is due to
    (A) synapsis.
    (B) fertilization.
    (C) mitosis.
    (D) DNA replication.
    (E) chiasmata.

15. During the first meiotic division (meiosis I),
    (A) homologous chromosomes separate.
    (B) the chromosome number becomes haploid.
    (C) crossing over between nonsister chromatids occurs.
    (D) paternal and maternal chromosomes assort randomly.
    (E) all of the above occur.

16. A cell with a diploid number of 6 could produce gametes with how many different combinations of maternal and paternal chromosomes?
    (A) 6
    (B) 8
    (C) 12
    (D) 64
    (E) 128

17. The DNA content of a diploid cell is measured in the G_1 phase. After meiosis I, the DNA content of one of the two cells produced would be
    (A) equal to that of the G_1 cell.
    (B) twice that of the G_1 cell.
    (C) one-half that of the G_1 cell.
    (D) one-fourth that of the G_1 cell.
    (E) impossible to estimate due to independent assortment of homologous chromosomes.

18. A synaptonemal complex would be found during
    (A) prophase I of meiosis.
    (B) fertilization or syngamy of gametes.
    (C) metaphase II of meiosis.
    (D) prophase of mitosis.
    (E) anaphase I of meiosis.

19. Meiosis II is similar to mitosis because
    (A) sister chromatids separate.
    (B) homologous chromosomes separate.
    (C) DNA replication precedes the division.
    (D) they both take the same amount of time.
    (E) haploid cells are produced.

20. Which of the following is *not* true of homologous chromosomes?
    (A) They behave independently in mitosis.
    (B) They synapse during the S phase of meiosis.
    (C) They travel together to the metaphase plate in prometaphase of meiosis I.
    (D) Each parent contributes one set of homologous chromosomes to an offspring.
    (E) Crossing over between nonsister chromatids of homologous chromosomes is indicated by the presence of chiasmata.

ANSWERS AND EXPLANATIONS

Multiple-Choice Questions

▌ **1. (E) is correct.** The probability that the woman will have a seventh child who is a daughter is ½. Since the probability that a sperm carrying an X chromosome and the probability that a sperm carrying a Y chromosome will fertilize an egg is equal—both 50%—fertilization is considered an independent event. The outcome of independent events is unaffected by what events occurred before or will occur after. Therefore, the probability that this woman's next child will be a girl is ½. Likewise, the probability that she will have a child that is a boy is also ½.

▌ **2. (A) is correct.** If the probability of allele *R* segregating into a gamete is ¼, and that of *S* segregating is ½, you can calculate the probability of two independent events occurring in a specific combination, order, or sequence by multiplying their probabilities. So in this case, you need to multiply ¼ by ½.

3. (D) is correct. Let's say that the yellow coat parent is C^YC^Y, and the homozygous brown coat parent is C^BC^B. Because the yellow coat parent can produce only gametes C^Y, and the brown coat parent can produce only gametes C^B, the F_1 generation will all have genotype C^YC^B. Crossing two members of this generation would give you a ratio of 1 yellow coat:2 gray coats:1 brown coat. This means that 25% of the offspring would have brown coats, 25% would have yellow coats, and 50% would have gray coats.

4. (A) is correct. All of the statements about meiosis are true except A. The spindle fibers attach during prophase, not metaphase.

5. (D) is correct. To find the answer to this problem, first look at the ratio of the offspring. It's 6:2, which can be reduced to 3:1. Next, you can quickly work through the crosses listed. You can immediately rule out answers A and B, because A would give you only offspring that exhibited the dominant traits, short hair and green eyes, and B would give you all offspring that had the recessive traits—long hair and blue eyes. If you look carefully at the remaining answers, you will want to choose the one that will give you all short-haired offspring, so you will need the dominant allele to be present in both parents. This rules out answer E. If you still cannot choose between C and D, write out what gametes the parents could produce, and then use a Punnett square to determine their offspring. By doing this, you can see that D is correct: the ratio of offspring is 12:4, or 3:1, which matches the ratio in the original question.

6. (A) is correct. If the boy is afflicted with hemophilia, then he must have inherited the recessive hemophilia gene from his mother. Sex-linked genes are usually located on the X chromosome. In order for the child to be a boy, he must have inherited a Y chromosome from his father. Because the gene causing hemophilia is located on the X chromosome, you can rule out answers B and C (it would not matter if the father possessed the allele for hemophilia because he can't pass on his X to a son). Therefore, you need to look for an answer choice that shows that the source of his X chromosome was a carrier of the allele—afflicted or not. This is answer A.

7. (E) is correct. While genes that are on the same chromosome tend to be inherited together, the process of crossing over enables "linked" genes to sort independently. Those that are linked but located farther apart on the chromosome will undergo crossing over more frequently than those located very close together on a chromosome simply because there are more sites between the two genes at which crossing over can take place.

8. (D) is correct. Autosomal dominant traits appear with equal frequency in both sexes, and they do not skip generations. These qualities are all exhibited by the trait that is illustrated in the pedigree. All three generations are affected with the trait; the sexes are affected roughly equally (four women and two men are affected).

9. (A) is correct. The father either has type A, B, or O blood. The mother, who has the phenotype blood type A, has the genotype I^Ai. In order to produce a son with genotype ii (the genotype of people with blood type O), she would need to reproduce with a man who had genotype I^Ai, genotype I^Bi, or genotype ii. Try writing out the Punnett square if you aren't confident of this.

10. (C) is correct. If the type O individual were to mate with an individual who was type AB, since I^A and I^B are both dominant over i, the genotype would be 1 I^A:1 I^B, and the phenotype would be a ratio of 1:1 offspring with type A or type B blood.

11. (B) is correct. The most likely reason for this 2:1 ratio in the offspring is that Y is lethal in homozygous form, and this caused the death of all of the *YY* individuals in the litter. The expected ratio of this cross would be 1 *YY*:2 *Yy*:1 *yy*. If you remove the *YY*, you get a ratio of 2 *Yy* (yellow mice, since the gene for yellow, *Y*, is dominant) to 1 *yy* (nonyellow mouse).

12. (D) is correct. The only process listed that does not lead directly to genetic recombination, or the recombining (scrambling) of genes in the offspring, is gene linkage. If genes are linked, they are located on the same chromosome and are more likely to segregate together into the same cell.

13. (C) is correct. Recall that a dihybrid cross between two heterozygotes produces a 9:3:3:1 offspring ratio. The question is asking for one of the heterozygotes (which would be one of the 3s in the above ratio). To come up with the answer, you could complete a Punnett square and see that the ratio of offspring produced is 9 warty green, 3 warty orange, 3 dull green, and 1 dull orange. The total number of offspring produced is 144; thus, you can deduce that that's 9 times 16 (the sum of $9 + 3 + 3 + 1$). So the number of dull green cucumbers must be equal to $9 \times 3 = 27$. The closest answer choice to 27 is 28.

14. (B) is correct. Fertilization restores the diploid number in a sexually reproducing organism. The two major events in the life cycle of sexually reproducing organisms are meiosis and fertilization.

15. (E) is correct. All of the responses are correct. Homologous chromosomes are a key conceptual point in Mendelian Genetics. If you missed this question, review more thoroughly the concept of homologous chromosomes.

16. (B) is correct. The number of possible gametes can be determined by using the formula 2^n where n equals the haploid number. With a diploid number of 6, thus a haploid number of 3, the answer is $2 \times 2 \times 2 = 8$.

17. (A) is correct. Recall that in G_1 the chromosomes have not replicated. After meiosis I the chromosomes have replicated, which doubles the amount of DNA. Next, the homologues separate in meiosis I. This separation reduces the amount of DNA by one-half and back to the original amount.

18. (A) is correct. The synaptonemal complex forms during prophase I and is the platform from which crossing over occurs.

19. (A) is correct. In meiosis II the most important event is the separation of sister chromatids. Mitosis involves only the separation of sister chromatids; homologues do not form in mitosis.

20. (B) is correct. Synapsis is the process of homologues moving to the tetrad position. This occurs in prophase I, not S phase as indicated in the question.

Answer from page 98: The karotype in Figure 4.1 is that of a male. Note the unpaired X and Y in the bottom right corner.

Molecular Genetics

Chapter 16: The Molecular Basis of Inheritance

YOU MUST KNOW

- The structure of DNA.
- The major steps to replication.
- The difference between replication, transcription, and translation.
- How DNA is packaged into a chromosome.

Concept 16.1 DNA is the genetic material

▎ Once chromosomes were known to carry genes, the next question became which of the two organic compounds that make chromosomes, DNA or protein, was the genetic material?

▪ In 1952 **Alfred Hershey and Martha Chase** answered this question utilizing *bacteriophages*—viruses that infect bacteria. Bacteriophages were excellent organisms for this study, in part because they are made of only two organic compounds, DNA and protein. Hershey and Chase used a radioactive isotope of phosphorus to tag the DNA in one culture of bacteriophages and radioactive sulfur to tag the protein in a second culture. Their results clearly showed that only the DNA entered bacteria infected by the virus; the radioactive protein never entered the cell. This research convinced scientists that DNA must be the genetic material.

▎ The next big question centered on the structure of DNA. Would the structure of DNA give any clues as to how it functioned as the genetic material?

▪ **James Watson and Francis Crick** were the first to solve the puzzle of the structure of DNA. Critical to their success was the work of Rosalind Franklin and Maurice Wilkins, both working in the field of X-ray crystallography.

▪ **X-ray crystallography** is a process used to visualize molecules three-dimensionally. X-rays are diffracted as they pass through the molecule, and they bounce back to produce patterns that can be interpreted through mathematical equations. Through this technique, a rough blueprint of the molecule was formed.

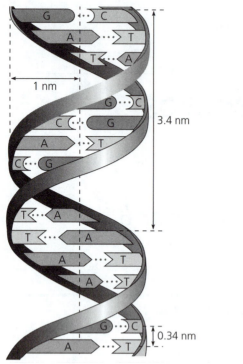

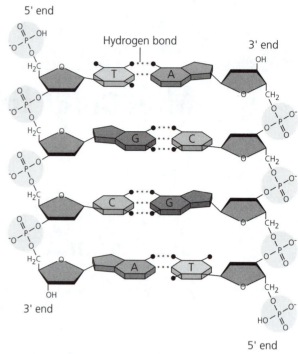

Key Features of DNA Structure   **Partial Chemical Structure**

Figure 5.1

■ Watson and Crick's model determined three major features of DNA. Find each major point by following Figure 5.1 as the model is explained.

1. DNA is a **double helix**, which can be described as a twisted ladder with rigid rungs. The side, or backbone, is made up of sugar-phosphate components, whereas the rungs are made up of pairs of nitrogenous bases.

2. The nitrogenous bases of DNA are adenine (A), thymine (T), guanine (G), and cytosine (C). In DNA, adenine pairs only with thymine, and guanine pairs only with cytosine.

3. Notice that the right side of the model is in the reading position, but the left side of the chain runs in the opposite, upside-down direction. The strands are termed **antiparallel**. The left side, the side in the reading position, runs 5′ to 3′ while the opposite strand runs 3′ to 5′. (Recall that the carbons are numbered, and you will see that the #5 carbon and #3 carbon and the resultant nucleotides are flipped relative to each other.) Nucleic acid strands are always antiparallel, whether it is DNA/DNA or DNA/RNA or RNA/RNA interactions.

Concept 16.2 Many proteins work together in DNA replication and repair

▌ **Replication** is the making of DNA from an existing DNA strand. DNA replication is *semiconservative*. This means that at the end of replication, each of the daughter molecules has one old strand, derived from the parent strand of DNA, and one strand that is newly synthesized. Study Figure 5.2 to see the pattern of semiconservative replication.

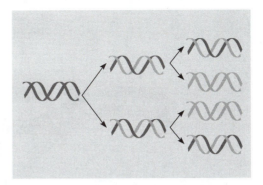

Figure 5.2   Semiconservative replication

> *Tip from the Readers:* Know the difference between *replication* (DNA to DNA), *transcription* (DNA to RNA), and *translation* (RNA to protein). In essay questions that use these terms, often 25% of the students confuse the processes!

▌ The replication of DNA includes six major points:

1. The replication of DNA begins at sites called the *origins of replication.*
2. Initiation proteins bind to the origin of replication and separate the two strands, forming a *replication bubble.* DNA replication then proceeds in both directions along the DNA strand until the molecule is copied.
3. A group of enzymes called **DNA polymerases** catalyzes the elongation of new DNA at the replication fork.
4. DNA polymerase adds nucleotides to the growing chain one by one, working in a 5′ to 3′ direction, matching adenine with thymine and guanine with cytosine.
5. Recall that the strands of DNA are antiparallel. This means that DNA replication occurs continuously along the 5′ to 3′ strand, which is called the **leading strand**. The strand that runs 3′ to 5′ is copied in series of segments and termed the **lagging strand**. Read steps 1–3 in Figure 5.3 to visualize this process.
6. The lagging strand is synthesized in separate pieces called **Okazaki fragments**, which are then sealed together by **DNA ligase** (step 4, Figure 5.3), forming a continuous DNA strand.

▌ There are several different factors contributing to the accuracy of DNA replication:

- The specificity of base pairing (A=T, G=C).
- **Mismatch repair**, in which special repair enzymes fix incorrectly paired nucleotides.
- **Nucleotide excision repair**, in which incorrectly placed nucleotides are excised or removed by enzymes termed **nucleases**, and the gap left over is filled in with the correct nucleotides.

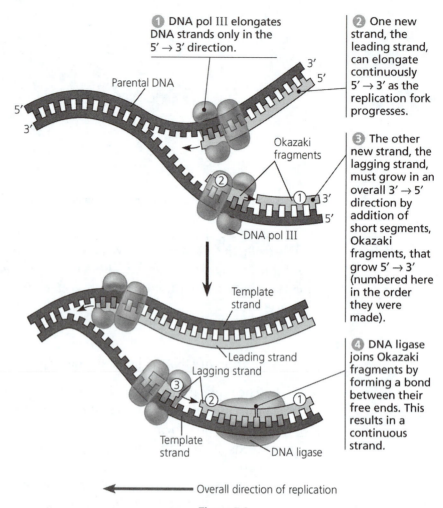

① DNA pol III elongates DNA strands only in the 5′ → 3′ direction.

② One new strand, the leading strand, can elongate continuously 5′ → 3′ as the replication fork progresses.

③ The other new strand, the lagging strand, must grow in an overall 3′ → 5′ direction by addition of short segments, Okazaki fragments, that grow 5′ → 3′ (numbered here in the order they were made).

④ DNA ligase joins Okazaki fragments by forming a bond between their free ends. This results in a continuous strand.

Parental DNA

Okazaki fragments

DNA pol III

Template strand

Leading strand

Lagging strand

Template strand

DNA ligase

Overall direction of replication

Figure 5.3

▌ The fact that DNA polymerase can add nucleotides only to the 3′ end of a molecule means that it would have no way to complete the 5′ end of the DNA molecule at the end of the chromosome. Every time the chromosome is replicated for mitosis, a small portion of the tip of the chromosome is removed. To avoid losing the terminal genes, the linear ends of eukaryotic chromosomes are "capped" with **telomeres**, short, repetitive nucleotide sequences that do not contain genes.

Concept 16.3 A chromosome consists of a DNA molecule packed together with proteins

▌ Know the general differences between the bacterial chromosome and those of eukaryotes.

  ▪ A bacterial chromosome is one double-stranded, circular DNA molecule associated with a small amount of protein.
  ▪ Eukaryotic chromosomes are linear DNA molecules associated with large amounts of protein.

▌ In eukaryotic cells, DNA and proteins are packed together as **chromatin**. Eukaryotic DNA shows four levels of packaging. Visualize each level of packaging by studying Figure 5.4 as you read.

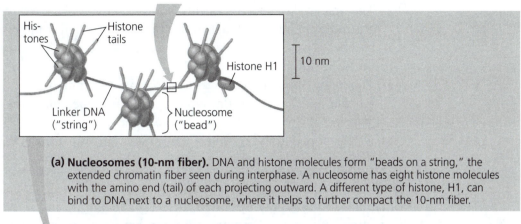

(a) Nucleosomes (10-nm fiber). DNA and histone molecules form "beads on a string," the extended chromatin fiber seen during interphase. A nucleosome has eight histone molecules with the amino end (tail) of each projecting outward. A different type of histone, H1, can bind to DNA next to a nucleosome, where it helps to further compact the 10-nm fiber.

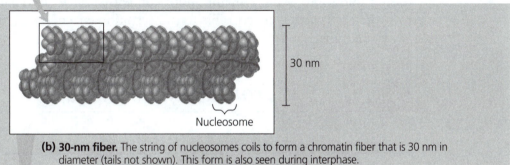

(b) 30-nm fiber. The string of nucleosomes coils to form a chromatin fiber that is 30 nm in diameter (tails not shown). This form is also seen during interphase.

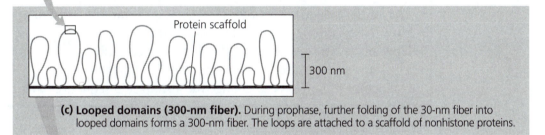

(c) Looped domains (300-nm fiber). During prophase, further folding of the 30-nm fiber into looped domains forms a 300-nm fiber. The loops are attached to a scaffold of nonhistone proteins.

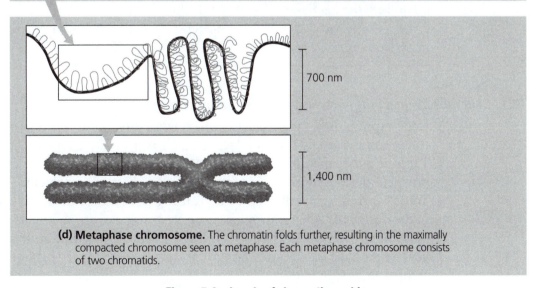

(d) Metaphase chromosome. The chromatin folds further, resulting in the maximally compacted chromosome seen at metaphase. Each metaphase chromosome consists of two chromatids.

Figure 5.4   Levels of chromatin packing

1. The first level of packing in chromosomes involves DNA wrapped around proteins called **histones**. This DNA/protein complex resembles beads on a string and is termed a **nucleosome**. Nucleosomes are the basic unit of DNA packing.
2. The string of nucleosomes folds to form a **30nm fiber**.
3. Further folding of the 30nm fiber results in **looped domains** of 300nm.
4. As the looped domains fold, a **metaphase chromosome** is formed.

▮ As DNA becomes more highly packaged, it becomes less accessible to transcription enzymes. This reduces gene expression. In interphase cells, most chromatin is in the highly extended form (*euchromatin*) and is available for transcription, but some remains more condensed (*heterochromatin*). Heterochromatin is largely inaccessible to transcription enzymes and, thus, generally is not transcribed. Barr bodies are heterochromatin.

Chapter 17: From Gene to Protein

YOU MUST KNOW

- The key terms gene expression, transcription, and translation.
- How to explain the process of transcription.
- How eukaryotic cells modify RNA after transcription.
- The steps to translation.
- How point mutations can change the amino acid sequence of a protein.

Tip from the Readers:
This is the central chapter for molecular genetics. It is one of the top five chapters you must know to perform well on the AP exam!

Concept 17.1 Genes specify proteins via transcription and translation

▮ **Gene expression** is the process by which DNA directs the synthesis of proteins (or, in some cases, RNAs).

▮ The **one gene–one polypeptide hypothesis** states that each gene codes for a polypeptide, which can be—or can constitute a part of—a protein.

▮ **Transcription** is the synthesis of RNA using DNA as a template. It takes place in the nucleus of eukaryotic cells.

▮ **Messenger RNA**, or **mRNA**, is produced during transcription. It carries the genetic message of DNA to the protein-making machinery of the cell in the cytoplasm, the *ribosome*.

▮ In eukaryotes, transcription results in pre-mRNA, which undergoes **RNA processing** to yield the final mRNA

- In prokaryotes, transcription results directly in mRNA, which is not processed. Transcription and translation can occur simultaneously.
- **Translation** is the production of a polypeptide chain using the mRNA transcript and occurs at the ribosomes.
- The instructions for building a polypeptide chain are written as a series of three-nucleotide groups; this is called a *triplet code*.
- During transcription, only one strand of the DNA is transcribed, and it is called the **template strand**. The mRNA that is produced is said to be *complementary* to the original DNA strand. The mRNA base triplets are called **codons**. They are written in the 5′ to 3′ direction.
- The genetic code is *redundant*, meaning that more than one codon codes for each of the 20 amino acids. The codons are read based on a consistent reading frame—the groups of three must be read in the correct groupings in order for translation to be successful.

Concept 17.2 Transcription is the DNA-directed synthesis of RNA

- **RNA polymerase** is an enzyme that separates the two DNA strands and connects the RNA nucleotides as they base-pair along the DNA template strand.
- The RNA polymerases can add RNA nucleotides only to the 3′ end of the strand, so RNA elongates in the 5′ to 3′ direction. As RNA nucleotides are added, remember that uracil replaces thymine when base pairing to adenine.
- The DNA sequence at which RNA polymerase attaches is called the **promoter**, while the DNA sequence that signals the end of transcription is called the **terminator**.
- A **transcription unit** is the entire stretch of DNA that is transcribed into an RNA molecule. A transcription unit may code for a polypeptide or an RNA, like transfer RNA or ribosomal RNA.
- There are three main stages of transcription:

  1. **Initiation:** In bacteria, RNA polymerase recognizes and binds to the promoter. In eukaryotes, RNA polymerase II, the specific RNA polymerase that transcribes mRNA, cannot bind to the promoter without supporting help from proteins known as transcription factors. **Transcription factors** assist the binding of RNA polymerase to the promoter and, thus, the initiation of transcription. The whole complex of RNA polymerase II and transcription factors is called a **transcription initiation complex** (see Figure 5.5).

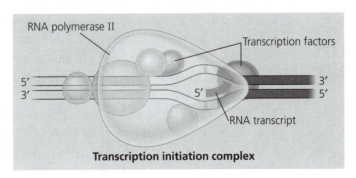

Transcription initiation complex

Figure 5.5

2. **Elongation:** RNA polymerase moves along the DNA, continuing to untwist the double helix. RNA nucleotides are continually added to the 3′ end of the growing chain. As the complex moves down the DNA strand, the double helix re-forms, with the new RNA molecule straggling away from the DNA template. Find these key steps in Figure 5.6.

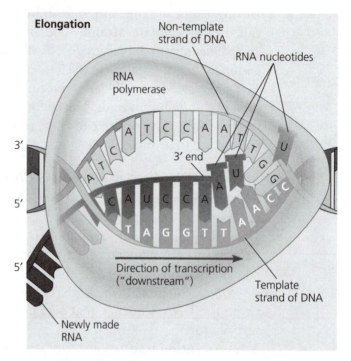

Figure 5.6   Elongation stage of transcription

3. **Termination:** After RNA polymerase transcribes a terminator sequence in the DNA, the RNA transcript is released, and the polymerase detaches.

Concept 17.3 Eukaryotic cells modify RNA after transcription

▌ In eukaryotes, there are a couple of key post-transcription modifications to RNA—the addition of a **5′ cap** and the addition of a **poly-A tail**.

▌ The 5′ cap and the poly-A tail facilitate the export of mRNA from the nucleus, help protect the mRNA from degradation by enzymes, and facilitate the attachment of the mRNA to the ribosome.

▌ **RNA splicing** also takes place in eukaryotic cells. In RNA splicing, large portions of the newly synthesized RNA strand are removed. The sections of the mRNA that are spliced out are called **introns**, and the sections that remain—and subsequently spliced together by a *spliceosome*—are called **exons**. Use Figure 5.7 to help you visualize exons and introns.

▌ One amazing thing about how spliceosomes work is the role of a special kind of RNA, termed small nuclear RNA (snRNA). snRNA plays a major role in catalyzing the excision of the introns and joining of exons. When RNA serves a catalytic role, the molecule is termed a **ribozyme**. For many years it was

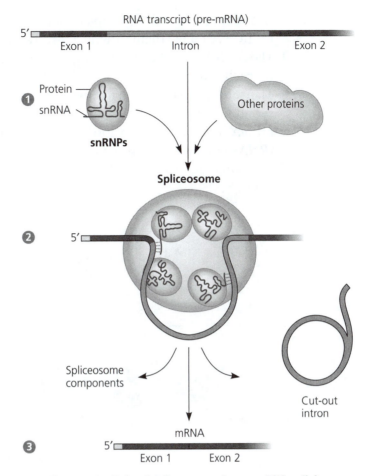

Figure 5.7   Role of spliceosomes in pre-mRNA splicing

thought that only proteins could be catalytic, but the discovery of ribozymes totally changed that idea!

▌ Another rethinking that has taken place came with the realization that we have fewer than 25,000 genes to make approximately 100,000 polypeptides. One gene can often make more than one polypeptide. An intron removed in the production of one polypeptide can be an exon in a second polypeptide made from the same gene! Alternative RNA splicing allows for different combinations of exons, resulting in more than one polypeptide per gene.

Concept 17.4 Translation is the RNA-directed synthesis of a polypeptide

▌ In addition to mRNA, two additional types of RNA play important roles in translation: transfer RNA (tRNA) and ribosomal RNA (rRNA).

▌ **tRNA** functions in transferring amino acids from a pool of amino acids in the cell's cytoplasm to a ribosome. The ribosome accepts the amino acid from tRNA and incorporates the amino acid into a growing polypeptide chain.

   ▪ Each type of tRNA is specific for a particular amino acid; at one end it loosely binds the amino acid, and at the other end it has a nucleotide triplet called an **anticodon**, which allows it to pair specifically with a complementary codon on the mRNA.

- A **codon** is an mRNA triplet. Since there are four different nucleotides (A,T,C,G), taking them three at a time results in 64 different codons.
- The mRNA is read codon by codon, and one amino acid is added to the chain for each codon read.
- The rules for base-pairing between the third base of a codon and the corresponding base of a tRNA anticodon are not as strict as those for DNA and mRNA codons. This relaxation of base-pairing rules is called **wobble**.
- **rRNA** complexes with proteins to form the two subunits that form ribosomes. Ribosomes have three binding sites for tRNA (locate each tRNA binding site in Figure 5.8).

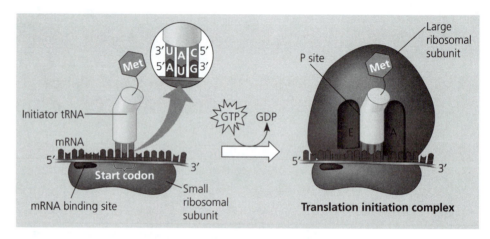

Figure 5.8   Initiation stage of translation

- A **P site**, which holds the tRNA that carries the growing polypeptide chain.
- An **A site**, which holds the tRNA that carries the amino acid that will be added to the chain next.
- An **E site**, which is the exit site for tRNA.

- Translation, like transcription, can be divided into three stages:

    1. **Initiation:** Organize initiation into these three steps. Use Figure 5.8 to find each step.
        A. A small ribosomal subunit binds to mRNA in such a way that the first codon of the mRNA strand, which is always AUG, is placed in the proper position.
        B. tRNA with anticodon UAC, which carries the amino acid methionine, hydrogen bonds to first codon (initiation factors are proteins that assist in holding all this together).
        C. Large subunit of ribosome attaches, allowing the tRNA with methionine to attach to the P site. Notice that the A site is now available to the tRNA that will bring the second amino acid.

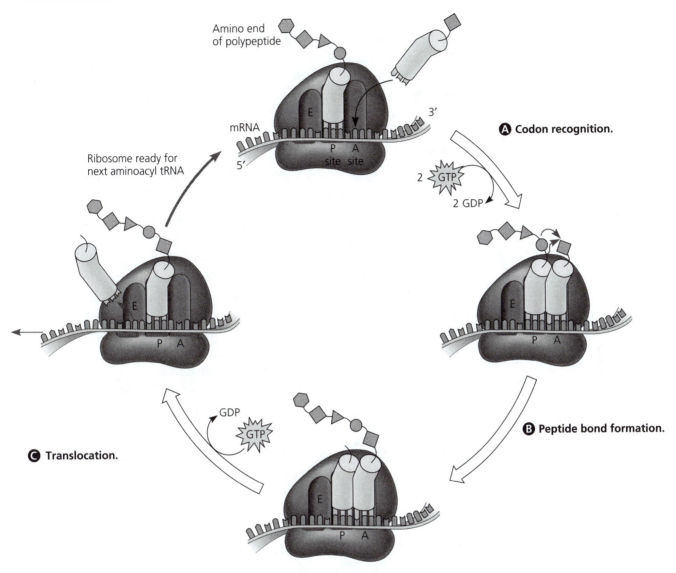

Figure 5.9 Protein synthesis

2. **Elongation:** Elongation also has three steps. Use Figure 5.9 to follow each of the three steps of elongation:
   A. *Codon recognition.* The codon in the A site is matched by the incoming tRNA anticodon.
   B. *Peptide bond formation.* The incoming amino acid in the A site forms a peptide bond with the existing chain of amino acids held in the P site. This bond is catalyzed by an rRNA and is an example of a ribozyme.
   C. *Translocation.* Translocation occurs when tRNA in the A site is moved to the P site, and the tRNA in the P site is moved to the E site, where it is released. Now the A site is clear and the process starts anew with codon recognition.

3. **Termination:** A stop codon in the mRNA is reached and translation stops. A protein called release factor binds to the stop codon, and the polypeptide is freed from the ribosome.

■ Polypeptides then fold to assume their specific conformation, and they are sometimes modified further to render them functional. The destination of a protein is often determined by the sequence of about 20 amino acids at the leading end of the polypeptide chain. The leading 20 or so amino acids is the **signal peptide** and serves as a sort of cellular zip code, directing proteins to their final destination.

Concept 17.5 Point mutations can affect protein structure and function

■ Mutations are alterations in the genetic material of the cell; **point mutations** are alterations of just one base pair of a gene. They come in two basic types:

1. **Base-pair substitution** refers to the replacement of one nucleotide and its complementary base pair in DNA with another pair of nucleotides.
   • **Missense mutations** are those substitutions that enable the codon to still code for an amino acid, although it might not be the correct one.
   • **Nonsense mutations** are those substitutions that change a regular amino acid codon into a stop codon, ceasing translation.
2. **Insertions** and **deletions** refer to the additions and losses of nucleotide pairs in a gene. If they interfere with the codon groupings, they can cause a **frameshift mutation**, which causes the mRNA to be read incorrectly.

■ **Mutagens** are substances or forces that interact with DNA in ways that cause mutations. X-rays and other forms of radiation are known mutagens, as are certain chemicals.

Chapter 18: Regulation of Gene Expression

YOU MUST KNOW

• The functions of the three parts of an operon.
• The role of repressor genes in operons.
• The impact of DNA methylation and histone acetylation on gene expression.
• The role of oncogenes, proto-oncogenes, and tumor suppressor genes in cancer.

Concept 18.1 Bacteria often respond to environmental change by regulating transcription

■ In bacteria, genes are often clustered into units called *operons*. Figure 5.10 shows a repressible operon with an inactive repressor. Locate each part of the operon and the regulatory gene as you read the accompanying text.

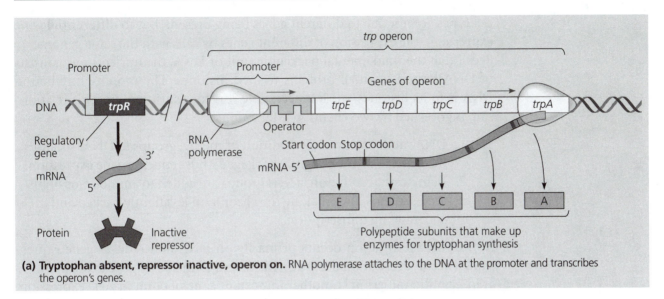

(a) **Tryptophan absent, repressor inactive, operon on.** RNA polymerase attaches to the DNA at the promoter and transcribes the operon's genes.

Figure 5.10   A repressible operon

- An **operon** consists of three parts:

   1. An **operator** that controls the access of RNA polymerase to the genes. The operator is found within the promoter site or between the promoter and the protein coding genes of the operon.
   2. The **promoter**, which is where RNA polymerase attaches.
   3. The **genes of the operon**. This is the entire stretch of DNA required for all the enzymes produced by the operon.

- Located some distance from the operon is a regulatory gene. **Regulatory genes** produce repressor proteins that may bind to the operator site. When a regulatory protein occupies the operator site, RNA polymerase is blocked from the genes of the operon. In this situation the operon is off.

- A **repressible operon** is normally on but can be inhibited. This type of operon is normally anabolic, building an essential organic molecule. The repressor protein produced by the regulatory gene is inactive. If the organic molecule being produced by the operon is provided to the cell, the molecule can act as a **corepressor** and bind to the repressor protein, activating it. The activated repressor protein binds to the operator site, shutting down the operon. This is the type of operon shown in Figure 5.10. The *lac* operon is inducible.

- An **inducible operon** is normally off but can be activated. This type of operon is normally catabolic, breaking down food molecules for energy. The repressor protein produced by the regulatory gene is active. To turn an inducible operon on, a specific small molecule, called an **inducer**, binds to and inactivates the repressor protein. With the repressor out of the operator site, RNA polymerase can access the genes of the operon.

Concept 18.2 Gene expression can be regulated at any stage

- The expression of eukaryotic genes can be turned off and on at any point along the pathway from gene to functional protein. Further, the differences between cell types are not due to different genes being present, but to **differential gene expression**, the expression of different genes by cells with the same genome.

- Recall that the fundamental packaging unit of DNA, the nucleosome, consists of DNA bound to small proteins termed histones. The more tightly bound DNA is to its histones, the less accessible it is for transcription. This relationship is governed by two chemical interactions:

  1. **DNA methylation** is the addition of methyl groups to DNA. It causes the DNA to be more tightly packaged, thus reducing gene expression.
  2. In **histone acetylation**, acetyl groups are added to amino acids of histone proteins, thus making the chromatin less tightly packed and encouraging transcription.

- Notice that methylation occurs primarily on DNA and reduces gene expression, whereas acetylation occurs on histones and increases gene expression.
- Transcription initiation is another important control point in gene expression. At this stage, DNA control elements that bind transcription factors (needed to initiate transcription) are involved in regulation. Notice in Figure 5.11 that DNA sequences far from the gene, termed **enhancer regions**, are bound to the promoter region by proteins termed **activators** and mediator proteins. The resulting **transcription initiation complex** greatly enhances gene expression.
- The control of gene expression may also occur prior to translation and just after translation, when proteins are processed.

Concept 18.3 Noncoding RNAs play multiple roles in controlling gene expression

- Since 1993 researchers have found that small molecules of single-stranded RNA can complex with proteins and influence gene expression. Two types of RNA, *micro RNAs (miRNA)* and *small interfering RNAs (siRNAs)*, can bind to mRNA and degrade the mRNA or bind to mRNA and block its translation.

Concept 18.4 A program of differential gene expression leads to the different cell types in a multicellular organism

- The zygote undergoes transformation through three interrelated processes

  1. **Cell division** is the series of mitotic divisions that increases the number of cells.
  2. **Cell differentiation** is the process by which cells become specialized in structure and function.
  3. **Morphogenesis** is the organization of cells into tissues and organs.

- What controls differentiation and morphogenesis?

  1. **Cytoplasmic determinants** are maternal substances in the egg that influence the course of early development. These are distributed unevenly in the early cells of the embryo and result in different effects.
  2. **Cell-cell signals** result from molecules, such as growth factors, produced by one cell influencing neighboring cells, a process called **induction** which causes cells to differentiate.

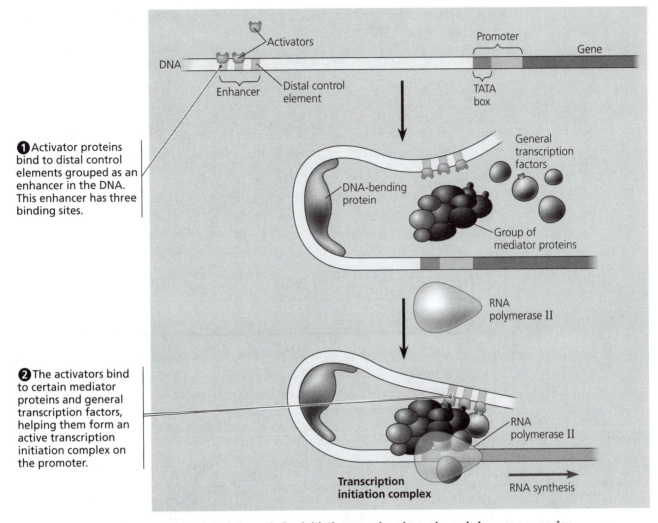

① Activator proteins bind to distal control elements grouped as an enhancer in the DNA. This enhancer has three binding sites.

② The activators bind to certain mediator proteins and general transcription factors, helping them form an active transcription initiation complex on the promoter.

Figure 5.11   Formation of transcription initiation complex plays a key role in gene expression

▌ **Determination** is the series of events that lead to observable differentiation of a cell. Differentiation is caused by cell-cell signals and is irreversible.

▌ **Pattern formation** sets up the body plan and is a result of cytoplasmic determinants and inductive signals. This is what determines head and tail, left and right, back and front. Uneven distribution of substances called **morphogens** plays a role in establishing these axes.

Concept 18.5 Cancer results from genetic changes that affect cell cycle control

▌ **Oncogenes** are cancer-causing genes; **proto-oncogenes** are genes that code for proteins that are responsible for normal cell growth. Proto-oncogenes become oncogenes when a mutation occurs that causes an increase in the product of the proto-oncogene, or an increase in the activity of each protein molecule produced by the gene.

▌ **Cancer** can also be caused by a mutation in a gene whose products normally inhibit cell division. These genes are called **tumor-suppressor genes**.

- An important tumor-suppressor gene is the **p53 gene**. This gene suppresses cancer in three ways:
  1. The p53 gene halts the cell cycle by binding to cyclin-dependent kinases. This allows time for DNA to be repaired before the resumption of cell division.
  2. The p53 gene turns on genes directly involved in DNA repair.
  3. When DNA damage is too great to repair, p53 activates suicide genes whose products cause cell death, a process termed **apoptosis**.
- The multistep model of cancer development is based on the idea that cancer results from the accumulation of mutations that occur throughout life. The longer we live, the more mutations that are accumulated and the more likely that cancer might develop.

Chapter 19: Viruses

YOU MUST KNOW

- The components of a virus.
- The differences between lytic and lysogenic cycles.

Concept 19.1 A virus consists of a nucleic acid surrounded by a protein coat

- Smaller than ribosomes, the tiniest viruses are about 20nm across.
- Genomes can be double- or single-stranded DNA or double- or single-stranded RNA.
- The **capsid** is a protein shell that surrounds the genetic material.
- Some viruses also have **viral envelopes** that surround the capsid and aid the viruses in infecting their hosts. (See Figure 5.12.)
- **Bacteriophages**, or **phages**, are viruses that infect bacterial cells.

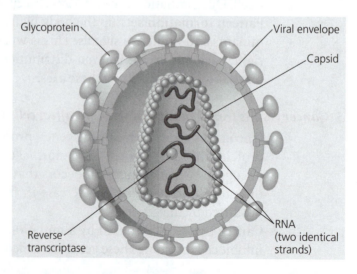

Figure 5.12   Structure of HIV. Note the components of a typical virus: nucleic acid, capsid. This virus also has an envelope, and reverse transcriptase.

Concept 19.2 Viruses reproduce only in host cells

▐ Viruses have a limited **host range**. This means they can infect only a very limited variety of hosts. *Example*: Human cold virus infects only cells of the upper respiratory tract.

▐ Viral reproduction occurs only in host cells. Two variations have been studied in bacteriophages. Study Figure 5.13 as you read the following:

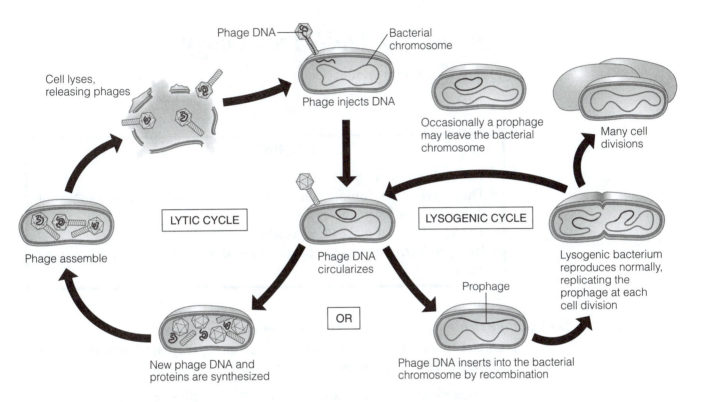

Figure 5.13   Infection of a bacterial cell by a bacteriophage. Note the two cycles

1. The **lytic cycle** ends in the death of the host cell by rupturing it (lysis). In this cycle, a bacteriophage injects its DNA into a host cell and takes over the host cell's machinery to synthesize new copies of the viral DNA as well as protein coats. These self-assemble, and the bacterial cell is lysed, releasing multiple copies of the virus.

2. In the **lysogenic cycle** the bacteriophage's DNA becomes incorporated into the host cell's DNA and is replicated along with the host cell's genome. The viral DNA is known as a **prophage**. Under certain conditions, the prophage will enter the lytic cycle, described above.

▐ **Retroviruses** are RNA viruses that use the enzyme **reverse transcriptase** to transcribe DNA from an RNA template. The new DNA then permanently integrates into a chromosome in the nucleus of an animal cell. The host transcribes the viral DNA into RNA that may be used to synthesize viral proteins or may be released from the host cell to infect more cells. *Example*: HIV is a retrovirus.

Concept 19.3 Viruses, viroids, and prions are formidable pathogens in animals and plants

▌ **Viroids** are circular RNA molecules several hundred nucleotides in length that infect plants. They cause errors in regulatory systems that control plant growth.

▌ **Prions** are misfolded, infectious proteins that cause the misfolding of normal proteins they contact in various animal species. *Examples* of diseases caused by prions include mad cow disease and, in humans, Creutzfeldt-Jakob.

Chapter 20: DNA Technology and Genomics

```
YOU MUST KNOW

• The terminology of biotechnology.
• The steps in gene cloning with special attention to the biotechnology tools
  that make cloning possible.
• The key ideas that make PCR possible.
• How gel electrophoresis can be used to separate DNA fragments or protein
  molecules.
```

Concept 20.1 DNA cloning yields multiple copies of a gene or other DNA segment

▌ The key to unlocking the concepts of biotechnology is to understand the terms. Know the following commonly used terms.

  ■ **Genetic engineering** is the process of manipulating genes and genomes.
  ■ **Biotechnology** is the process of manipulating organisms or their components for the purpose of making useful products.
  ■ **Recombinant DNA** is DNA that has been artificially made, using DNA from different sources—and often different species. An example is the introduction of a human gene into an *E. coli* bacterium.
  ■ **Gene cloning** is the process by which scientists can produce multiple copies of specific segments of DNA that they can then work with in the lab.
  ■ **Restriction enzymes** are used to cut strands of DNA at specific locations (called **restriction sites**). They are derived from bacteria.
  ■ When a DNA molecule is cut by restriction enzymes, the result will always be a set of **restriction fragments**, which will have at least one single-stranded end, called a **sticky end**. Sticky ends can form hydrogen bonds with complementary single-stranded pieces of DNA. These unions can be sealed with the enzyme **DNA ligase**.

In this example, a human gene is inserted into a plasmid from *E. coli*. The plasmid contains the amp^R gene, which makes *E. coli* cells resistant to the antibiotic ampicillin.

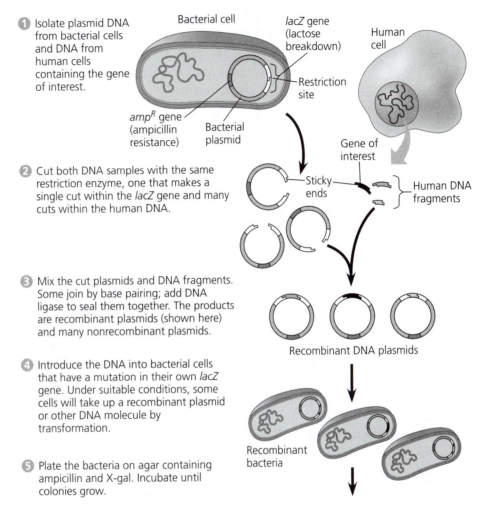

1 Isolate plasmid DNA from bacterial cells and DNA from human cells containing the gene of interest.

2 Cut both DNA samples with the same restriction enzyme, one that makes a single cut within the *lacZ* gene and many cuts within the human DNA.

3 Mix the cut plasmids and DNA fragments. Some join by base pairing; add DNA ligase to seal them together. The products are recombinant plasmids (shown here) and many nonrecombinant plasmids.

4 Introduce the DNA into bacterial cells that have a mutation in their own *lacZ* gene. Under suitable conditions, some cells will take up a recombinant plasmid or other DNA molecule by transformation.

5 Plate the bacteria on agar containing ampicillin and X-gal. Incubate until colonies grow.

Figure 5.14   Cloning a human gene in a bacterial plasmid

▎ The cloning of genes generally occurs in five steps. Follow each step in Figure 5.14. The overall picture for each step is given below.

1. *Identify and isolate the gene of interest and a **cloning vector**.* The vector is the plasmid (usually bacterial) that will carry the DNA sequence to be cloned.

2. *Cut both the gene of interest and the vector with the same restriction enzyme.* This gives the plasmid and the human gene matching sticky ends.

3. *Join the two pieces of DNA.* Form recombinant plasmids by mixing the plasmids with the DNA fragments. The human DNA fragments can be sealed into the plasmid using DNA ligase.

4. *Get the vector carrying the gene of interest into a host cell.* The plasmids are taken up by the bacterium by *transformation*. The process of transformation is a key part of Lab 6.

5. *Select for cells that have been transformed.* The bacterial cells carrying the clones must be identified or selected. This can be done by linking the gene of interest to an antibiotic resistance gene or a *reporter gene* such as green fluorescent protein. In AP Lab 6, we used an ampicillin-resistant plasmid. Any bacterial cells that did not pick up the plasmid by transformation will be killed when grown on agar with the antibiotic ampicillin.

▮ The next problem is finding the gene of interest among the many colonies present after transformation. A process known as **nucleic acid hybridization** can be used to find the gene. If we know at least part of the nucleotide sequence of the gene of interest, we can synthesize a probe complementary to it. For example, if the known sequence is G-G-C-T-A-A, then we would synthesize the complementary probe C-C-G-A-T-T. If we make the probe radioactive or fluorescent, the probe will be easy to track, taking us to the proper gene of interest.

▮ The process just described leads to a genomic library. A **genomic library** is a set of thousands of recombinant plasmid clones, each of which has a piece of the original genome being studied. A **cDNA library** is made up of complementary DNA made from mRNA transcribed by reverse transcriptase. This technique rids the gene of introns but may not contain every gene in the organism.

▮ **PCR** (polymerase chain reaction) is a method used to greatly amplify a particular piece of DNA without the use of cells. PCR is used to amplify DNA when the source is impure or scanty (as it would be at a crime scene). Figure 5.15 shows the basic steps of the PCR procedure.

Concept 20.2 DNA technology allows us to study the sequence, expression, and function of a gene

▮ **Gel electrophoresis** is a lab technique that is used to separate macromolecules, primarily DNA and proteins, on the basis of their size and charge with the use of an electrical current. In separating DNA, the negative charges on phosphates in the molecule cause DNA to move toward the positive pole. The gel allows smaller molecules to move more easily than larger fragments of DNA. The DNA fragments are separated by size. Follow Figure 5.16 as the specific steps are shown. Because gel electrophoresis is a required AP Lab, pay special attention to the concepts explained in the figure.

▮ **Southern blotting** combines gel electrophoresis and nucleic acid hybridization, allowing researchers to find a specific human gene. This technique is specific enough to find and note the difference between alleles. For example, it can distinguish a normal hemoglobin gene from a sickle cell gene.

▮ Genome-wide studies of gene expression are made possible by the use of **DNA microarray assays**. DNA microarray chips work as follows:

1. Small amounts of single-stranded DNA fragments representing different genes are fixed to a glass slide in a tight grid, termed a *DNA chip*.
2. All mRNA molecules from the cell being tested are isolated and converted to cDNA by reverse transcriptase, then tagged with a fluorescent dye.

3. The cDNA bonds to the ssDNA on the chip, indicating which genes are "on" in the cell (actively producing mRNA). The dye alerts the researcher. This enables researchers, for example, to see differences in gene expression between breast cancer tumors and noncancerous breast tissue.

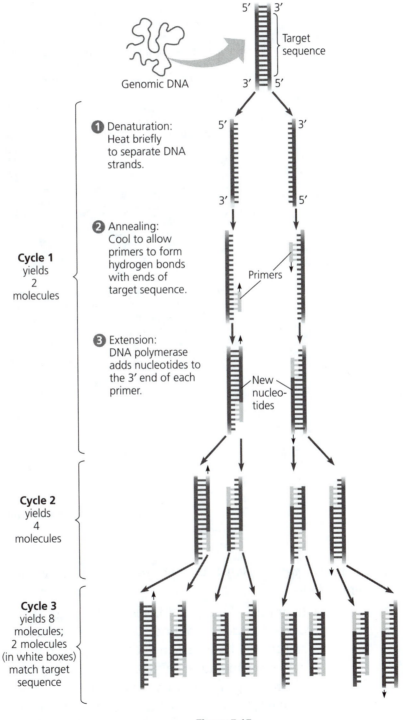

Figure 5.15

Gel Electrophoresis

APPLICATION Gel electrophoresis is used for separating nucleic acids or proteins that differ in size, electrical charge, or other physical properties. DNA molecules are separated by gel electrophoresis in restriction fragment analysis of both cloned genes (see Figure 20.10) and genomic DNA (see Figure 20.11).

TECHNIQUE Gel electrophoresis separates macromolecules on the basis of their rate of movement through a polymeric gel in an electric field: The distance a DNA molecule travels is inversely proportional to its length. A mixture of DNA molecules, usually fragments produced by restriction enzyme digestion (cutting) or PCR amplification, is separated into bands. Each band contains thousands of molecules of the same length.

1 Each sample, a mixture of DNA molecules, is placed in a separate well near one end of a thin slab of gel. The gel is set into a small plastic support and immersed in an aqueous solution in a tray with electrodes at each end.

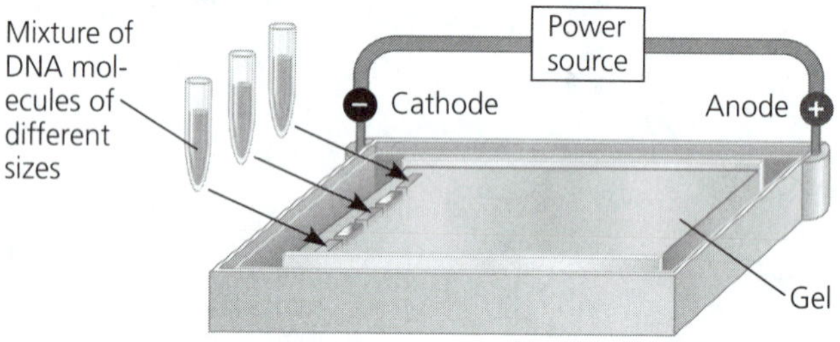

Mixture of DNA molecules of different sizes

Power source

Cathode   Anode

Gel

2 When the current is turned on, the negatively charged DNA molecules move toward the positive electrode, with shorter molecules moving faster than longer ones. Bands are shown here in blue, but on an actual gel, the bands would not be visible at this time.

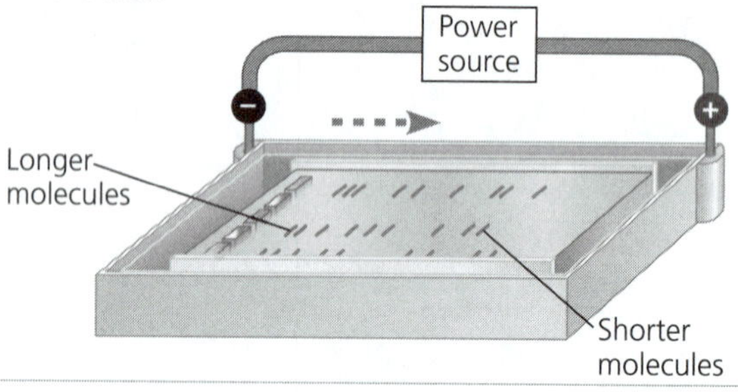

Power source

Longer molecules

Shorter molecules

Figure 5.16   Gel electrophoresis

Concept 20.3 Cloning organisms may lead to production of stem cells for research and other applications

▌ In animal cloning the nucleus of an egg is removed and replaced with the diploid nucleus of a body cell, a process termed *nuclear transplantation.* The ability of a body cell to successfully form a clone decreases with embryonic development and cell differentiation.

▌ The major goal of most animal cloning is reproduction, but not for humans. In humans, the major goal is the production of **stem cells**. A stem cell can both reproduce itself indefinitely and, under the proper conditions, produce other specialized cells. Stem cells have enormous potential for medical applications.

Concept 20.4 The practical applications of DNA technology affect our lives in many ways

There are many different uses for DNA technology, some of which are as follows:

1. **Diagnosis of disease:**

   - If the sequence of a particular virus's DNA or RNA is known, PCR can be used to amplify patients' blood samples to detect even small traces of the virus.
   - Different alleles have different DNA sequences. These differing sequences can be found using restriction enzymes that yield different lengths of DNA fragments, or **restriction fragment length polymorphisms.** The difference in banding patterns after electrophoresis allows for diagnosis of the disease, or even a carrier of the disease.

2. **Gene therapy:** The alteration of an afflicted individual's genes. Gene therapy holds great potential for treating disorders traceable to a single defective gene, such as cystic fibrosis.

3. **The production of pharmaceuticals:** Gene splicing and cloning can be used to produce large amounts of particular proteins in the lab.

4. **Forensic applications:** DNA samples taken from the blood, skin cells, or hair of alleged criminal suspects can be compared to DNA collected from the crime scene. DNA fingerprints (electrophoretic bands that are unique to each individual) can be compared and used to identify persons at that crime scene.

5. **Environmental cleanup:** Scientists engineer metabolic capabilities into microorganisms, which are then used to treat environmental problems, such as removing heavy metals from toxic mining sites.

6. **Agricultural applications:** Certain genes that produce desirable traits have been inserted into crop plants to increase their productivity or efficiency. An organism that has acquired by artificial means one or more genes from another species or variety is termed a **genetically modified organism** (GM organism). Currently, a debate is in progress over the safety of GM organisms.

Multiple-Choice Questions

1. Which of the following is NOT a potential control mechanism for regulation of gene expression in eukaryotic organisms?
   (A) The degradation of RNA
   (B) The transport of mRNA from the nucleus
   (C) The lactose operon
   (D) Transcription
   (E) Gene amplification

2. Which of the following exists as DNA surrounded by a protein coat?
   (A) a retrovirus
   (B) a virus
   (C) a eukaryote
   (D) a prokaryote
   (E) ampicillin

3. A goat can produce milk containing the same polymers present in the silk produced by spiders when particular genes from a spider are inserted into the goat's genome. Which of the following reasons describes why this is possible?
   (A) Goats and spiders share a common ancestor and, thus, produce similar protein excretions.
   (B) The opposite is true, too–when genes from a goat are inserted into a spider's genome, the spider produces goats' milk instead of silk.
   (C) The proteins in goats' milk and spiders' silk have the same amino acid sequence.
   (D) The processes of transcription and translation in the cells of spiders and goats are fundamentally similar.
   (E) The processes of transcription and translation in the cells of spiders and goats produce exactly the same proteins anyway.

4. Restriction enzymes are generally used in the laboratory for which of the following reasons?
   (A) Restricting the replication of DNA
   (B) Restricting the transcription of DNA
   (C) Restricting the translation of mRNA
   (D) Cutting DNA molecules at specific locations
   (E) Cutting DNA into manageable sizes for manipulation

Directions: The group of questions below consists of five lettered choices followed by a list of numbered phrases or sentences. For each numbered phrase or sentence, select the one choice that is most closely related to it. Each choice may be used once, more than once, or not at all.

Questions 5–9
   (A) Transcription
   (B) Translation
   (C) Transposon
   (D) DNA methylation
   (E) Histone acetylation

5. A mobile segment of DNA that travels from one location on a chromosome to another, one element of genetic change

6. The addition of groups to certain bases of DNA after DNA synthesis, this is thought to be an important control mechanism for gene expression

7. The synthesis of polypeptides from the genetic information coded in mRNA

8. The synthesis of RNA from a DNA template

9. The attachment of groups to particular amino acids of specific proteins, this is thought to be an important control mechanism for gene expression

10. The figure at the top of page 141 shows which of the following processes?
    (A) the lytic cycle of a phage
    (B) the lysogenic cycle of a phage
    (C) transcription
    (D) translation
    (E) DNA replication

11. The actions of which of the following enzymes are responsible for ensuring that chromosomes do not decrease in length with every round of replication?
    (A) telomerase
    (B) DNA ligase
    (C) DNA polymerase
    (D) helicase
    (E) primase

12. PCR (polymerase chain reaction) makes gene cloning possible because it enables lab technicians to do which of the following very quickly?
    (A) isolate gene-source DNA
    (B) insert DNA into an appropriate vector
    (C) introduce the cloning vector into a host cell
    (D) amplify DNA samples
    (E) identify clones carrying the gene of interest

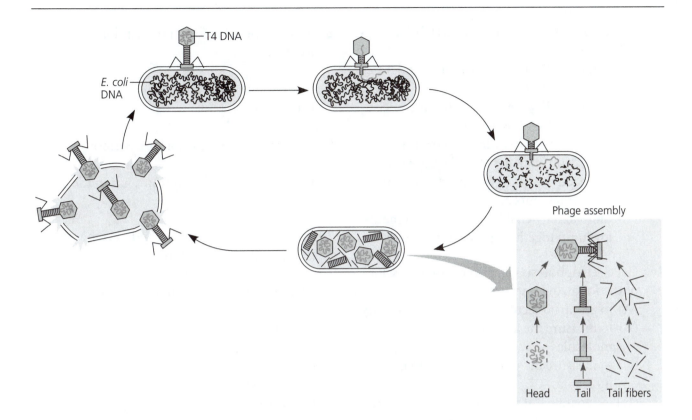

E. coli DNA

T4 DNA

Phage assembly

Head   Tail   Tail fibers

Questions 13–16 refer to an experiment that was performed to separate DNA fragments from three samples radioactively labeled with ^{32}P. The fragments were then separated using gel electrophoresis. The visualized bands are depicted below:

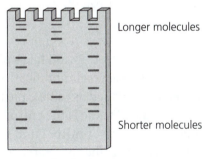

Longer molecules

Shorter molecules

Completed gel

13. When the electric field was applied, the fragments of DNA in each of the three samples migrated to different locations along the gel because
    (A) the fragments differed in their levels of radioactivity.
    (B) the fragments differed in their charges—some were positively charged, whereas others were negatively charged.
    (C) the fragments differed in size.
    (D) the fragments differed in polarity.
    (E) the fragments differed in solubility.

14. How many sites on DNA were cut by the particular restriction enzyme used in Sample 1 (the leftmost sample)?
    (A) 5
    (B) 6
    (C) 7
    (D) 8
    (E) 9

15. The DNA in this experiment was labeled with ^{32}P because
    (A) without ^{32}P, the DNA would not migrate through the gel.
    (B) without ^{32}P, we would not be able to visualize the DNA fragments.
    (C) ^{32}P is required in order to allow the restriction enzymes to make their cuts.
    (D) radioactivity limits the interference of scrap fragments of DNA.
    (E) radioactivity enables the DNA fragments to clump together and produce bands.

16. Gel electrophoresis can also be used for which of the following purposes?
    (A) to group molecules based on their polarity
    (B) to measure the acidity of certain large molecules
    (C) to measure the polarity of certain large molecules
    (D) to separate out the proteins in a mixture
    (E) to measure the amount of protein in a mixture of substances

17. In genetic engineering, DNA ligase is used for which of the following purposes?
    (A) to act as a probe for locating cloned genes
    (B) to create breaks in DNA in order to allow foreign DNA fragments to be inserted
    (C) to seal up nicks created in newly created recombinant DNA
    (D) to ensure that "sticky ends" of like DNA fragments do not re-anneal
    (E) in Southern blotting

Directions: The group of questions below consists of five lettered choices followed by a list of numbered phrases or sentences. For each numbered phrase or sentence, select the one choice that is most closely related to it. Each choice may be used once, more than once, or not at all.

Questions 18–22
    (A) tRNA
    (B) mRNA
    (C) poly-A tail
    (D) RNA polymerase
    (E) rRNA

18. An example of a post-transcriptional modification

19. Binds to the promoter on DNA to initiate transcription

20. Along with proteins, comprises ribosomes

21. Loosely binds to free amino acids in the cytoplasm

22. Travels out of the nucleus and into the cytoplasm to participate in translation

23. All of the following nitrogenous bases are included in DNA EXCEPT
    (A) adenine.
    (B) cytosine.
    (C) guanine.
    (D) thymine.
    (E) uracil.

24. The expression of genes can be controlled at all the following stages of protein synthesis EXCEPT
    (A) initiation of transcription.
    (B) RNA processing.
    (C) DNA unpacking.
    (D) degradation of protein.
    (E) protein folding.

25. After eukaryotic transcription takes place, mRNA undergoes several modifications before leaving the nucleus to take part in translation. One of these is the cutting out of nonessential sections of mRNA and the subsequent splicing together of stretches of mRNA necessary for the final functional molecule. Which of the following mRNA sections are spliced together into the finished mRNA molecule?
    (A) introns
    (B) exons
    (C) genes
    (D) coding sequences
    (E) ribozymes

26. In the process of eukaryotic translation, the term *wobble* refers to
    (A) the tendency of the two ribosome subunits to come closer to one another and to separate at different points in translation.
    (B) the tendency of the amino acid loosely attached to the tRNA to move back and forth before finally attaching to the polypeptide chain.
    (C) the fact that the genetic code is redundant.
    (D) the fact that the anticodon and codon bind very loosely.
    (E) the fact that the third nucleotide of a tRNA can form hydrogen bonds with more than one kind of base in the third position of a codon.

27. Which of the following is an example of a missense mutation?
    (A) A nucleotide and its partner are replaced with an "incorrect" pair of nucleotides, which destroys the function of the final protein.
    (B) A nucleotide pair is added into a gene, destroying the reading frame of the genetic message.
    (C) A nucleotide pair is lost from the gene, destroying the reading frame of the genetic message.
    (D) A frameshift mutation occurs, ultimately causing the production of nonfunctional proteins.
    (E) A nucleotide pair substitution occurs, which causes the codon to code for an amino acid that may not be the "correct" one, although translation continues.

28. Which of the following is an example of a nonsense mutation?
    (A) A chemical change occurs in just one base pair of a gene, and it has no effect on the final protein.
    (B) Part of the gene breaks off and travels to a distant location on the chromosome, inserting itself there.
    (C) A substitution occurs, which changes a regular amino acid codon into a stop codon, causing translation to cease.
    (D) A substitution occurs, which changes a regular amino acid codon into a start codon, and translation begins again, creating two unfinished polypeptides.
    (E) A nucleotide pair substitution occurs, which causes the codon to code for an amino acid that may not be the "correct" one, although translation continues.

29. At the end of DNA replication, each of the daughter molecules has one old strand, derived from the parent strand of DNA, and one strand that is newly synthesized. This explains why DNA replication is described as
    (A) conservative.
    (B) largely conservative.
    (C) nonconservative.
    (D) semiconservative.
    (E) unconservative.

30. The segment of DNA shown below has restriction sites I and II, which create restriction fragments a, b, and c. Which of the following gels produced by electrophoresis would represent the separation and identity of these fragments?

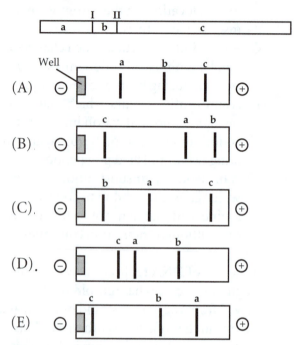

31. Which of the following is a difficulty in getting prokaryotic cells to express eukaryotic genes?
    (A) The signals that control gene expression are different and prokaryotic promoter regions must be added to the vector.
    (B) The genetic code differs because prokaryotes substitute the base uracil for thymine.
    (C) Prokaryotic cells cannot transcribe introns because their genes do not have them.
    (D) The ribosomes of prokaryotes are not large enough to handle long eukaryotic genes.
    (E) The RNA splicing enzymes of bacteria work differently from those of eukaryotes.

32. Complementary DNA does not create as complete a library of genes as the shotgun approach because
    (A) it has eliminated introns from the genes.
    (B) a cell produces mRNA for only a small portion of its genes.
    (C) the shotgun approach produces more restriction fragments.
    (D) cDNA is not as easily integrated into plasmids.
    (E) restriction enzymes are not used to create cDNA.

33. You have affixed the chromosomes from a cell onto a microscope slide. Which of the following would *not* make a good radioactively labeled probe to help map a particular gene to one of those chromosomes? (Assume DNA of chromosomes and probes is single stranded.)
    (A) cDNA made from the mRNA transcribed from the gene
    (B) a portion of the amino acid sequence of that protein
    (C) mRNA transcribed from the gene
    (D) a piece of the restriction fragment on which the gene is located
    (E) a sequence of nucleotide bases determined from the genetic code needed to produce a known sequence of amino acids found in the protein product of the gene.

34. The human genome appears to have only one-third more genes than the simple nematode, *C. elegans*. Which of the following best explains how the more complex humans can have relatively few genes?
    (A) The unusually long introns in human genes are involved in regulation of gene expression.
    (B) More than one polypeptide can be produced from a gene by alternative splicing.
    (C) Human genes code for many more types of domains.
    (D) The human genome has a high proportion of noncoding DNA.
    (E) The large number of SNPs (single nucleotide polymorphisms) in the human genome provides for a great deal of genetic variability.

Free-Response Question

1. *Genes are located on chromosomes and are the basic unit of heredity that is passed on from parent to child, through generations.*

    (a) **Explain** how a chromosome mutation could occur and why mutations are detrimental to the organism in which they take place.
    (b) **Explain** why it is that—although there are very few genes located on the Y chromosome—human males may suffer from having just one copy of the X chromosome, whereas females have two.

ANSWERS AND EXPLANATIONS

Multiple-Choice Questions

▌ 1. **(C) is correct.** Operons are gene expression mechanisms of bacteria. They are not found in eukaryotic cells.

▌ 2. **(B) is correct.** Viruses are made up of nucleic acid surrounded by a protein coat. The answer cannot be *A* because retroviruses consist of RNA. They reproduce by injecting their genetic material into a host cell and using the cell's replicative machinery to replicate their DNA and proteins. The new viruses leave the cell to infect more cells, sometimes killing the host cell in the process.

▌ 3. **(D) is correct.** The process of genetic engineering is possible because the processes of transcription and translation are so similar in all eukaryotic cells. Once the spider gene or genes that were responsible for coding for the silk proteins were isolated and then inserted into a bacterial plasmid (which would serve as the vector), the cloning vector would be taken up by the goat's cells, and the goat's cells' transcription/translation machinery would begin the process of producing the spider protein, along with its own proteins.

▌ 4. **(D) is correct.** Restriction enzymes can be used to cut DNA at specific locations, and this enables researchers to perform recombinant DNA techniques. When specific restriction enzymes are added to the DNA, they produce cuts in the sugar-phosphate backbone and create "sticky ends," which can bind to DNA fragments from a different source to produce recombinant DNA. DNA ligase is then added to seal the strands together permanently.

5. (C) is correct. Transposons are also called transposable genetic elements, and they are pieces of DNA that can move from location to location in a chromosome—or a genome. Transposons are also called "jumping genes," and most of them are capable of moving to many different target sites in the genome.

6. (D) is correct. One of the two important ways that the cell has of controlling gene expression is through DNA methylation. In DNA methylation, methyl groups are attached to certain DNA bases after DNA is synthesized. This appears to be responsible for the long-term inactivation of genes.

7. (B) is correct. The process by which genetic information flows from mRNA to protein is called translation. Translation occurs in the cytoplasm of the cell, at ribosomes. A molecule of mRNA is moved through the ribosome, and codons are translated into amino acids one at a time. Transfer RNAs add their associated amino acids onto a growing polypeptide as its anticodon pairs with a codon on the mRNA and then departs from the ribosome to bind more free amino acids.

8. (A) is correct. In transcription, RNA is synthesized using the genetic information encoded by DNA. Transcription occurs in the nucleus of the cell. The double-stranded DNA helix unwinds, allowing enzymes and proteins to synthesize a new complementary single-stranded mRNA molecule from the template strand of DNA.

9. (E) is correct. In histone acetylation, acetyl groups are attached to certain amino acids of histones. Deacetylation is the process by which they are removed. Acetylation makes the histones change shape so they are less tightly bound to DNA, and this allows the proteins involved in transcription to move in and begin the process. Therefore, acetylation is one way for the cell to initiate transcription and to control the expression of its genes.

10. (A) is correct. This art portrays the lytic cycle of phage reproduction. In the lytic cycle, the phage first attaches to the cell surface and injects its DNA into the cell. It then hydrolyzes the host cell's DNA and uses the cell's machinery to produce phage proteins and to replicate its genome. The phage proteins are then assembled in the cell until the host cell lyses (breaks open), and the new phages are released to infect other cells. In the lysogenic cycle, the phage genome becomes incorporated into the host cell's DNA without destroying the host cell.

11. (A) is correct. The enzyme responsible for adding nucleotides to the replicating DNA strand, DNA polymerase can add nucleotides only to the 3′ end of a molecule. This means that it would have no way to complete the 5′ end of the molecule; thus, the linear chromosomes of eukaryotes use an enzyme called telomerase, which catalyzes the ends of the molecules (called telomeres).

12. (D) is correct. PCR, the polymerase chain reaction, is a technique by which any piece of DNA can be copied many times without the use of cells. The DNA is heated to separate its strands and then cooled to allow primers to attach to the single strands. DNA polymerase is added, which begins to add nucleotides to the 3′ end of each primer on the two strands. With each turn of the cycle, the amount of DNA is multiplied by two.

■ **13. (C) is correct.** The fragments of DNA separated out from one another along the gel once the electric field was applied because they differ in size. For DNA in gel electrophoresis, how far a molecule travels through a gel (while the current is applied) is inversely proportional to its size. The larger a fragment is, the more slowly it will migrate.

■ **14. (D) is correct.** The restriction enzyme used to cut the DNA that was placed into the first well of the gel must have cut the DNA at eight sites, because it produced nine DNA fragments. The number of fragments produced is always one more than the number of restriction sites cut.

■ **15. (B) is correct.** Radioactivity is conferred to the DNA fragments so we can visualize them once they have ceased migrating through the gel. This can be done by applying a piece of film to the gel—the radioactivity exposes the film to form an image that corresponds to the bands of DNA shown.

■ **16. (D) is correct.** Gel electrophoresis is a method used to separate macromolecules (DNA, protein, most types of macromolecules) based on their rate of movement through a gel once an electric field has been applied. The rate of their movement will be inversely proportional to their size.

■ **17. (C) is correct.** In genetic engineering (the manipulation of genes for practical purposes), DNA ligase is an enzyme that is used to seal the strands of newly recombinant DNA (DNA that is spliced together from two different sources) by catalyzing the formation of phosphodiester bonds.

■ **18. (C) is correct.** The addition of a poly-A tail after transcription is one example of post-transcriptional modifications that the mRNA undergoes. This poly-A tail inhibits the degradation of the newly synthesized mRNA strand and is thought also to help ribosomes attach to it. Another important modification that mRNA undergoes is the addition of a 5′ cap. The 5′ cap helps protect mRNA from degradation and also acts as the point of attachment for the ribosomes, just prior to translation.

■ **19. (D) is correct.** RNA polymerase is the most prominent enzyme involved in the transcription of DNA to make mRNA. It is responsible for binding to the promoter sequence on the template DNA, prying the two DNA strands apart, and hooking the RNA nucleotides together as they base-pair along the DNA template. RNA polymerases add nucleotides to the 3′ end of the growing chain until a terminator sequence is reached—it transcribes entire transcription units.

■ **20. (E) is correct.** Ribosomal RNA (rRNA), together with proteins, makes up ribosomes. Ribosomes, the sites of protein synthesis, are composed of two subunits, the large and the small subunit. The large subunit of the ribosome contains the A, P, and E sites, which shuttle through the tRNA and mRNA during translation.

■ **21. (A) is correct.** tRNA, or transfer RNA, interprets the genetic message coded in mRNA. It transfers amino acids taken from the cytoplasmic pool to a ribosome, which adds the specific amino acid brought to it by tRNA to the end of a growing polypeptide chain. Each type of tRNA binds loosely to a specific amino acid at one end; its other end contains an anticodon, which base-pairs with a complementary codon on the mRNA strand.

22. (B) is correct. mRNA, also known as messenger RNA, is a type of RNA that is synthesized from DNA and attaches to ribosomes in the cytoplasm to specify the primary structure of a protein. Since mRNA is the product of transcription, which occurs in the nucleus, it must travel out of the nucleus and into the cytoplasm in order to participate in translation.

23. (E) is correct. The base uracil is found in RNA but not in DNA. The bases in DNA are cytosine, guanine, thymine, and adenine, whereas the bases found in RNA are cytosine, guanine, adenine, and uracil. In DNA, cytosine is capable of forming three hydrogen bonds with guanine, and thymine and adenine form two hydrogen bonds between them—the bases form the "rungs" of the double helix ladder, and sugar-phosphate groups form the rails of the ladder.

24. (E) is correct. Protein folding is the mechanism by which the polypeptide assumes its functional conformation. It is the only answer listed that does not describe a stage in the pathway from gene to protein that is involved in controlling gene expression. At almost all of the stages in this pathway, the cell has some mechanism for controlling the expression of its genes or the amount of gene product produced.

25. (B) is correct. In the modification of mRNA that occurs after transcription, a process called RNA splicing occurs. In this process, noncoding regions of nucleic acid that are situated between coding regions are cut out. These noncoding regions are called introns. The remaining regions are called exons, and these are spliced together to form the final mRNA product. When you think of exons, think expressed—because they are actually translated into proteins, whereas introns are not.

26. (E) is correct. In eukaryotic translation, the term *wobble* refers to the fact that more than one tRNA exists for every mRNA codon that specifies for an amino acid. Some tRNAs have codons that can recognize two or more different codons because of wobble—wherein the base-pairing rules are relaxed, and the third nucleotide of a tRNA can form hydrogen bonds with more than one kind of base in the third position of a codon.

27. (E) is correct. A missense mutation is a base-pair substitution (the replacement of a nucleotide and its partner in the cDNA strand with a different pair of nucleotides) that still enables the codon to code for an amino acid. The amino acid may or may not ultimately contribute to a functional protein, but a missense mutation is one where an amino acid is still chosen to be added to the polypeptide chain, and translation will continue.

28. (C) is correct. In a nonsense mutation, a base-pair substitution takes place—one base-pair is replaced by another—and the point mutation changes the codon for a regular amino acid into a stop codon. This makes translation end prematurely, and this results in a shortened and usually nonfunctional protein.

29. (D) is correct. DNA replication is semiconservative. Each new daughter molecule contains one newly synthesized strand, and one strand that used to belong to the parent double helix DNA.

30. (B) is correct. In gel electrophoresis the smallest fragments travel furthest; the largest fragments are closest to the well. Fragment b is shortest, followed by a and c. We would expect a gel with c closest to the well, then a, with b at the far end.

31. (A) is correct. Several of the other possible answers have errors that would be instructive to note. Answer *B* is false because the genetic code is near universal; answer *C* is false because introns are not part of the inserted gene, as they have been removed; answer *D* is false because ribosomes are not governed by the length of the gene; answer *E* is false because prokaryotes do not have splicing enzymes to take out introns. Answer *A* is correct because the signals that control gene expressions and promoter regions are different between prokaryotes and eukaryotes.

32. (B) is correct. At any given time only a part of a cell's genes are on, with only those mRNA transcripts in the cytoplasm. Reverse transcriptase is used to copy the mRNA to DNA, yielding cDNA for insertion into the vector. Only the genes on at the time the cell is sampled will be present in this library.

33. (B) is correct. In order to be a probe, it would have to contain complementary nucleotides, not amino acids. It is surprising the number of students who confuse the matching of nucleotides with nucleic acids (DNA or RNA) and amino acids with proteins.

34. (B) is correct. The number of genes in the human genome has been revised sharply downward since the Human Genome Project. Less than a decade ago the number of human genes was at 100,000; now it is 20,000! How can only 20,000 genes be enough for a human? The answer is alternative gene splicing yielding multiple proteins per gene.

Free-Response Question

(a) The reason that it is detrimental to an organism to have an abnormal chromosome number is that genes, which are located on chromosomes, code for proteins, which have specific functions in the cell. If an organism has two copies of a particular gene, then this gene will be transcribed twice, creating twice the usual gene product. This alters the relative amounts, or doses, of interacting products in the cell, and this can cause serious developmental problems. Likewise, if a gene is missing from a chromosome, it will not be transcribed, and its corresponding protein will not be produced. If that protein has an important cellular function, the organism will be seriously affected.

There are many ways by which chromosomes can be altered to cause problems for the cell. Among these is nondisjunction—when during mitosis or meiosis the chromosomes fail to separate properly, and one cell ends up with two copies of a chromosome while the other gets no copies. This results in a condition called aneuploidy. Smaller chromosomal mutations are deletions (in which part of the chromosome breaks off and is lost), inversions (in which a chromosome segment is reversed within a chromosome), duplications (in which a chromosome segment is repeated in a chromosome), and translocation (in which part of a chromosome is moved from one chromosome to another).

(b) When fertilization occurs and a sperm carrying a Y chromosome penetrates the egg first, a male zygote with one X and one Y chromosome is produced. If a sperm carrying an X chromosome penetrates the egg first, a female zygote with XX is produced. Although it seems as though the female zygote would have the advantage of having twice the cell product as the male, due to its double dose of genes located on the large X versus the small Y, this is not the case. The reason for this is that, in every cell of the female human body, one of the X chromosomes is inactivated. The mechanisms for this are not understood, but the X chromosome that is inactivated condenses into a structure called a Barr body, which then associates with the nuclear envelope. As a Barr body, most of the X chromosome's genes are not expressed—although some of them do remain active. As a result of this, females are a mosaic consisting of cells with the X chromosome from their mother activated and cells with the X chromosome from their father activated in about a 50–50 ratio. This is also the reason sex-linked disorders are usually not expressed in females. Though one of the X chromosomes may be incapable of producing a crucial gene product, this mosaic effect ensures that the other half of the somatic cells produce sufficient amounts of the protein in question.

This response demonstrates knowledge of the following terms and processes:

chromosome	*inversion*
gene	*duplication*
gene product	*translocation*
doses	*X and Y chromosomes*
transcription	*zygote*
nondisjunction	*X chromosome inactivation*
aneuploidy	*Barr body*
deletion	*somatic cell*

Moreover, the response describes the following processes in a clear, concise, and organized way:

> *—the function of genes and why their loss or duplication affects the cell*
> *—the many types of chromosome mutations that can occur*
> *—the difference between X and Y chromosomes*
> *—the process and result of X chromosome inactivation*

Mechanisms of Evolution

Chapter 22: Descent with Modification: A Darwinian View of Life

---

YOU MUST KNOW

- How Lamarck's view of the mechanism of evolution differed from Darwin's.
- Several examples of evidence for evolution.
- The difference between structures that are homologous and those that are analogous, and how this relates to evolution.
- The role of adaptations, variation, time, reproductive success, and heritability in evolution.

Concept 22.1 **The Darwinian revolution challenged traditional views of a young Earth inhabited by unchanging species**

▌ Historical Setting

- **Scala naturae** (Aristotle, 384–322 B.C.): Life-forms could be arranged on a scale of increasing complexity.
- **Old Testament:** Perfect species were individually designed by God.
- **Carolus Linnaeus** (1707–1778): Grouped similar species into increasingly general categories reflecting what he considered the pattern of their creation.
    - Developed **taxonomy,** the branch of biology dedicated to the naming and classification of all forms of life.
    - Developed **binomial nomenclature,** a two-part naming system that includes the organism's genus and species.
- **Georges Cuvier** (1769–1832): French geologist opposed to idea of evolution
    - Advocated **catastrophism,** the principle that events in the past occurred suddenly and by different mechanisms than those occurring today.
    - This explained boundaries between strata and location of different species.

- **Charles Lyell** (1797–1875): English geologist
  - Developed principle of **uniformitarianism,** the idea that the geologic processes that have shaped the planet have not changed over the course of Earth's history.
  - *Importance: **The earth must be very old**.*
  - Lyell's *Principles of Geology* was studied by Darwin during his journeys.
- **Jean-Baptiste de Lamarck** (1744–1829): Developed an early theory of evolution based on two principles:
  - **Use and disuse** is the idea that parts of the body that are used extensively become larger and stronger, while those that are not used deteriorate.
  - **Inheritance of acquired characteristics** assumes that characteristics acquired during an organism's lifetime could be passed on to the next generation. *Example*: A weightlifter's child could be born with a more muscular anatomy.
  - *Importance:* Lamarck recognized that species evolve, though his explanatory mechanism was flawed.

Concept 22.2 Descent with modification by natural selection explains the adaptations of organisms and the unity and diversity of life

- **Charles Darwin's** voyage on the HMS *Beagle* from 1831 to 1836 was the impetus for the development of his theory of evolution by natural selection.
- Darwin's mechanism for evolution was natural selection. Recall that Larmarck's mechanism was the inheritance of acquired characteristics.
- **Natural selection** explains how adaptations arise.

  - **Adaptations:** Characteristics that enhance organisms' ability to survive and reproduce in specific environments. *Example*: Desert foxes have large ears which radiate heat. Arctic foxes have small ears which conserve body heat.
  - Natural selection is a process in which individuals that have certain *heritable characteristics* survive and reproduce at a higher rate than other individuals.
  - Over time, natural selection can increase the match between organisms and their environment.
  - If an environment changes, or if individuals move to a new environment, natural selection may result in adaptation to these new conditions, sometimes giving rise to new species in the process.

- **Artificial selection** is the process by which species are modified by humans. *Example*: Selective breeding for milk or meat production; development of dog breeds.
- *Individuals do not evolve. **Populations** evolve.*

Concept 22.3 Evolution is supported by an overwhelming amount of scientific evidence

ORGANIZE YOUR THOUGHTS

EVIDENCE FOR EVOLUTION is seen in the following ways:

1. Direct observations
2. The fossil record
3. Homology
4. Biogeography

▌ Evidence for Evolution

1. **Direct Observations of Evolutionary Change**
   1. Intense predation of wild guppies results in more drably colored males.
   2. Evolution of drug-resistant viruses and antibiotic-resistant bacteria
2. **The Fossil Record:** *Fossils provide evidence for the theory of evolution.*
   - Fossils are remains or traces of organisms from the past. They are found in sedimentary rock.
   - **Paleontology** is the study of fossils.
   - Fossils show evolutionary changes have occurred over time and the origin of major new groups of organisms.
3. **Homology and Convergent Evolution**
   - **Homology:** Characteristics in related species can have an underlying similarity even though they have very different functions.
   - **Homologous structures** are anatomical signs of evolution. *Examples:* Forelimbs of mammals that are now used for a variety of purposes, such as flying in bats or swimming in whales, but were present and used in a common ancestor for walking.
   - **Embryonic homologies:** Comparison of early stages of animal development reveals many anatomical homologies in embryos that are not visible in adult organisms. *Examples:* All vertebrate embryos have a post-anal tail and pharyngeal pouches.
   - **Vestigial organs** are structures of marginal, if any, importance to the organism. They are remnants of structures that served important functions in the organisms' ancestors. *Example:* Remnants of the pelvis and leg bones are found in some snakes.
   - **Molecular homologies** are shared characteristics on the molecular level. *Examples:* All life forms use the same genetic language of DNA and RNA. Amino acid sequences coding for hemoglobin in primate species shows great similarity, thus indicating a common ancestor.
   - **Convergent evolution** explains why distantly related species can resemble one another. Convergent evolution has taken place when two organisms developed similarities as they adapted to similar environmental challenges—not because they evolved from a common ancestor. The likenesses that result from convergent evolution are considered **analogous** rather than homologous.

- **Similar problem, similar solution** *Examples*:
  1. The torpedo-shapes of a penguin, dolphin, and shark are the solution to movement through an aqueous environment.
  2. Sugar-gliders (marsupial mammals) and flying squirrels (eutherian mammals) occupy similar niches in their respective habitats.

STUDY TIP: Homologous structures show evidence of relatedness (whale fin/bat wing).

Analogous structures are similar solutions to similar problems but do *not* indicate close relatedness (bird wing/bat wing).

---

4. **Biogeography:** The geographic distribution of species.
   - Species in a discrete geographic area tend to be more closely related to each other than to species in distant geographic areas. *Example:* In South America, desert animals are more closely related to local animals in other habitats than they are to the desert animals of Asia. This reflects evolution, not creation.
   - **Continental drift** and the break-up of *Pangaea* can explain the similarity of species on continents that are distant today.
   - **Endemic species** are found at that certain geographic location and nowhere else.
     - *Example:* Marine iguanas are endemic to the Galapagos.
   - Darwin's theory of evolution through natural selection explains the succession of forms in the fossil record. Transitional fossils have been found that link ancient organisms to modern species, just as Darwin's theory predicts.

ORGANIZE YOUR THOUGHTS

1. Evolution is change in species over time.
2. Heritable variations exist within a population.
3. These variations can result in differential reproductive success.
4. Over generations, this can result in changes in the genetic composition of the population.

And remember . . . Individuals do not evolve! **Populations** *evolve.*

Chapter 23: The Evolution of Populations

YOU MUST KNOW

- How mutation and sexual reproduction each produce genetic variation.
- The conditions for Hardy-Weinberg Equilibrium.
- How to use the Hardy-Weinberg equation to calculate allelic frequencies and to test whether a population is evolving.

Concept 23.1 Mutation and sexual reproduction produce the genetic variation that makes evolution possible

- **Microevolution** is change in the allele frequencies of a population over generations. This is evolution on its smallest scale.
- **Mutations** are the only source of *new* genes and *new* alleles.

  - Only mutations in cell lines that produce gametes can be passed to offspring.

- **Point mutations** are changes in one base in a gene. They can have significant impact on phenotype, as in sickle-cell disease.
- **Chromosomal mutations** delete, disrupt, duplicate, or rearrange many loci at once. They are almost certain to be harmful, but not always.
- However, **most of the genetic variations** within a population are due to the sexual recombination of alleles that already exist in a population.
- Sexual reproduction rearranges alleles into new combinations in every generation. Recall there are *three mechanisms* for this shuffling of alleles:

  1. **Crossing over** during Prophase I of meiosis.
  2. **Independent assortment** of chromosomes during meiosis (2^{23} different combinations possible in the formation of human gametes!)
  3. **Fertilization** ($2^{23} \times 2^{23}$ different possible combinations for human sperm and egg)

Concept 23.2 The Hardy-Weinberg equation can be used to test whether a population is evolving

- **Population genetics** is the study of how populations change genetically over time.
- **Population:** A group of individuals of the same species that live in the same area and interbreed, producing fertile offspring.
- **Gene pool:** All of the alleles at all loci in all the members of a population.
- In diploid species, each individual has two alleles for a particular gene, and the individual may be either heterozygous or homozygous.
- If all members of a population are homozygous for the same allele, the allele is said to be **fixed.** Only one allele exists at that particular locus in the population.

> **ORGANIZE YOUR THOUGHTS**
>
> Five Conditions for Hardy-Weinberg Equilibrium
>
> 1. No mutations.
> 2. Random mating.
> 3. No natural selection.
> 4. The population size must be extremely large. (No genetic drift.)
> 5. No gene flow. (Emigration, immigration, transfer of pollen, etc.)

- The greater the number of fixed alleles, the lower the species' diversity.
- The **Hardy-Weinberg theorem** is used to describe a population that is *not* evolving. It states that the frequencies of alleles and genes in a population's gene pool will remain constant over the course of generations unless they are acted upon by forces *other* than Mendelian segregation and the recombination of alleles. The population is at **Hardy-Weinberg equilibrium.**
- However, it is unlikely that all the conditions for Hardy-Weinberg equilibrium will be met. Allelic frequencies change. Populations evolve. This can be tested by applying the **Hardy-Weinberg equation.**

> **TRY THIS FOR PRACTICE!**
>
> Suppose in a plant population that red flowers (R) are dominant to white flowers (r). In a population of 500 individuals, 25% show the recessive phenotype. How many individuals would you expect to be homozygous dominant and heterozygous for this trait? (**Answer below in bold. Cover this solution and try it first!**)
>
> 1. Let p = frequency of the dominant allele (R) and q = frequency of the recessive allele (r)
> 2. q^2 = frequency of the homozygous recessive =25% = .25. Since q^2 = .25, q = .5
> 3. Now, p + q =1, so p = .5
> 4. Homozygous dominant individuals are RR or p^2 = .25, and will represent (.25)(500) = **125 individuals.**
> 5. The heterozygous individuals are calculated from 2 pq = (2) (.5) (.5) = .5, and in a population of 500 individuals will be (.5)(500) = **250 individuals.**

Concept 23.3 Natural selection, genetic drift, and gene flow can alter allele frequencies in a population

- **Mutations can alter gene frequency but are rare.**
- **The three major factors** that alter allelic frequencies and bring about most evolutionary change are *natural selection, genetic drift,* and *gene flow.*

  1. **Natural selection** results in alleles being passed to the next generation in proportions different from their relative frequencies in the present generation. Individuals with variations that are better suited to their

environment tend to produce more offspring than those with variations that are less suited.

2. **Genetic drift** is the unpredictable fluctuation in allelic frequencies from one generation to the next. The smaller the population, the greater the chance is for genetic drift. This is a *random, nonadaptive* change in allelic frequencies. Two examples follow.

   A. **Founder effect:** A few individuals become isolated from a larger population and establish a new population whose gene pool is not reflective of the source population.

   B. **Bottleneck effect:** A sudden change in the environment (for example, an earthquake, flood, or fire) drastically reduces the size of a population. The few survivors that pass through the restrictive bottleneck may have a gene pool that no longer reflects the original population's gene pool. *Example:* The population of California condors was reduced to nine individuals. This represents a bottlenecking event.

3. **Gene flow** occurs when a population gains or loses alleles by genetic additions to and/or subtractions from the population. This results from the movement of fertile individuals or gametes. Gene flow tends to reduce the genetic differences between populations, thus making populations more similar.

Concept 23.4 Natural selection is the only mechanism that consistently causes adaptive evolution

▌ **Relative fitness** refers to the contribution an organism makes to the gene pool of the next generation relative to the contributions of other members. Fitness does *not* indicate strength or size. It is measured only by reproductive success.

▌ Natural selection acts more directly on the phenotype and indirectly on the genotype and can alter the frequency distribution of heritable traits in three ways (see Figure 6.1).

1. **Directional Selection:** Individuals with one extreme of a phenotypic range are favored, shifting the curve toward this extreme. *Example:* Large black bears survived periods of extreme cold better than smaller ones, and so became more common during glacial periods.

2. **Disruptive selection** occurs when conditions favor individuals on both extremes of a phenotypic range rather than individuals with intermediate phenotypes. *Example:* A population has individuals with either large beaks or small beaks, but few with the intermediate beak size. Apparently the intermediate beak size is not efficient in cracking either the large or small seeds that are common.

3. **Stabilizing selection** acts against both extreme phenotypes and favors intermediate variants. *Example:* Birth weights of most humans lie in a narrow range, as those babies who are very large or very small have higher mortality.

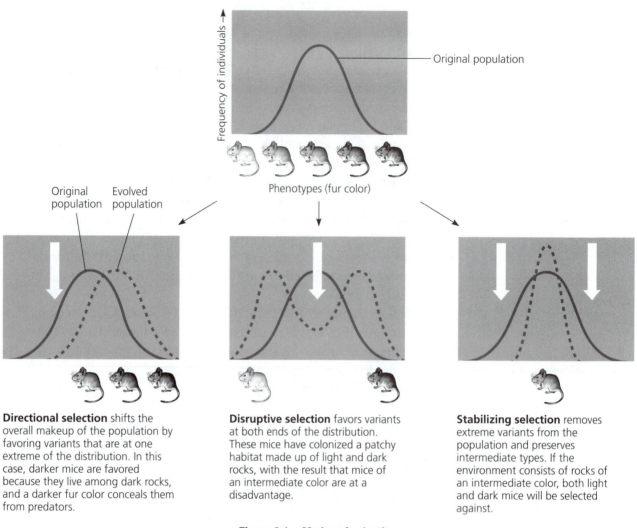

Directional selection shifts the overall makeup of the population by favoring variants that are at one extreme of the distribution. In this case, darker mice are favored because they live among dark rocks, and a darker fur color conceals them from predators.

Disruptive selection favors variants at both ends of the distribution. These mice have colonized a patchy habitat made up of light and dark rocks, with the result that mice of an intermediate color are at a disadvantage.

Stabilizing selection removes extreme variants from the population and preserves intermediate types. If the environment consists of rocks of an intermediate color, both light and dark mice will be selected against.

Figure 6.1   Modes of selection

How is genetic variation preserved in a population?

- **Diploidy:** Because most eukaryotes are diploid, they are capable of hiding genetic variation (recessive alleles) from selection.
- **Heterozygote advantage:** Individuals who are heterozygous at a certain locus have an advantage for survival. *Example:* In sickle-cell disease, individuals homozygous for normal hemoglobin are more susceptible to malaria, whereas homozygous recessive individuals suffer from the complications of sickle-cell disease. Heterozygotes benefit from protection from malaria and do not have sickle-cell disease, so the mutant allele remains relatively common.

Why natural selection cannot produce perfect organisms:

1. Selection can only edit existing variations.
2. Evolution is limited by historical constraints.
3. Adaptations are often compromises.
4. Chance, natural selection, and the environment interact.

Chapter 24: The Origin of Species

YOU MUST KNOW

- The difference between microevolution and macroevolution.
- The biological concept of species.
- Prezygotic and postzygotic barriers that maintain reproductive isolation in natural populations.
- How allopatric and sympatric speciation are similar and different.
- How an autopolyploid or an allopolyploid chromosomal change can lead to sympatric speciation.
- How punctuated equilibrium and gradualism describe two different tempos of speciation.

Concept 24.1 *The biological species concept emphasizes reproductive isolation*

- **Speciation** is the process by which new species arise.
- **Microevolution** is change in the genetic makeup of a population from generation to generation. It refers to adaptations that are confined to a single gene pool.
- **Macroevolution** refers to evolutionary change above the species level, such as the appearance of feathers and other such novelties, used to define higher taxa.
- The **biological species concept** defines a species as a population or group of populations whose members have the potential to interbreed in nature and produce viable, fertile offspring but are unable to produce viable, fertile offspring with members of other populations.
- **Reproductive isolation** is defined as the existence of biological barriers that impede members of two species from producing viable, fertile hybrids. A population may gain or lose alleles by **gene flow,** genetic additions to and subtractions from a population resulting from the movement of fertile individuals or gametes.
- **Prezygotic and postzygotic** are two types of barriers that prevent members of different species from producing offspring that can also successfully reproduce. Examples of prezygotic barriers, those that prevent mating or hinder fertilization, include the following:

    1. **Habitat isolation:** Two species can live in the same geographic area but not in the same habitat; this will prevent them from mating.
    2. **Behavioral isolation:** Some species use certain signals or types of behavior to attract mates, and these signals are unique to their species. Members of other species do not respond to the signals, thus mating does not occur.
    3. **Temporal isolation:** Species may breed at different times of day, different seasons, or different years, and this can prevent them from mating.
    4. **Mechanical isolation:** Species may be anatomically incompatible.
    5. **Gametic isolation:** Even if the gametes of two species do meet, they might be unable to fuse to form a zygote.

Examples of postzygotic barriers, those that prevent a fertilized egg from developing into a fertile adult, include the following:

1. **Reduced hybrid viability:** When a zygote *is* formed, genetic incompatibility may cause development to cease.
2. **Reduced hybrid fertility:** Even if the two species produce a viable offspring, reproductive isolation is still occurring if the offspring is sterile and can't reproduce.
3. **Hybrid breakdown:** Sometimes two species mate and produce viable, fertile hybrids; however, when the hybrids mate, their offspring are weak or sterile.

Concept 24.2 *Speciation can take place with or without geographic separation*

▌ The two main types of speciation are **allopatric speciation,** in which a population forms a new species because it is geographically isolated from the parent population, and **sympatric speciation,** in which a small part of a population becomes a new population without being geographically separated from the parent population.

▌ Some **geologic events or processes** that can fragment a population include the emergence of a mountain range, the formation of a land bridge, or evaporation in a large lake that produces several small lakes.

▌ Small, newly isolated populations undergo allopatric speciation more frequently because they are more likely to have their gene pools significantly altered. Allopatric speciation is confirmed when individuals from the new population are unable to mate successfully with individuals from the parent population.

▌ One mechanism that can lead to **sympatric speciation** in plants is the formation of **autopolyploid** plants through nondisjunction in meiosis. For example, these plants may have a $4n$ chromosome number instead of the normal $2n$ number. They cannot breed with diploid members and produce fertile offspring.

▌ **Polyploid speciation** occurs in animals, but it is not common. Instead, in animals, sympatric speciation can result from part of the population switching to a new habitat, food source, or other resource.

▌ **Adaptive radiation** occurs when many new species arise from a single common ancestor. Adaptive radiation typically occurs when a few organisms make their way to new, distant areas or when environmental changes cause numerous extinctions, opening up ecological niches for the survivors.

Concept 24.3 *Speciation can occur rapidly or slowly, and it can result from changes in few or many genes*

▌ **Gradualism** proposes that species descended from a common ancestor and gradually diverge more and more in morphology as they acquire unique adaptations.

▌ **Punctuated equilibrium** is a term used to describe periods of apparent stasis punctuated by sudden change observed in the fossil record.

> *A Tip from the Readers:*
> Remember this: Individuals do not evolve! They do not "struggle to survive." They cannot change their genetic makeup in response to a catastrophe. The individual lives or dies. Those that live reproduce and pass on adaptive heritable variations. INDIVIDUALS DO NOT EVOLVE! ONLY POPULATIONS CAN EVOLVE!

Chapter 25: The History of Life on Earth

YOU MUST KNOW

- The age of the Earth and when prokaryotic and eukaryotic life emerged.
- Characteristics of the early planet and its atmosphere.
- How Miller and Urey tested the Oparin-Haldane hypothesis and what they learned.
- Methods used to date fossils and rocks.
- Evidence for endosymbiosis.
- How continental drift can explain the current distribution of species.

Concept 25.1 Conditions on early Earth made the origin of life possible

▌ Current theory about how life arose consists of four main stages:

   1. Small organic molecules were synthesized.
   2. These small molecules joined into macromolecules, such as proteins and nucleic acids.
   3. All these molecules were packaged into **protobionts,** membrane-containing droplets, whose internal chemistry differed from that of the external environment.
   4. Self-replicating molecules emerged that made inheritance possible.

▌ Earth was formed about **4.6 billion years ago,** and life on Earth emerged about **3.8 to 3.9 billion years ago.** For the first three-quarters of Earth's history, all of its living organisms were microscopic and primarily unicellular.

▌ Hypothetical early conditions of Earth have been simulated in laboratories, and organic molecules have been produced.

   ■ **Oparin** and **Haldane** hypothesized that the early atmosphere, thick with water vapor, nitrogen, carbon dioxide methane, ammonia, hydrogen, and hydrogen sulfide, provided with energy from lightning and UV radiation, could have formed organic compounds, a primitive "soup" from which life arose.

   ■ **Miller** and **Urey** tested this hypothesis and produced a variety of amino acids.

▌ It is hypothesized that **self-replicating RNA** (not DNA) was the **first genetic material.**

Concept 25.2 The fossil record documents the history of life on Earth

❚ The **fossil record** is the sequence in which fossils appear in the layers of sedimentary rock that constitute Earth's surface. **Paleontologists** study the fossil record. Fossils, which may be remnants of dead organisms or impressions they left behind, are most often found in sedimentary rock formed from layers of minerals settling out of water. The fossil record is incomplete because it favors organisms that existed for a long time, were relatively abundant and widespread, and had shells or hard bony skeletons.

❚ Rocks and fossils are dated several ways:

  ▪ **Relative dating** uses the order of rock strata to determine the relative age of fossils.

  ▪ **Radiometric dating** uses the decay of radioactive isotopes to determine the age of the rocks or fossils. It is based on the rate of decay, or **half-life** of the isotope.

Concept 25.3 Key events in life's history include the origins of single-celled and multicelled organisms and the colonization of land

❚ The earliest living organisms were **prokaryotes.**

❚ About 2.7 billion years ago, **oxygen** began to accumulate in Earth's atmosphere as a result of photosynthesis.

❚ **Eukaryotes** appeared about 2.1 billion years ago.

  ▪ **The endosymbiotic hypothesis** proposes that mitochondria and plastids (chloroplasts) were formerly small prokaryotes that began living within larger cells.

  ▪ **Evidence** for this hypothesis includes

    1. Both organelles have enzymes and transport systems homologous to those found in the plasma membranes of living prokaryotes.
    2. Both replicate by a splitting process similar to prokaryotes.
    3. Both contain a single, circular DNA molecule, not associated with histone proteins.
    4. Both have their own ribosomes which can translate their DNA into proteins.

❚ **Multicellular eukaryotes** evolved about 1.2 billion years ago.

❚ **The colonization of land** occurred about 500 million years ago, when **plants, fungi, and animals** began to appear on Earth.

Concept 25.4 The rise and fall of dominant groups reflect continental drift, mass extinctions, and adaptive radiations

❚ **Continental drift** is the movement of Earth's continents on great plates that float on the hot, underlying mantle. The San Andreas Fault marks where two plates are sliding past each other. Where plates have collided, mountains are uplifted.

- Continental drift alters the habitats in which organisms live and promotes allopatric speciation on a grand scale.
- Continental drift can help explain the disjunct geographic distribution of certain species, such as a fossil freshwater reptile found in both Brazil and Ghana in west Africa, today widely separated by ocean.
- Continental drift can explain why no eutherian (placental) mammals are indigenous to Australia.
- **Mass extinctions,** loss of large numbers of species in a short period, have resulted from global environmental changes that have caused the rate of extinction to increase dramatically.
- By removing large numbers of species, a mass extinction can drastically alter a complex ecological community. Evolutionary lineages disappear forever. *Example:* The dinosaurs were lost in a mass extinction 65 million years ago.
- **Adaptive radiations** are periods of evolutionary change in which groups of organisms form many new species whose adaptations allow them to fill different ecological niches. *Example:* The Galapagos finch species are the result of an adaptive radiation.

Concept 25.5 *Major changes in body form can result from changes in the sequence and regulation of developmental genes*

- Evolutionary novelty can arise when structures that originally played one role gradually acquire a different one. Structures that evolve in one context but become co-opted for another function are sometimes called **exaptations.** For example, it is possible that feathers of modern birds were co-opted for flight after functioning in some other capacity, such as thermoregulation.
- **"Evo-devo"** is a field of study in which evolutionary biology and developmental biology converge. This field is illuminating how slight genetic divergences can be magnified into major morphological differences between species.
- **Heterochrony** is an evolutionary change in the rate or timing of developmental events. Changing relative rates of growth even slightly can change the adult form of organisms substantially, thus contributing to the potential for evolutionary change.
- **Homeotic genes** are master regulatory genes that determine the location and organization of body parts.
- *Hox* genes are one class of homeotic genes. Changes in *Hox* genes and in the genes that regulate them can have a profound effect on morphology, thus contributing to the potential for evolutionary change. An example is seen in the variable expression of a *Hox gene* in a fish fin bud and a chicken leg bud resulting in a skeletal extension in the chicken.

Multiple-Choice Questions

1. The condition in which there are barriers to inbreeding between individuals of the same species separated by a portion of a mountain range is referred to as
   (A) minute variations.
   (B) geographic isolation.
   (C) infertility.
   (D) reproductive isolation.
   (E) differential breeding capacity.

2. Which of the following statements best expresses the concept of punctuated equilibrium?
   (A) Minute changes in the genome of individuals eventually lead to the evolution of a population.
   (B) The five conditions of Hardy-Weinberg equilibrium will prevent populations from evolving quickly.
   (C) Evolution occurs in rapid bursts of change alternating with long periods in which species remain relatively unchanged.
   (D) Profound change over the course of geologic history is the result of an accumulation of slow, continuous processes.
   (E) When two species compete for a single resource in the same environment, one of them will gradually become extinct.

3. In a particular bird species, individuals with average-sized wings survive severe storms more successfully than other birds in the same population with longer or shorter wings. This illustrates
   (A) the founder effect.
   (B) stabilizing selection.
   (C) artificial selection.
   (D) gene flow.
   (E) diversifying selection.

4. All of the following statements are part of Darwin's theory of evolution EXCEPT
   (A) The most prominent contribution to evolution is made by the process of genetic mutation.
   (B) Natural selection is the force behind evolution.
   (C) Natural selection occurs as a result of the differing reproductive success of individuals in a population.
   (D) The driving force of evolution is the adaptation of a population of organisms to their environment.
   (E) More individuals are born in a population than will survive to reproduce.

5. In a certain group of rabbits, the presence of yellow fur is the result of a homozygous recessive condition in the biochemical pathway producing hair pigment. If the frequency of the allele for this condition is 0.10, which of the following is closest to the frequency of the dominant allele in this population? (Assume that the population is in Hardy-Weinberg equilibrium.)
   (A) 0.01
   (B) 0.20
   (C) 0.40
   (D) 0.90
   (E) 1.0

Directions: The group of questions below consists of five lettered choices followed by a list of numbered phrases or sentences. For each numbered phrase or sentence, select the one choice that is most closely related to it. Each choice may be used once, more than once, or not at all.

Questions 6–10
 (A) Artificial selection
 (B) Homology
 (C) Gene pool
 (D) The founder effect
 (E) The bottleneck effect

6. Leads to new species with certain traits desired by humans

7. Can result in a new island population with a limited gene pool

8. One result of evolution from a common ancestor

9. A result of drastic reduction in population size due to a sudden change in the environment

10. Constitutes all of the alleles in a population

11. All of the following are examples of prezygotic barriers EXCEPT
 (A) habitat isolation.
 (B) behavioral isolation.
 (C) temporal isolation.
 (D) mechanical isolation.
 (E) hybrid breakdown.

12. Species that are found only in one particular geographic location are said to be
 (A) behaviorally evolved.
 (B) endemic.
 (C) speciated.
 (D) undergoing behavioral isolation.
 (E) undergoing mechanical isolation.

13. The allele that causes sickle-cell disease is found with greater frequency in Africa, where malaria is more of a threat, than in the United States. Which genetic phenomena most likely contributes to the difference in frequency?
 (A) heterozygote advantage
 (B) heterozygote protection theory
 (C) balanced polymorphism
 (D) frequency-dependent selection
 (E) neutral variations

14. In a population of squirrels, the allele that causes bushy tail (B) is dominant, while the allele that causes bald tail is recessive (b). If 64% of the squirrels have a bushy tail, what is the frequency of the dominant allele?
 (A) 0.8
 (B) 0.6
 (C) 0.4
 (D) 0.36
 (E) 0.2

15. Which of the following factors would *not* contribute to allopatric speciation?
 (A) A population becomes geographically isolated from the parent population.
 (B) The separated population is small, and genetic drift occurs.
 (C) The isolated population is exposed to different selection pressures than the ancestral population.
 (D) Gene flow between the two populations is minimal or does not occur.
 (E) The different environments of the two populations create different mutations.

16. A marsupial living in Australia has evolved to eat tree leaves, be diurnal, and raise its young until they are of reproductive age. A grazing placental mammal has also evolved to eat tree leaves, be diurnal, and raise its young until they are of reproductive age. This is an example of which of the following types of evolution?
    (A) divergent evolution
    (B) species-specific evolution
    (C) convergent evolution
    (D) neutral evolution
    (E) sibling evolution

17. Which of the following can lead to sympatric speciation?
    (A) migration of a small number of individuals
    (B) natural disaster that cuts off contact between members of a population
    (C) a newly formed river separates segments of a population
    (D) autopolyploidy
    (E) bottleneck effect

18. Mitochondria and plastids contain DNA and ribosomes and make some, but not all, of their proteins. Some of their proteins are coded for by nuclear DNA and produced in the cytoplasm. What may explain this division of labor?
    (A) Over the course of evolution, some of the original endosymbiont's genes were transferred to the host cell's nucleus.
    (B) The host cell's genome always included genes for making mitochondrial and plastid proteins.
    (C) These organelles do not have sufficient resources to make all their proteins and rely on the help of the cell.
    (D) Some mitochondria and plastid genes were contributed by early bacterial prokaryotes that shared genes with other primitive cells.
    (E) The smaller prokaryotic ribosomes in these organelles cannot produce the eukaryotic proteins required for their functions.

19. Banded iron formations in marine sediments provide evidence of
    (A) the first prokaryotes around 3.5 billion years ago.
    (B) oxidized iron layers in terrestrial rocks.
    (C) the accumulation of oxygen in the seas from the photosynthesis of cyanobacteria.
    (D) the evolution of photosynthetic archaea near deep-sea vents.
    (E) the crashing of meteorites onto Earth, possibly transporting abiotically produced organic molecules from space.

20. Which of the following constitutes the smallest unit capable of evolution?
    (A) an individual
    (B) a group
    (C) a population
    (D) a clade
    (E) a community

Free-Response Question

> *1. Microevolution is the change in the gene pool from one generation to the next.*
>
> (a) **Describe** three ways in which microevolution can take place.
> (b) **Describe** the difference between microevolution and macroevolution.

ANSWERS AND EXPLANATIONS

Multiple-Choice Questions

▌ **1. (B) is correct.** When two members of the same species are prevented from breeding by a geographic barrier such as a mountain range or river, the fact that they live in different ponds, or any other physical obstruction, the two individuals are said to be geographically isolated.

▌ **2. (C) is correct.** The concept of punctuated equilibrium was put forth recently by Niles Eldredge and Steven J. Gould. It explains periods of apparently little charge (stasis) "punctuated" by sudden changes observed in the fossil record.

▌ **3. (B) is correct.** Since selection favors the average-sized wings and reduces the frequency of two extremes, this is an example of stabilizing selection. If you missed this, study the graphs in Figure 6.1 again.

▌ **4. (A) is correct.** All the answers are parts of Darwin's theory of evolution except A—genetic mutation is not the most important factor contributing to the process of evolution. In fact, when Darwin was developing his ideas, no one knew about genes. Darwin's theory centered on his observation of changing populations rather than molecular evidence.

▌ **5. (D) is correct.** If the frequency of the recessive allele is 0.10, then we know that the frequency of the other allele is 1 – 0.10, which is equal to 0.90. The Hardy-Weinberg equation states that, if a population contains just two alleles for a given trait, and we know the frequency of one of the alleles, we can calculate the frequency of the other using the equation $p + q = 1$. If we designate the frequency of the occurrence of the recessive allele as q, and use a value of 0.1, we can rearrange the equation to read $p + 0.10 = 1$. Then, $1 - 0.10 = 0.9$, which is equal to p, or the frequency of the other allele—in this case the dominant one.

▌ **6. (A) is correct.** Artificial selection is the selective breeding of domesticated plants and animals in order to modify them to better suit the needs of humans. Humans have been practicing artificial selection for thousands of generations, and many of the common foods we eat are a result of this process.

▌ **7. (D) is correct.** The founder effect occurs when a few individuals from a population colonize a new, isolated habitat. The smaller the number of individuals who start this new population, the more limited will be the starting gene pool for the population—and the less the new gene pool will resemble that of the parent population.

▌ **8. (B) is correct.** Homology is the result of descent from a common ancestor. It can be described as the underlying structural or molecular similarities (even in structures that are no longer used for the same function) that exist in organisms as a result of common ancestry.

9. (E) is correct. Bottleneck effects are often the result of a natural disaster such as a flood, drought, fire, or anything that destroys most members of a population. The gene pool of the surviving members of the population may not resemble the gene pool of that of the parent population—some genes will be overrepresented, and some will be underrepresented. Bottlenecking reduces the genetic variability in a population because of the loss of alleles.

10. (C) is correct. The gene pool is the collection of all the alleles that exist in a population—a population is defined as a group of individuals living in a certain geographic location that are capable of interbreeding.

11. (E) is correct. Prezygotic barriers are those that prevent or hinder the mating of two species, or they prevent fertilization even if two species mate. All the answers are examples of prezygotic barriers except the last one. Hybrid breakdown is an example of a postzygotic barrier (postzygotic barriers are those that prevent hybrid zygotes from developing into viable adults) in which the second generation of offspring from two species is either weak or sterile.

12. (B) is correct. Species that are found in only one geographic location are said to be endemic. Some examples of endemic species are kangaroos (endemic to Australia) and blue-footed boobies (endemic to the Galápagos Islands).

13. (A) is correct. The allele that causes sickle-cell disease also prevents severe malarial infections. Individuals who have two copies of the sickle-cell allele will come down with the disease and most likely die. Those who are heterozygous, however, have an advantage. Because they have one normal allele, they will not contract full-blown sickle-cell disease; however they will have an increased protection against malaria. Individuals with two normal alleles will not have this same protection and are more likely to die from malaria.

14. (C) is correct. Did you select B? You were too hasty and forgot a very important concept: the 64% of the population that show the dominant trait includes heterozygotes, so you cannot simply take the square root of .64. You must subtract .64 from 1.00 to obtain .36 = q^2. Taking the square root of q^2 yields q = 0.6, so p = .4.

15. (E) is correct. Recall that mutations are not "created" in response to some change in the environment or need. All the other responses are factors that contribute to allopatric speciation. Understand them!

16. (C) is correct. In convergent evolution, species from different evolutionary branches appear alike as a result of undergoing evolution in very similar ecological roles and environments. Similarity between species that have undergone convergent evolution is known as analogy, and structures they share are analogous (not homologous) structures. (Remember, similar problem, similar solution. However, convergence does **not** indicate a common ancestor.) Homologous structures are those that are shared in two species as a result of those species' having a common ancestor.

17. (D) is correct. One mechanism that can lead to sympatric speciation, which is the formation of a small new population within the parent population in plants, is the formation of autopolyploids through nondisjunction in meiosis. These plants have 4*n* chromosomes, instead of the normal 2*n* number, and they are unable to breed with members of the parent population—though they are still able to breed with other tetraploids.

 18. **(A) is correct.** For this question recall the endosymbiotic hypothesis proposes that mitochondria and plastids were engulfed by other cells and have their own genomes. It is most likely that over time some of their genes were incorporated into the host cell's genome. Hint: Review the endosymbiotic hypothesis.

 19. **(C) is correct.** It is important to recall that the early atmosphere did not have oxygen, and that the evolution of photosynthesis by cyanobacteria resulted in the accumulation of oxygen in the atmosphere. The first life forms were not able to do photosynthesis.

 20. **(C) is correct.** The smallest unit capable of evolution is the population. Individuals cannot undergo evolution because they exist for only one generation, and evolution is the changing and refinement of a group's gene pool to best fit the group's environment.

Free-Response Question

(a) Microevolution is the change in the allelic frequencies of a population that occurs from one generation to the next, and three factors that contribute to microevolution are genetic drift (including both the bottleneck effect and the founder effect), natural selection, and gene flow.

The term *genetic drift* refers to any change in a population's allele frequencies due to chance. Two examples of what this "chance" can consist of are the bottleneck effect and the founder effect. The bottleneck effect occurs after a natural disaster such as a violent storm or fire causes a drastic reduction in the size of the population. Such an event leaves only a few individuals to continue to produce offspring, so bottlenecking usually reduces the genetic variability in a population because some alleles are lost from the gene pool and others are overrepresented. Similarly, the founder effect occurs when a few members of a population colonize an isolated location that isn't accessed by members of the parent population. The smaller the number of founders, the more limited the variability of the genes in the new population.

Gene flow refers to genetic exchange due to the migration of individuals or gametes between populations. This can take place in the course of one generation to the next, so this is another valid contributor to microevolution.

Natural selection is another route by which microevolution can take place. This term refers to the differing reproductive success of all the individuals in a population. Those who are best suited to their environment will survive to pass on their alleles to the next generation. Natural selection is the only one of these factors that is adaptive.

(b) Macroevolution is the process of the evolution of new taxonomic groups (meaning, new species, families, or kingdoms) through evolution. It differs from microevolution in that microevolution refers only to changes that occur in populations from generation to generation—no new species or other taxonomic groups need arise in the course of microevolution. Microevolution occurs on a small scale, whereas macroevolution concerns the "bigger picture."

This is a good free-response answer because it shows knowledge of the following key terms that you will be expected to know for the exam:

microevolution	*founder effect*
gene pool	*natural selection*
population	*gene flow*
genetic drift	*macroevolution*
bottleneck effect	*taxonomic groups*

It also shows that the student understands the following processes: microevolution, genetic drift, including the bottleneck effect and the founder effect, natural selection, gene flow, and macroevolution. Note that the student did not slip into clichés like "survival of the fittest" and neither should you!

The Evolutionary History of Biodiversity

Chapter 26: Phylogeny and the Tree of Life

YOU MUST KNOW

- The taxonomic categories and how they indicate relatedness.
- How systematics is used to develop phylogenetic trees.
- The three domains of life including their similarities and their differences.

Concept 26.1 Phylogenies show evolutionary relationships

▍ **Phylogeny** is the evolutionary history of a species or a group of related species. It is constructed by using evidence from **systematics,** a discipline that focuses on classifying organisms and their evolutionary relationships. Its tools include fossils, morphology, genes, and molecular evidence.

▍ **Taxonomy** is an ordered division of organisms into categories based on a set of characteristics used to assess similarities and differences.

▍ **Binomial nomenclature** uses a two-part naming system that consists of the **genus** to which the species belongs, as well as the organisms' **species** within the genus, such as *Canis familiaris*, the scientific name of the common dog. This system was developed by **Carolus Linnaeus.**

▍ The hierarchical classification of organisms consists of the following levels, beginning with the most general or inclusive: **domain, kingdom, phylum, class, order, family, genus, and species.** Each categorization at any level is called a **taxon.**

STUDY TIP: Use a mnemonic device, such as "King Phillip climbed over the fence and got shot," to remember the categories in order. Know that the degree of relatedness increases with each successive level down (in other words, the organisms share more common traits).

■ Systematists use branching diagrams called **phylogenetic trees** to depict hypotheses about evolutionary relationships. The branches of such trees reflect the hierarchical classifications of groups nested within more inclusive groups. (See Figure 7.1.)

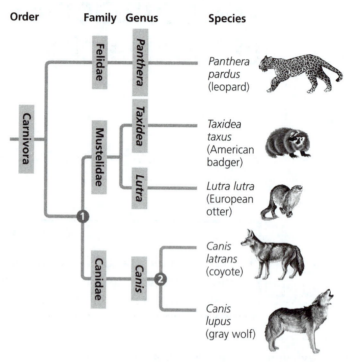

Figure 7.1

Concept 26.2 *Phylogenies are inferred from morphological and molecular data*

■ **Homologous structures** are similarities due to shared ancestry, such as the bones of a whale's flipper and a tiger's paw.

■ **Convergent evolution** has taken place when two organisms developed similarities as they adapted to similar environmental challenges—not because they evolved from a common ancestor. *Example:* The streamlined bodies of a tuna and a dolphin show convergent evolution.

■ The likenesses that result from convergent evolution are considered **analogous** rather than homologous. They do not indicate relatedness, but rather similar solutions to similar problems. *Example:* The wing of a butterfly is analogous to the wing of a bat. Both are adaptations for flight.

■ **Molecular systematics** uses DNA and other molecular data to determine evolutionary relationships. The more alike the DNA sequences of two organisms, the more closely related they are evolutionarily.

Concept 26.3 *Shared characters are used to construct phylogenetic trees*

■ A **cladogram** depicts patterns of shared characteristics among taxa and forms the basis of a **phylogenetic tree.**

■ A **clade,** within a tree, is defined as a group of species that includes an ancestral species and all its descendants.

Concept 26.4 An organism's evolutionary history is documented in its genome

▌ The rate of evolution of DNA sequences varies from one part of the genome to another; therefore, comparing these different sequences helps us to investigate relationships between groups of organisms that diverged a long time ago.

- DNA that codes for ribosomal RNA changes relatively slowly and is useful for investigating relationships between taxa that diverged hundreds of millions of years ago.
- DNA that codes for mitochondrial DNA evolves rapidly and can be used to explore recent evolutionary events.

Concept 26.5 Molecular clocks help track evolutionary time

▌ **Molecular clocks** are methods used to measure the absolute time of evolutionary change based on the observation that some genes and other regions of the genome appear to evolve at constant rates.

Concept 26.6 New information continues to revise our understanding of the tree of life

▌ Taxonomy is in flux! When your authors were in high school, we were taught there were two kingdoms, plants and animals; then in our college courses, we were introduced to five kingdoms: Monera, Protista, Plantae, Fungi, and Animalia.

▌ Now biologists have adopted a **three-domain** system, which consists of the domains Bacteria, Archaea, and Eukarya. This system arose from the finding that there are two distinct lineages of prokaryotes.

▌ The domains Bacteria and Archaea contain *prokaryotic* organisms, and Eukarya contains *eukaryotic* organisms. As we gain more tools for analysis, earlier ideas about evolutionary relatedness are changed, and so taxonomy, too, continues to evolve. That said, let's look at the three domains. Principal differences between the groups are simplified and presented below.

A Comparison of the Three Domains of Life

Characteristic	Bacteria	Archaea	Eukarya
Nuclear Envelope	no	no	yes
Membrane-enclosed organelles	no	no	yes
Introns	no	yes	yes
Histone proteins assoc. with DNA	no	yes	yes
Circular chromosome	yes	yes	no

Chapter 27: Bacteria and Archaea

```
┌─────────────────────────────────────────────────────────────────┐
│                         YOU MUST KNOW                             │
│  • The key ways in which prokaryotes differ from eukaryotes with  │
│    respect to genome, membrane-bound organelles, size, and        │
│    reproduction.                                                  │
│  • Mechanisms that contribute to genetic diversity in prokaryotes,│
│    including transformation, conjugation, transduction, and       │
│    mutation.                                                      │
└─────────────────────────────────────────────────────────────────┘
```

Concept 27.1 *Structural and functional adaptations contribute to prokaryotic success*

▌ Life is divided into three domains: **Archaea, Bacteria, and Eukarya.** Both domain Bacteria and domain Archaea are made up of prokaryotes.

▌ The most common shapes of prokaryotes are spheres, rods, and helices; most are about 1–5 μm in size. This is perhaps 1/10 the size of a typical eukaryotic cell.

▌ Prokaryotes have *no true nuclei* or internal compartmentalization. The DNA is concentrated in a nucleoid region and has little associated protein. Relative to eukaryotes, prokaryotes have simple, *small genomes.*

▌ In addition to their one major chromosome, prokaryotic cells may also possess smaller, circular, independent pieces of DNA called **plasmids.**

▌ Prokaryotes reproduce through an asexual process called **binary fission,** and they continually synthesize DNA.

▌ Outside their cell membranes, most prokaryotes possess a cell wall that contains **peptidoglycans. Gram-positive** bacteria have simpler walls with more peptidoglycans, whereas **gram-negative** cells have walls that are structurally more complex.

▌ Prokaryotes use appendages called **pili** that adhere to each other or to surrounding surfaces. About half of the prokaryotes are **motile,** because they possess whiplike **flagella.**

  ▪ Because the flagellum of a bacterium is structurally different from the eukaryotic flagellum, this is another example of analogous structures.

Concept 27.2 *Rapid reproduction, mutation, and genetic recombination promote genetic diversity in prokaryotes*

▌ Three mechanisms by which bacteria can transfer genetic material between each other are

  1. **Transformation,** in which a prokaryote takes up DNA from its environment.
  2. **Conjugation,** in which genes are directly transferred from one prokaryote to another.
  3. **Transduction,** in which viruses transfer genes between prokaryotes.

▌ The major source of genetic variation in prokaryotes is **mutation.**

Concept 27.3 A great diversity of nutritional and metabolic adaptations has evolved in prokaryotes

- Prokaryotes can be placed in four groups according to how they take in carbon and how they obtain energy:

  - **Photoautotrophs** are photosynthetic, and they use the power of sunlight to convert carbon dioxide into organic compounds.
  - **Chemoautotrophs** also use carbon dioxide as their source of carbon, but they get energy from oxidizing inorganic substances.
  - **Photoheterotrophs** use light to make ATP but must obtain their carbon from an outside source already fixed in organic compounds.
  - **Chemoheterotrophs** get both carbon and energy from organic compounds.

- **Obligate aerobes** cannot grow without oxygen because they need oxygen for cellular respiration.
- **Obligate anaerobes** are poisoned by oxygen. Some use fermentation, whereas others extract chemical energy by another form of anaerobic respiration.
- **Facultative anaerobes** use oxygen if it is available; when oxygen is not available, they undergo fermentation.
- Some prokaryotes can use atmospheric nitrogen as a direct source of nitrogen in a process called **nitrogen fixation.** They convert N_2 to NH_4^+. You might want to review the nitrogen cycle from Chapter 54.

Concept 27.4 Molecular systematics is illuminating prokaryotic phylogeny

- The first prokaryotes that were classified in the domain Archaea are known as **extremophiles** and live in extreme environments such as geysers:

  1. **Extreme halophiles** live in saline environments (highly concentrated with salt).
  2. **Extreme thermophiles** live in very hot environments.

- Other Archaea do not live in extreme environments. **Methanogens** use carbon dioxide to oxidize H_2 and produce methane as a waste product.

Concept 27.5 Prokaryotes play crucial roles in the biosphere

- Many prokaryotes are **decomposers,** breaking down dead corpses, vegetation, and waste products.
- Many prokaryotes are **symbiotic,** meaning that they form relationships with other species:

  1. **Mutualism:** Both symbiotic organisms benefit.
  2. **Commensalism:** One organism benefits, whereas the other is neither helped nor harmed.
  3. **Parasitism:** One organism benefits at the expense of the other.

Concept 27.6 Prokaryotes have both harmful and beneficial impacts on humans

- Some prokaryotes are **pathogenic** and cause illness by producing poisons.

- **Antibiotics** are chemicals that can kill prokaryotes. They are *not* effective against viruses. Many bacterial plasmids contain resistance genes to different antibiotics.
- Prokaryotes are used by humans in many diverse ways:

    1. **Bioremediation,** removing pollutants from soil, air, or water. This includes treating sewage, cleaning up oil spills, and precipitating radioactive materials.
    2. Symbionts in the gut, manufacturing vitamins, and digesting foods.
    3. Gene cloning and producing transgenic organisms.
    4. Production of cheese and yogurt and other products.

Chapter 28: Protists

YOU MUST KNOW

- Protista is no longer considered a kingdom! This probably contradicts what you learned in your introductory biology class.
- How chloroplasts and mitochondria evolved through endosymbiosis.

Concept 28.1 Most eukaryotes are single-cell organisms

- **Protist** is now a term used to refer to eukaryotes that are neither plants, animals, nor fungi. Biologists no longer consider *Protista* a kingdom because it is paraphyletic. Study Figure 7.2 to review what this means. Some protists are more closely related to plants, fungi, or animals than to other protists.

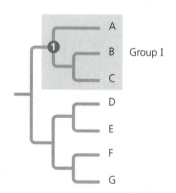

(a) Monophyletic group (clade). Group I, consisting of three species (A, B, C) and their common ancestor ❶, is a clade, also called a monophyletic group. A monophyletic group consists of an ancestral species and *all* of its descendants.

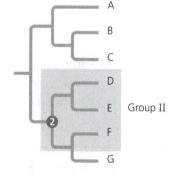

(b) Paraphyletic group. Group II is paraphyletic, meaning that it consists of an ancestral species ❷ and some of its descendants (species D, E, F) but not all of them (missing species G).

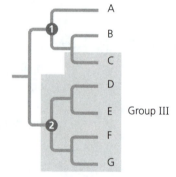

(c) Polyphyletic group. Group III is polyphyletic, meaning that it includes descendants of two or more common ancestors. In this case, species D, E, F, and G share common ancestor ❷, but species C has a different ancestor: ❶.

Figure 7.2

■ Protists vary in structure and function more than any other group of eukaryotes. Here are some general commonalities, but even these are not true for all groups:

1. Most are unicellular.
2. Most use aerobic metabolism and have mitochondria.

■ According to current theory, mitochondria and chloroplasts evolved through **endosymbiosis.** They were originally unicellular organisms engulfed by other cells that ultimately became organelles in the host cell.

■ Protists can be divided into three categories: photosynthetic (plant-like) algae, ingestive (animal-like) protozoans, and absorptive (fungus-like) organisms.

■ Most protists are aquatic and are important constituents of plankton. Many other protists live as symbionts in other organisms.

■ Protists are such a diverse group that their classification continues to undergo revision. Despite this, you may recognize these protists:

1. *Giardia intestinalis* (causes "hiker's diarrhea"; always treat your water!)
2. *Trichomonas vaginalis* (sexually transmitted infection)
3. *Trypanosoma sp.* (sleeping sickness and Chagas' disease)
4. *Euglena* (remember seeing the tiny flagellated green cell with a red eyespot in Bio. I?)
5. Dinoflagellates (blooms cause "red tides"; many are bioluminescent)
6. *Plasmodium* (causative agent of malaria)
7. Ciliates (*Paramecium* and *Stentor* are examples; micro- and macronuclei)
8. Amoeba (move by pseudopodia)
9. Diatoms (two-part glass-like wall made of silica)
10. Golden algae
11. Brown algae (kelp)
12. Oomycetes (water molds and their relatives; includes causative agent of potato blight)
13. Red algae (multicellular; some found at great depths; sushi wraps)
14. Green algae (*Clamydomonas, Ulva, Volvox*; this group is the closest relative of land plants)
15. Slime molds

Chapter 29: Plant Diversity I: How Plants Colonized Land

```
┌─────────────────────────────────────────────────────────────┐
│                      YOU MUST KNOW                          │
│                                                             │
│  • Why land plants are thought to have evolved from green   │
│    algae.                                                   │
│  • Some of the disadvantages and advantages of life on land.│
│  • That plants have a unique life cycle termed alternation  │
│    of generations with a gametophyte generation and a       │
│    sporophyte generation.                                   │
│  • The role of antheridia and archegonia in gametophytes.   │
│  • The major characteristics of bryophytes.                 │
│  • The major characteristics of seedless vascular plants.   │
└─────────────────────────────────────────────────────────────┘
```

Concept 29.1 Land plants evolved from green algae

▮ Land plants evolved from green algae more than 500 million years ago. Plants have enabled other life forms to survive on land. Plants supply oxygen and are the ultimate provider of most of the food eaten or absorbed by animals and fungi.

▮ The evolution of land plants from the green algae group known as the charophytes includes the following five lines of evidence:

1. They produce cellulose for cell walls in the same, unique fashion.
2. The peroxisomes of these two groups, but no others, have enzymes that reduce the effects of photorespiration.
3. The structure of their sperm is closely related.
4. They both produce cell plates in the same way during cell division.
5. Genetic evidence including analysis of nuclear and chloroplast genes indicates the two groups are closely related.

▮ Movement onto land had both advantages and challenges:

■ Advantages included increased sunlight unfiltered by water, more carbon dioxide in the atmosphere than the water, soils rich in nutrients, and fewer predators.

■ Challenges included a lack of water, desiccation, and a lack of structural support against gravity.

▮ All land plants have a life cycle that consists of two multicellular stages, called **alternation of generations.**

▮ The two stages are the **gametophyte** stage (in which the plant cells are haploid) and the **sporophyte** stage (in which the plant cells are diploid).

■ In the gametophyte stage, the gametes are produced. During fertilization, egg and sperm fuse to form a diploid zygote—the sporophyte—which divides mitotically. The sporophyte produces spores by meiosis.

- The zygote develops within the tissues of the female parent, deriving nutrients from it. For this reason land plants are sometimes referred to as **embryophytes.**

- Another key feature of plants is the production of gametes in multicellular organs called **gametangia.** The female gametangia are the **archegonia,** which produces a single egg. The male gametangia are the **antheridia,** which produces many sperm.

Concept 29.2 Mosses and other nonvascular plants have their life cycles dominated by gametophytes

- Bryophytes include three phyla: mosses, liverworts, and hornworts. Currently, debate continues over whether the three phyla share a common ancestor (in which case this group would be a single clade).

- The bryophytes are considered **nonvascular** (no xylem or phloem tissue), even though some mosses do have simple vascular tissue. The lack of vascular tissue accounts for the small size of bryophytes.

- Unlike vascular plants, in all three bryophyte phyla the gametophytes are the dominant stage of the life cycle. The archegonia and antheridia are found on the gametophytes. Bryophytes require water for the sperm to swim to the egg during fertilization. Follow Figure 7.3 for the basic steps to the life cycle of a typical moss.

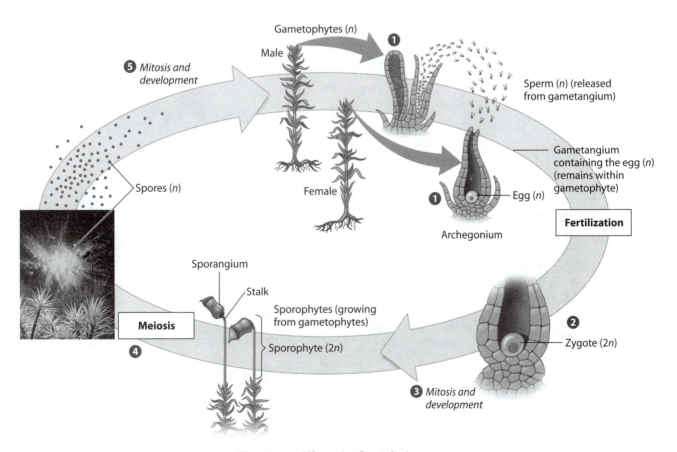

Figure 7.3   Life cycle of a typical moss

■ Bryophytes have the smallest, simplest sporophytes of all plant groups. Even though the sporophytes are photosynthetic when young, they must absorb water, sugars, and other nutrients from parental gametophytes. Notice from Figure 7.3 that meiosis occurs as the mechanism from sporophyte to gametophyte.

Concept 29.3 Ferns and other seedless vascular plants were the first plants to grow tall

■ The evolution of vascular tissues allowed vascular plants to grow taller than bryophytes, gaining access to sunlight. Ferns and other seedless vascular plants, however, still require a film of water for the sperm to reach the egg.

■ The seedless vascular plant life cycle of alternation of generations is dominated by the **sporophyte** stage. Meiosis occurs in the sporophyte stage in structures termed **sporangia,** producing haploid **spores.** The haploid spores may grow into gametophytes, where swimming sperm fertilize eggs, yielding the diploid zygote which will grow into the sporophyte generation. Through this alternation between generations, the life cycle name originated. Use Figure 7.4 to follow the key steps to a typical fern life cycle.

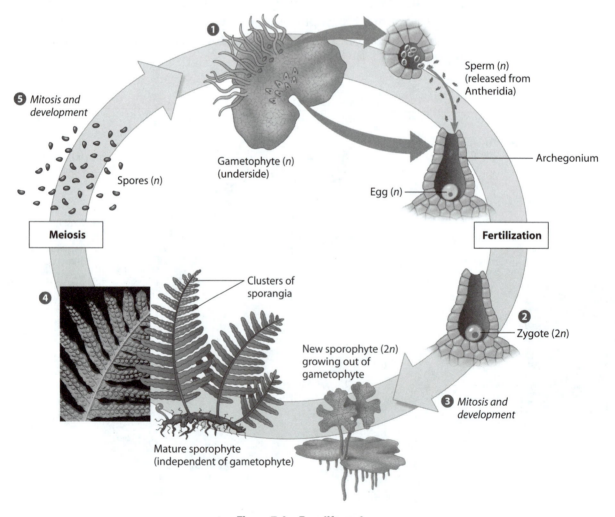

Figure 7.4   Fern life cycle

- Biologists recognize two clades of living seedless vascular plants. The phylum *Lycophyta* includes club mosses (not real mosses; recall that they are nonvascular) and quillworts. The other clade is the ferns, horsetails, and whisk ferns, or the *Pterophytes*.
- Seedless vascular plants formed forests of great heights in the Carboniferous period, eventually forming deposits of coal that we use for fuel today.

Chapter 30: Plant Diversity II: The Evolution of Seed Plants

YOU MUST KNOW

- Key adaptations to life on land unique to seed plants.
- The evolutionary significance of seeds and pollen.
- The role of flowers and fruits in angiosperm reproduction.
- The role of stamens and carpels in angiosperm reproduction.

Concept 30.1 Seeds and pollen grains are key adaptations for life on land

- **Seeds** are plant embryos packaged with a food supply in a protective coat.
- Five crucial adaptations led to the success of seed plants:

   1. *Reduced gametophytes.* Gametophytes in seed plants are mostly microscopic and entirely dependent on the sporophyte for food and protection. This protects the delicate antheridia and archegonia, increasing reproductive success.
   2. *Heterospory.* **Heterospory** means the production of two types of spores. **Megaspores** produce female gametophytes, which produce the eggs. **Microspores** produce male gametophytes, which contain sperm nuclei.
   3. *Ovules and the production of eggs.* The megasporangium, megaspore, and the protective tissue around them make an ovule. The ovule increases protection of the egg and the developing zygote, thus increasing reproductive fitness.
   4. *Pollen and production of sperm.* A pollen grain is a male gametophyte, containing two sperm nuclei. The pollen grain has a waterproof coating, allowing for transfer by the wind. Until pollen, water was required for sperm transfer. *The evolution of pollen was a key adaptation to land.*
   5. *Seeds.* Seeds have several advantages over spores. Seeds are multicellular, with several layers of protective tissue, safeguarding the embryo. Unlike spores, seeds have a supply of stored energy which allows the seed to wait for good germination conditions and use stored energy to finance the early growth of the embryo.

Concept 30.2 Gymnosperms bear "naked" seeds, typically on cones

▌ Gymnosperms are plants that have "naked" seeds that are not enclosed in ovaries. Their seeds are often exposed on modified leaves that form cones. To compare, angiosperms (flowering plants) have seeds enclosed in fruits, which are mature ovaries. Gymnosperms do not have fruits.

▌ Four phyla of plants are considered gymnosperms, but the most ecologically significant group is the conifers (*Coniferophyta*), which include pines, spruces, firs, and redwoods.

▌ Use Figure 7.5 to review the life cycle of a pine. Notice the evolutionary advances listed in Figure 7.5 and the important role they play in the life cycle.

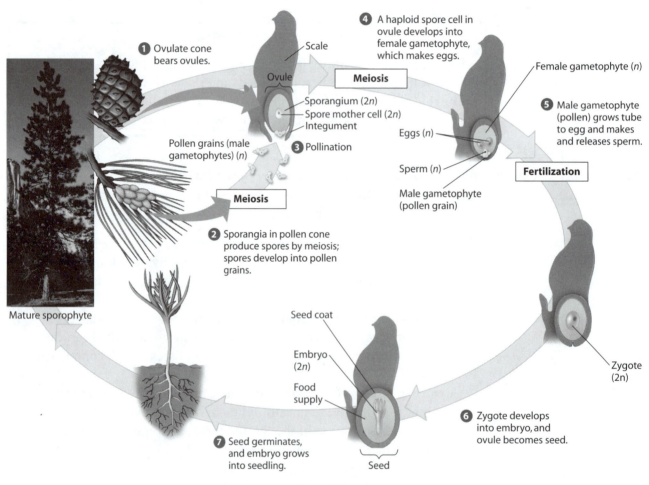

Figure 7.5

Concept 30.3 The reproductive adaptations of angiosperms include flowers and fruits

▌ **Angiosperms** are seed plants that produce the reproductive structures called flowers and fruits. Flowering plants are classified in the phylum **Anthophyta.** Today, angiosperms account for about 250,000 species, about 90% of all plant species.

■ The major reproductive adaptation of the angiosperm is the **flower,** which consists of four floral organs: sepals, petals, stamens, and carpels. Use Figure 7.6 to review the major parts of the flower.

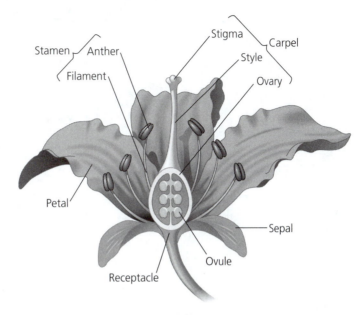

Figure 7.6

  ▪ **Stamens** are the male reproductive structure, producing microspores in the anthers that develop into pollen grains.
  ▪ **Carpels** are the female reproductive structure, producing megaspores and their products—female gametophytes with eggs.

■ **Fruits** are mature ovaries of the plant. As seeds develop from ovules after fertilization, the wall of the ovary thickens to become the fruit. Fruits help disperse the seeds of angiosperms.
■ The life cycle of the angiosperm is a refined version of the alternation of generations that all plants undergo. Use Figure 7.7 to review the life cycle. Notice the evolutionary advances listed in Figure 7.5 and the important role they play in the life cycle. More specifics of angiosperm reproduction are covered in Chapter 38.
■ Angiosperms have traditionally been divided into monocots and eudicots:

  ▪ **Monocots** (about 70,000 species) have one cotyledon in the seed, parallel leaf venation, and flowering parts in multiples of threes. Examples include orchids, lilies, and grasses.
  ▪ **Eudicots** (about 170,000 species) have two cotyledon in the seed, net leaf venation, and flowering parts usually in multiples of fours or fives. Examples include roses, peas, beans, and oaks.

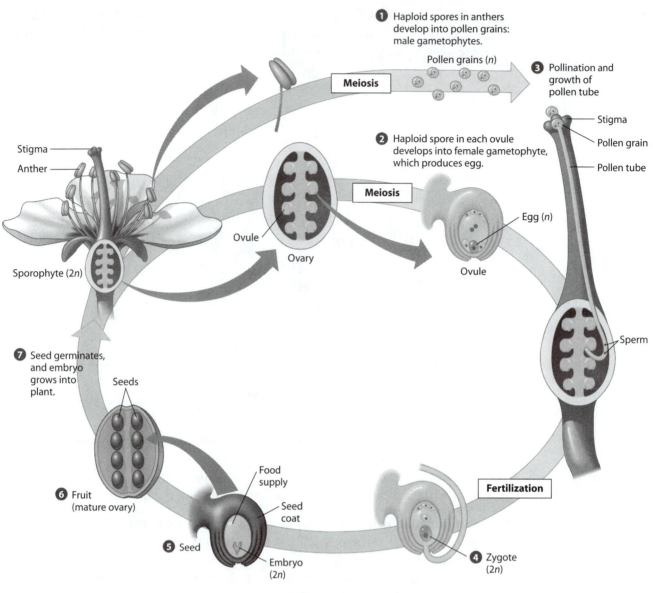

① Haploid spores in anthers develop into pollen grains: male gametophytes.

Pollen grains (n)

Meiosis

③ Pollination and growth of pollen tube

Stigma
Pollen grain
Pollen tube

Stigma
Anther

② Haploid spore in each ovule develops into female gametophyte, which produces egg.

Meiosis

Egg (n)

Ovule
Ovary

Ovule

Sporophyte (2n)

Sperm

⑦ Seed germinates, and embryo grows into plant.

Seeds

Food supply

Fertilization

⑥ Fruit (mature ovary)

Seed coat

⑤ Seed

Embryo (2n)

④ Zygote (2n)

Figure 7.7

Chapter 31: Fungi

YOU MUST KNOW

- The characteristics of fungi.
- Important ecological roles of fungi in mycorrhizal associations, and as decomposers and parasitic plant pathogens.

Concept 31.1 Fungi are heterotrophs that feed by absorption

▮ Fungi are eukaryotes with these characteristics:

1. **Multicellular heterotrophs** that obtain nutrients by **absorption.** Fungi secrete hydrolytic enzymes, digest food outside their bodies, and absorb the small molecules.
2. The cell walls of fungi are made of *chitin.*
3. Bodies composed of filaments called **hyphae** that are entwined to form a mass, the **mycelium.**
4. Most fungi are multicellular with hyphae divided into cells by cross walls called **septa.** *Coenocytic* fungi lack septa and consist of a continuous cytoplasmic mass containing hundreds of nuclei.
5. Fungi reproduce by spores.
6. Modes of nutrition include decomposers, parasites, and mutualists.

Concept 31.4 Fungi have radiated into a diverse set of lineages

▮ There are **five phyla of fungi.** Several phyla and examples follow:

1. **Zygomycota** (zygote fungi) are terrestrial, and include fast-growing molds, parasites, and commensal symbionts. A common zygomycete is bread mold, *Rhizopus.*
2. **Ascomycota** (sac fungi) produce sexual spores in sac-like structures called *asci;* ascomycetes include yeast, *Neurospora* (remember the nutritional mutants studied by Beadle and Tatum to investigate "one gene, one enzyme"? See Chapter 17.) and *Sordaria,* the subject of AP Lab 3 on Meiosis.
3. **Basidiomycota** (club fungi) include mushrooms and shelf fungi and are important decomposers of organic material.

Concept 31.5 Fungi play key roles in nutrient cycling, ecological interactions, and human welfare

▮ Fungi are important decomposers of organic material, including cellulose and lignin.

▮ **Mycorrhizal fungi** are found in association with plant roots and may improve delivery of minerals to the plant, while being supplied with organic nutrients. This is a fine example of mutualism.

▮ **Lichens** are symbiotic associations of photosynthetic microorganisms (algae) embedded in a network of fungal hyphae. They are very hardy organisms that are pioneers on rock and soil surfaces.

▮ Thirty percent of known species of fungi are parasites. Many are plant pathogens (e.g., Dutch elm disease, chestnut blight, dogwood anthracnose, wheat rust, and ergot). Some infect animals (e.g., ringworm, athlete's foot, and *Candida*).

Chapter 32: An Introduction to Animal Diversity

YOU MUST KNOW

- The characteristics of animals.
- The stages of animal development.
- How to sort the animal phyla based on symmetry, development of a body cavity, and the fate of the blastopore.

Concept 32.1 *Animals are multicellular, heterotrophic eukaryotes with tissues that develop from embryonic layers*

▍ Animals have the following characteristics:

1. They are multicellular heterotrophs.
2. Most have muscle and nervous tissue.
3. Most reproduce sexually, with a flagellated sperm and a large egg uniting to form a diploid *zygote*. The diploid stage dominates the life cycle.

▍ Study Figure 7.8. As you do, note and learn each of the following:

- **Zygote**—Fertilized egg
- **Cleavages**—Successive mitotic cell divisions without cell growth between cycles
- **Blastula**—A hollow ball of cells surrounding a cavity called the *blastocoel*
- **Gastrula**—As the blastula is "punched in," the embryonic tissue layers will form
- **Ectoderm**—The outer tissue layer
- **Endoderm**—The inner tissue layer
- **Blastopore**—The opening into the gastrula; it will become the mouth in *protostomes* and the anus in *deuterostomes*
- **Archenteron**—The blind pouch formed by gastrulation

▍ Some animals have **larvae,** an immature form distinct from the adult stage that will undergo **metamorphosis.**

▍ Animals share *Hox genes*, a unique homeobox-containing family of genes that play important roles in development.

Concept 32.3 *Animals can be characterized by "body plans"*

▍ **No symmetry:** the sponges

▍ **Radial symmetry** occurs in jellyfish and other organisms, in which any cut through the central axis of the organism would produce mirror images.

▍ **Bilateral symmetry** occurs in lobsters, humans, and many other organisms. These animals have a right side and a left side, and a single cut would divide the animal into two mirror image halves. There is also a *dorsal* (back) side, *ventral* (belly) side, and *anterior* (head) and *posterior* (tail) ends.

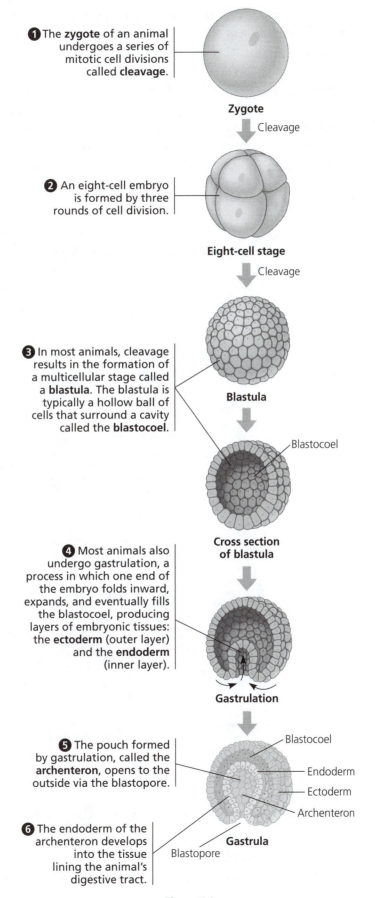

1 The **zygote** of an animal undergoes a series of mitotic cell divisions called **cleavage**.

Zygote

Cleavage

2 An eight-cell embryo is formed by three rounds of cell division.

Eight-cell stage

Cleavage

3 In most animals, cleavage results in the formation of a multicellular stage called a **blastula**. The blastula is typically a hollow ball of cells that surround a cavity called the **blastocoel**.

Blastula

Blastocoel

Cross section of blastula

4 Most animals also undergo gastrulation, a process in which one end of the embryo folds inward, expands, and eventually fills the blastocoel, producing layers of embryonic tissues: the **ectoderm** (outer layer) and the **endoderm** (inner layer).

Gastrulation

Blastocoel

Endoderm

Ectoderm

Archenteron

5 The pouch formed by gastrulation, called the **archenteron**, opens to the outside via the blastopore.

6 The endoderm of the archenteron develops into the tissue lining the animal's digestive tract.

Gastrula

Blastopore

Figure 7.8

- **Cephalization** is the concentration of sensory equipment at one end (usually the anterior, or head end) of the organism.
- **Acoelomates,** such as flatworms, have no cavities between their alimentary canal and the outer wall of their bodies.
- **Pseudocoelomates** are triploblastic animals (animals with three tissue layers) with a cavity formed from the mesoderm and endoderm.
- **Coelomates** possess a **true coelom.** This is a body cavity filled with fluid, and this space separates an animal's digestive tract from the outer body wall. The coelom forms from tissue derived from mesoderm only.

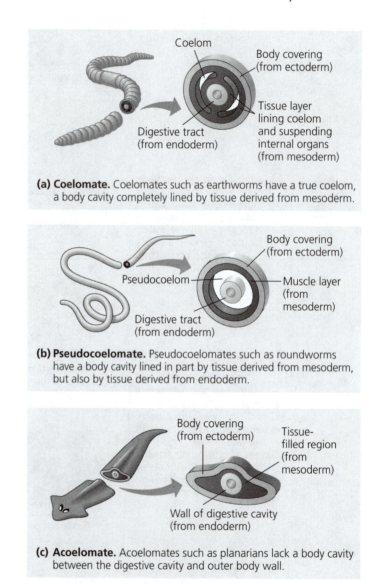

(a) Coelomate. Coelomates such as earthworms have a true coelom, a body cavity completely lined by tissue derived from mesoderm.

(b) Pseudocoelomate. Pseudocoelomates such as roundworms have a body cavity lined in part by tissue derived from mesoderm, but also by tissue derived from endoderm.

(c) Acoelomate. Acoelomates such as planarians lack a body cavity between the digestive cavity and outer body wall.

Figure 7.9

- *Functions of the body cavity* (see Figure 7.9) include
    1. Cushion suspended organs
    2. Act as a hydrostatic skeleton
    3. Enable internal organs to grow and move independently

■ **Protostomes** and **deuterostomes** differ in three major ways. Study Figure 7.10 to summarize these.

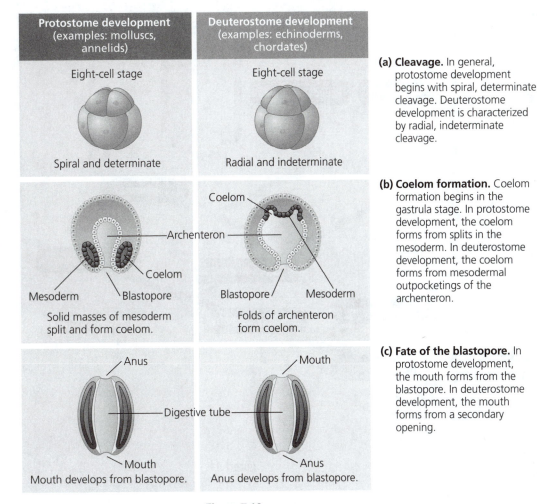

Protostome development
(examples: molluscs, annelids)

Eight-cell stage

Spiral and determinate

Deuterostome development
(examples: echinoderms, chordates)

Eight-cell stage

Radial and indeterminate

Key

Ectoderm

Mesoderm

Endoderm

Archenteron

Coelom

Mesoderm   Blastopore

Solid masses of mesoderm split and form coelom.

Coelom

Blastopore   Mesoderm

Folds of archenteron form coelom.

Anus

Digestive tube

Mouth

Mouth develops from blastopore.

Mouth

Anus

Anus develops from blastopore.

(a) Cleavage. In general, protostome development begins with spiral, determinate cleavage. Deuterostome development is characterized by radial, indeterminate cleavage.

(b) Coelom formation. Coelom formation begins in the gastrula stage. In protostome development, the coelom forms from splits in the mesoderm. In deuterostome development, the coelom forms from mesodermal outpocketings of the archenteron.

(c) Fate of the blastopore. In protostome development, the mouth forms from the blastopore. In deuterostome development, the mouth forms from a secondary opening.

Figure 7.10

STUDY TIP: Focus on Figure 7.11 as you consider:
Animals in which phylum

1. Lack symmetry and true tissues?
2. Show radial symmetry and are diploblastic?
3. Have three tissue layers but lack a body cavity?
4. Show bilateral symmetry and have a pseudocoelom?
5. Have a true coelom and are protostomes?
6. Have a true coelom and are deuterostomes?

If you can group the animal phyla based on the characteristics above, you are ready for the most common type of animal questions you will see on the AP Biology Exam!

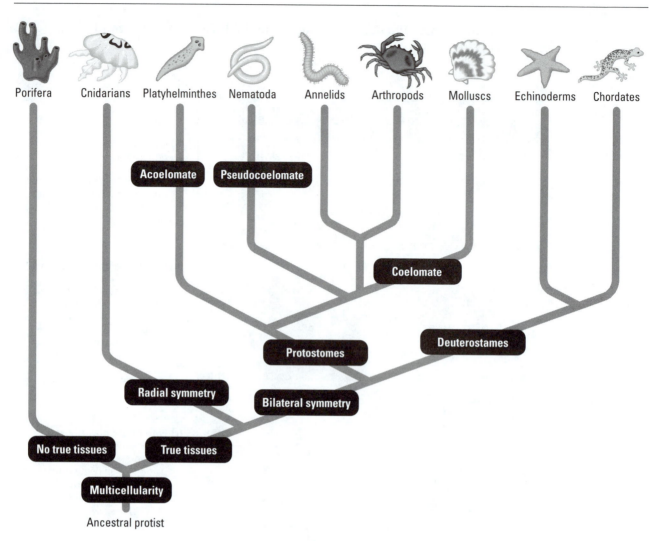

Porifera Cnidarians Platyhelminthes Nematoda Annelids Arthropods Molluscs Echinoderms Chordates

Acoelomate

Pseudocoelomate

Coelomate

Protostomes

Deuterostames

Radial symmetry

Bilateral symmetry

No true tissues

True tissues

Multicellularity

Ancestral protist

Figure 7.11

Chapter 33: Invertebrates

YOU MUST KNOW

- The traits from Figure 7.11 which are used to divide the animals into groups.
- Examples and unique traits for each phylum discussed.
- The evolution of systems for gas exchange, respiration, excretion, circulation, and nervous control.

Invertebrates are animals that lack backbones. Ninety-five percent of all known species are invertebrates. Before we begin our look at them, remember that taxonomy is in flux due to new molecular data. An outline of animal diversity and classification follows.

1. **Subkingdom Parazoa** (lacking true tissues)

   - The **sponges,** once placed in a single phylum, Porifera, are now in the phyla Calcarea and Silicea. Let's focus on the main characteristics of all sponges.
   - Sponges are *sessile* (anchored to the substrate) and lack true tissues. They are diploblastic (their body consists of only two layers of cells).
   - Sponges have no nerves or muscles. The body of a sponge looks like a sac with holes in it. Water is drawn through the pores into the **spongocoel** and flows out through the **osculum** through the movement of flagellated **choanocytes.**
   - **Spicules** comprise a skeletal framework.

2. **Subkingdom Eumetazoa** (animals with true tissues)

   - **A. Radially symmetrical animals**
     All animals except the Parazoa belong to the clade Eumetazoa; they are animals with true tissues.
     1. Members of **Phylum Cnidaria** exist in *polyp* (vase-like; think of a sea anemone) and *medusa* form (think of a jellyfish).
     2. They have radial symmetry, a central digestive compartment known as a gastrovascular cavity, and *cnidocytes* (cells that function in defense and the capture of prey). Examples of cnidarians are hydras, jellyfish, and corals.

   - **B. Bilaterally symmetrical animals**
     1. **Acoelomates** (animals without a body cavity)
        Phylum **Platyhelminthes** (flatworms) Examples: Planaria, tapeworms, flukes
        1. Flattened bodies with *cephalization* (sense organs at the anterior end)
        2. Excretion by *flame bulbs* and *protonephridia*
        3. No specialized organs for gas exchange or circulation
        4. Gastrovascular cavity with a single opening
     2. **Pseudocoelomates**
        Phylum **Nematoda** (roundworms) Examples: *Ascaris, C. elegans, Trichinella spiralis*, pinworms, hookworms
        1. Cylindrical bodies, with a tough *cuticle*
        2. Complete alimentary canal; no circulatory system
     3. **Coelomates**
        - **Phylum Mollusca** (soft-bodied animals) Examples: slugs, clams, snails, squids, and octopuses
          1. Muscular *foot* for movement, a *visceral mass* containing most of the organs, and a *mantle*, which secretes a shell.

2. Open circulatory system. This means that the circulatory fluid (hemolymph) is not always contained within vessels but sometimes circulates through body sinuses called the *hemocoel.*
3. Excretion through *nephridia.*
4. Classes include Polyplacophora (chitons), Gastropoda (snails and slugs), Bivalvia (clams and oysters), and Cephalopoda (squids and octopuses).

• **Phylum Annelida** (Segmented worms) Examples: earthworms, leeches
1. Internal and external segmentation
2. Excretion by *metanephridia* in each segment
3. Closed digestive system with specialized regions (the crop, gizzard, esophagus, and intestine)
4. Brainlike *central ganglia* with a ventral nerve cord
5. Closed circulatory system

• **Phylum Arthropoda** (jointed-legged animals) Examples: insects, arachnids, millipedes, centipedes, crustaceans
1. *Exoskeleton* of *chitin*, which must be *molted* to grow
2. Jointed appendages
3. Open circulatory system
4. Various organs for gas exchange including *gills, book lungs, tracheal systems*
5. Ventral nerve cords
6. Insects undergo metamorphosis during development.
   a. **Incomplete metamorphosis:** egg, nymph, adult (ex. grasshoppers)
   b. **Complete metamorphosis:** egg, larva, pupa, adult (ex. butterflies)

4. **Deuterostomia**
   • The clade Deuterostomia contains a diverse array of organisms, from sea stars to chordates. All have radial cleavage and share common developmental processes. There are two main phyla:
   • Phylum **Echinodermata** (spiny-skinned animals) Examples: sea stars, brittle stars, sea urchins, sand dollars, sea lilies, and sea cucumbers
     1. Larvae have bilateral symmetry; adults radiate from the center, often as five spokes.
     2. Have a thin skin covering an *exoskeleton.*
     3. Have a *water vascular system.* This is a network of internal canals that branch into *tube feet* used for moving, feeding, and gas exchange.
   • **Phylum Chordata** includes two invertebrate subphyla as well as all vertebrates.

Review the table below to learn the most important characteristics of these phyla.

Phylum		Description
Porifera (sponges)		Lack true tissues; have choanocytes (collar cells—unique flagellated cells that ingest bacteria and tiny food particles)
Cnidaria (hydras, jellies, sea anemones, corals)		Unique stinging structures (cnidae), each housed in a specialized cell (cnidocyte); gastrovascular cavity (digestive compartment with a single opening)
Platyhelminthes (flatworms)		Dorsoventrally flattened, unsegmented acoelomates; gastrovascular cavity or no digestive tract
Rotifera (rotifers)		Pseudocoelomates with alimentary canal (digestive tube with mouth and anus); jaws (trophi) in pharynx; head with ciliated crown
Lophophorates: Ectoprocta, Phoronida, Brachiopoda		Coelomates with lophophores (feeding structures bearing ciliated tentacles)
Nemertea (proboscis worms)		Unique anterior proboscis surrounded by fluid-filled sac; alimentary canal; closed circulatory system
Mollusca (clams, snails, squids)		Coelomates with three main body parts (muscular foot, visceral mass, mantle); coelom reduced; most have hard shell made of calcium carbonate
Annelida (segmented worms)		Coelomates with body wall and internal organs (except digestive tract) segmented
Nematoda (roundworms)		Cylindrical, unsegmented pseudocoelomates with tapered ends; no circulatory system
Arthropoda (crustaceans, insects, spiders)		Coelomates with segmented body, jointed appendages, and exoskeleton made of protein and chitin
Echinodermata (sea stars, sea urchins)		Coelomates with secondary radial anatomy (larvae bilateral; adults radial); unique water vascular system; endoskeleton
Chordata (lancelets, tunicates, vertebrates)		Coelomates with notochord; dorsal, hollow nerve cord; pharyngeal slits; muscular, post-anal tail

Metazoa

Eumetazoa

Bilateria

Deuterostomia

Chapter 34: Vertebrates

Concept 34.1 Chordates have a notochord and a dorsal, hollow nerve cord

▌ **Vertebrates** derive their name from the **vertebrae,** a series of bones which make up the backbone. In the majority of vertebrates, vertebrae enclose the spinal cord and have assumed the roles of the notochord.

▌ Vertebrates are members of the phylum **Chordata.**

▌ There are four characteristics of all chordates. Note that some of these are only present during embryonic development (see Figure 7.12).

1. A **notochord**—a long, flexible rod that appears during embryonic development between the digestive tube and the dorsal nerve cord. This is *not* the spinal cord or the vertebral column! Students commonly confuse all of these.

2. A **dorsal, hollow nerve cord**—formed from a plate of ectoderm that rolls into a hollow tube. Recall that other phyla you have studied, such as annelids and arthropods, had a *ventral* nerve cord.

3. **Pharyngeal clefts**—grooves that separate a series of pouches along the sides of the pharynx. In most chordates (but not tetrapods, the critters with four legs), the clefts develop into slits that allow water to enter and exit the mouth without going through the digestive tract.

4. A muscular **tail** posterior to the anus.

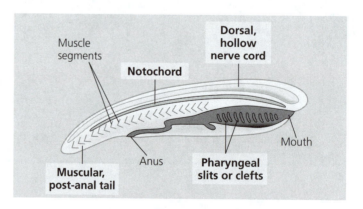

Figure 7.12

Concept 34.3 *Vertebrates are craniates that have a backbone*

As you review this section, use Figure 7.13 to help you organize these groups.

- **Lampreys** are the oldest lineage of vertebrates. They are jawless parasitic fish with a skeleton of cartilage.

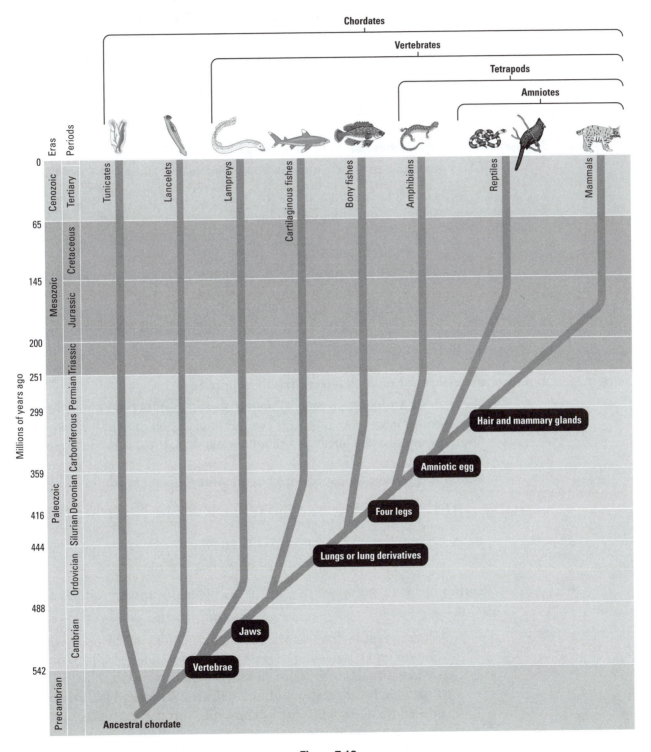

Figure 7.13

Concept 34.4 Gnathostomes are vertebrates that have jaws

▌ The jaws of vertebrates evolved from the modification of skeletal parts that had once supported the pharyngeal (gill) slits.

    1. **Class Chondrichthyes** have flexible endoskeletons composed of cartilage, possess streamlined bodies, are denser than water, and will sink if they stop swimming. Some examples are sharks and rays (such as stingrays).

    2. **Class Osteichthyes** are the bony fishes; these are the most numerous of all vertebrate groups. They have a bony endoskeleton, are covered in scales, and possess a swim bladder for buoyancy control. Some examples are trout and salmon.

Concept 34.5 Tetrapods are gnathostomes that have limbs and feet

▌ **Class Amphibia** members include frogs, toads, salamanders, and newts.

    1. Gas exchange can occur across their thin, moist skin, though many members have lungs.

    2. Some, such as frogs, have an aquatic larval stage with gills and metamorphosis to an adult stage with lungs.

    3. They have external fertilization and external development in an aquatic environment. Eggs lack a shell.

    4. As larvae, amphibians have a two-chambered heart (one atrium, one ventricle), while the adult stage generally has three chambers (two atria, one ventricle).

Concept 34.6 Amniotes are tetrapods that have a terrestrially adapted egg

▌ The clade of **Amniotes** consists of mammals and reptiles (including birds).

▌ The **amniotic egg** was an important evolutionary development for life on land. Amniotic eggs have a shell that retains water and, thus, can be laid in a dry environment.

▌ Amniotic eggs have four specialized **extraembryonic membranes:**

    1. The **amnion** encloses a fluid compartment that bathes the embryo and absorbs shock.

    2. The **chorion, allantois,** and **yolk sac** function in gas exchange, waste storage, and transfer of stored nutrients.

1. **Reptiles** include turtles, lizards and snakes, alligators and crocodiles, and the extinct dinosaurs.

    1. Modern reptiles have **scales,** containing keratin, which are an adaptation for terrestrial living because they reduce water loss. They obtain oxygen through their lungs, not their skin.

    2. Reproductive adaptations of internal fertilization and the amniote shelled egg allow reproduction on land. Nitrogenous waste is produced as *uric acid.*

3. Most reptiles are *ectothermic.* They regulate body temperature through behavioral adaptations rather than by metabolism.
4. Most reptiles have a heart with two atria and a ventricle, which has a partial septum.

2. **Birds** all have wings, a body covering of feathers, and many adaptations that facilitate flight.

1. They lay **amniotic eggs** and have keratin-containing **scales** on their legs—both of which are reptilian characteristics.
2. Most birds' bodies are constructed for flight, with light, hollow bones; relatively few organs; wings; and feathers.
3. **Feathers** and, in some cases, a layer of fat insulate birds and help them maintain internal temperature.
4. Birds have a **four-chambered heart** and a high rate of metabolism.
5. Birds are **endotherms** and maintain a warm, consistent body temperature.

Concept 34.7 Mammals are amniotes that have hair and produce milk

▌ **Mammals** share certain characteristics.

1. All possess **mammary glands** that produce milk.
2. Mammals have a body covering of **hair.**
3. They have a four-chambered heart and are **endothermic.**
4. Mammals have internal fertilization, and most are born rather than hatched.
5. They have **proportionally larger brains** than other vertebrates, and all have **teeth** of differing size and shape.

▌ Mammals can be placed into three groups:

1. **Monotremes** are egg-laying mammals that have hair and produce milk. Examples are platypuses and spiny anteaters.
2. **Marsupials** are born early in development and complete embryonic development in a marsupium (pouch) while nursing. Examples include kangaroos and opossums.
3. **Placental mammals (eutherians)** have a longer period of pregnancy; they complete their development in the uterus. Examples include mice, dogs, and humans.

▌ Humans belong to the order **Primates,** along with monkeys and gorillas. Some characteristics common to all **primates** include opposable thumbs, large brains and short jaws, forward-looking eyes, flat nails, well-developed parental care, and complex social behavior.

▌ Some features of **human evolution** are increased brain volume, shortening of the jaw, bipedal posture, reduced size-difference between the sexes, and certain important changes in family structure.

For Additional Review

Compare the land adaptations of plants and animals, including how each manages the uptake of nutrients and water as well as the excretion of wastes.

Multiple-Choice Questions

1. Which group thrives in extreme heat and acidic environments?
   (A) Bryophytes
   (B) Protists
   (C) Archaea
   (D) Fungi
   (E) Deuterostomia

2. Which of the following groups is best characterized as being eukaryotic, multicellular, heterotrophic, and without a cell wall?
   (A) Plantae
   (B) Animalia
   (C) Fungi
   (D) Viruses
   (E) Monera

3. A biologist captures an aquatic organism from a freshwater pond. It has a segmented body and a brainlike pair of central ganglia. It is also a bottom dweller that burrows into the pond floor. Most likely this organism is
   (A) a protist.
   (B) an annelid.
   (C) an arachnid.
   (D) a flatworm
   (E) a reptile.

4. In which of the following pairs are the organisms most closely related taxonomically?
   (A) squid; snails
   (B) mushrooms; tulips
   (C) clams; lobsters
   (D) sponges; hydras
   (E) lancelets; roundworms

Directions: The group of questions below consists of five lettered choices followed by a list of numbered phrases or sentences. For each numbered phrase or sentence, select the one choice that is most closely related to it. Each choice may be used once, more than once, or not at all.

Questions 5–9
   (A) Chordata
   (B) Arthropoda
   (C) Annelida
   (D) Platyhelminthes
   (E) Porifera

5. Acoelomates; lack a body cavity, flattened bodies, gastrovascular cavity with a single opening

6. Bodies segmented internally and externally, ventral nerve cord, respiration occurs through the skin

7. Possess a notochord; a dorsal, hollow nerve cord; pharyngeal clefts; and a post-anal tail at some point during development

8. Segmented coelomates that have exoskeletons of chiton and jointed appendages

9. Sessile, possessing no nerves or muscles, have a central cavity lined with choanocytes, are filter feeders

10. Systematists categorize all living creatures into what three domains?
    (A) Bacteria, Euglena, Eukarya
    (B) Bacteria, Archaea, Eukarya
    (C) Archaea, Plantae, Eukarya
    (D) Protista, Plantae, Eukarya
    (E) Prokaryota, Eukarya, Plantae

11. The cell walls of bacteria contain which of the following materials?
    (A) peptidoglycans
    (B) polynucleotides
    (C) pyrimidines
    (D) disaccharides
    (E) ribose

12. The most common method of locomotion in prokaryotes occurs via the movement of
    (A) claws.
    (B) contractions of the vascular cavity.
    (C) flagella.
    (D) tentacles.
    (E) repeated alteration of cell shape.

13. Which of the following compounds is initially *only* synthesized by an autotroph?
    (A) cholesterol
    (B) glycerol
    (C) polypeptides
    (D) ribonucleic acids
    (E) glucose

14. Which of the following is a symbiotic relationship in which both organisms benefit?
    (A) parasitism
    (B) commensalism
    (C) mutualism
    (D) obligate
    (E) heterotrophic

15. One example of a commensal symbiotic relationship is
    (A) a legume plant, which houses prokaryotic bacteria that fix nitrogen.
    (B) a pig which has roundworms in its intestine.
    (C) the vagina of a human, which hosts fermenting bacteria that produce acids.
    (D) cattle egrets feeding on insects flushed out by grazing cattle.
    (E) animal cells hosting a virus.

16. Which of the following constitutes the kingdom containing the widest array of organisms?
    (A) Animalia
    (B) Plantae
    (C) Fungi
    (D) Protista
    (E) Bacteria

17. From an evolution perspective, mitochondria and chloroplasts are the result of what process?
    (A) endosymbiosis
    (B) conjugation
    (C) transformation
    (D) ectosymbiosis
    (E) compartmentalization

18. Adaptations for terrestrial life seen in all plants are
    (A) chlorophylls *a* and *b*.
    (B) cell walls of cellulose and lignin.
    (C) vascular tissue and stomata.
    (D) alternation of generations.
    (E) sporopollenin, protection, and nourishment of embryo by gametophyte.

Directions: The group of questions below consists of five lettered choices followed by a list of numbered phrases or sentences. For each numbered phrase or sentence, select the one choice that is most closely related to it. Each choice may be used once, more than once, or not at all.

Questions 19–22
    (A) Bryophytes
    (B) Pteridophytes
    (C) Gymnosperms
    (D) Angiosperms
    (E) Fungi

19. Contain the mosses, lack vascular tissue

20. Contain the flowering plants, possess vascular tissue

21. Contain the ferns, are the seedless plants, and have vascular tissue

22. Contain the conifers, have seeds and vascular tissue

23. In alternation of generations in land plants, which of the following represents the haploid stage?
(A) zygote
(B) gametophyte
(C) sporophyte
(D) spore
(E) sporangia

24. All of the following are adaptations of plants to terrestrial life EXCEPT
(A) stomata.
(B) cuticle.
(C) xylem.
(D) phloem.
(E) photosynthesis.

25. Which plants have xylem and phloem?
(A) all plants
(B) all bryophytes
(C) bryophytes, ferns, gymnosperms, and angiosperms
(D) only the vascular plants with seeds
(E) all vascular plants, which include ferns, gymnosperms, and angiosperms

26. Which of the following is the dominant stage of the life cycle for seedless vascular plants?
(A) gametophyte
(B) sporophyte
(C) zygote
(D) heterosporous
(E) homosporous

27. A student is collecting samples of plants from a field in Connecticut. He picks one from the ground to study it more closely. In the process, he notices that the leaves show veins that run parallel from end to end. This plant is mostly likely a
(A) fern.
(B) conifer.
(C) monocot.
(D) eudicot.
(E) liverwort.

28. The tiny individual filaments that comprise the body of fungi are known as
(A) hyphae.
(B) mycelium.
(C) chitin.
(D) mycorrhizae.
(E) basidium.

29. Which one of the following is not a characteristic of fungi?
(A) heterotrophs that absorb nutrients
(B) mycelium made up of hyphae
(C) cell wall of chitin
(D) some members that are autotrophic
(E) reproduction by means of spores

30. Which of these are unicellular fungi that live in damp environments?
(A) mushrooms
(B) yeast
(C) lichens
(D) slime molds
(E) protozoans

31. Which of these is an organism with symbiotic associations of photosynthetic microorganisms in a network of fungal hyphae?
(A) mushrooms
(B) yeast
(C) lichens
(D) slime molds
(E) protozoans

32. Which of these is a mutualistic association of plant roots and fungi?
    (A) mushrooms
    (B) mycorrhizae
    (C) lichens
    (D) hyphae
    (E) sporangia

33. Place the following groups of plants in order beginning with those that first appeared on Earth:
    (A) moss, angiosperms, ferns, gymnosperms
    (B) liverworts, ferns, gymnosperms, angiosperms
    (C) moss, ferns, angiosperms, gymosperms
    (D) moss, liverworts, tracheophytes, bryophytes
    (E) seed plants, cone-bearing plants, bryophytes

34. All of the following terms reflect characteristics of almost all animals EXCEPT
    (A) multicellularity.
    (B) heterotrophic nutrition.
    (C) homeobox genes.
    (D) eukaryotic.
    (E) diurnal.

35. The body plans of sea anemones exhibit
    (A) bilateral symmetry.
    (B) radial symmetry.
    (C) dorsal and ventral symmetry.
    (D) anterior and posterior symmetry.
    (E) no symmetry.

36. Which of the following is descriptive of the embryonic development of most protostomes?
    (A) radial and determinate cleavage
    (B) spiral and determinate cleavage
    (C) spiral and indeterminate cleavage
    (D) radial and indeterminate cleavage
    (E) blastopore becomes the anus

37. Evolution of which feature enabled vertebrates to reproduce successfully on land?
    (A) the amniotic egg
    (B) quadruped locomotion
    (C) body hair
    (D) opposable thumbs
    (E) cloaca

38. Which of the following animals is characterized by a relatively short period of pregnancy followed by a period of nursing as its offspring completes development?
    (A) placental mammals
    (B) marsupial mammals
    (C) monotremes
    (D) therapsids
    (E) carnites

39. A body cavity that is not completely lined by tissue derived from the mesoderm is called a(n) and is seen in which phylum?
    (A) coelom/Chordata
    (B) coelom/Nematoda
    (C) pseudocoelom/Nematoda
    (D) pseudocoelom/Annelida
    (E) coelom/Echinodermata

40. Which three vertebrate classes have internal fertilization?
    (A) Birds, Mammalia, Osteichthyes
    (B) Amphibia, Reptilia, Birds
    (C) Reptilia, Birds, Mammalia
    (D) Chondrichthyes, Osteichthyes, Amphibia
    (E) Osteichthyes, Reptilia, Mammalia

Free-Response Question

1. Systematists are scientists who study evolutionary relationships between organisms; they use scientific evidence to construct hypothetical phylogenies that show these relationships. Systematists have replaced the five-kingdom system with a more accurate three-domain system, containing the Archaea, the Bacteria, and the Eukarya.

   (a) Describe how this scheme for classification differs from the old five-kingdom one.

   (b) Describe three types of evidence that scientists used to develop this three-domain system.

ANSWERS AND EXPLANATIONS

Multiple-Choice Questions

1. **(C) is correct.** The prokaryotes are split into two domains, Archaea and Bacteria. Many Archaea are known as extremophiles; they are often found in extreme (i.e., salty or very hot) environments, such as the geysers of Yellowstone National Park.

2. **(B) is correct.** Animals are heterotrophic—they are not capable of fixing carbon and must obtain it from phototropic organisms. They belong to the Eukarya, and are multicellular. The fungi are also eukaryotic, multicellular heterotrophs but have a cell wall made of chitin.

3. **(B) is correct.** Annelids are characterized by internally and externally segmented bodies; they are 1 mm to 3 m in length and live in freshwater habitats, the soil, and the sea. Earthworms are a common annelid. The flatworm would have been your second best choice, but they are Platyhelminthes and lack segmentation. Planaria, tapeworms, and flukes are all flatworms.

4. **(A) is correct.** The two organisms most closely related are the squids and snails. These two are members of the phylum Mollusca. The other choices do not include members of the same phylum.

5. **(D) is correct.** Platyhelminthes are flatworms—acoelomates that have gastrovascular cavities.

6. **(C) is correct.** Phylum Annelida contains organisms with segmented body plans that inhabit freshwater or damp terrestrial habitats. They have closed circulatory systems, and respiration occurs through the skin.

7. **(A) is correct.** The chordates are mostly vertebrates, although there are two groups of invertebrate chordates. The chordates are grouped according to the presence of a notochord; a dorsal, hollow nerve cord; pharyngeal clefts; and a post-anal tail. Many of these features exist only during embryonic development.

8. **(B) is correct.** Including the lobsters, spiders, and other related organisms, the arthropods are characterized by having an exoskeleton (which is sometimes shed in a process called molting), jointed appendages, and segmentation.

9. **(E) is correct.** Porifera are the simplest invertebrates. They are sessile and have no muscles or nerve cells. They resemble a sac with holes in it, and they draw water into their central cavity (spongocoel), filtering out food particles.

■ **10. (B) is correct.** The three domains into which all the living organisms are placed by systematists are Bacteria, Archaea, and Eukarya. Domains are one taxonomic level above kingdoms. Prokaryotes make up Archaea and Bacteria, whereas eukaryotes make up Eukarya.

■ **11. (A) is correct.** Peptidoglycans are polymers of modified sugars that are linked by short polypeptides that differ from species to species. Fungi have cell walls of chitin, whereas plants have cell walls of cellulose.

■ **12. (C) is correct.** The flagella account for most movements in prokaryotes. Prokaryotic flagella are much smaller than eukaryotic flagella and operate under different mechanisms.

■ **13. (E) is correct.** Heterotrophs need at least one organic compound in order to make others. Chemotrophs get their energy from chemicals in the environment. Autotrophs can make organic compounds such as glucose from CO_2.

■ **14. (C) is correct.** In mutualistic symbiosis, both organisms benefit from the association; in commensalistic symbiosis, one organism benefits, while the other is neither helped nor harmed; and in parasitic symbiosis, one organism benefits, while the other is harmed.

■ **15. (D) is correct.** Commensal association sometimes involves one species (the cattle egret) inadvertently helped out by another (the grazing cattle). Because the birds increase their feeding rate when following the cattle, they clearly benefit with no known benefit to the cattle.

■ **16. (D) is correct.** Protista is by far the most varied kingdom. Most of the protists are unicellular, but there are multicellular protists, too. Protists can be divided into protozoa (animal-like protists), algae (plant-like protists), and fungus-like protists.

■ **17. (A) is correct.** Endosymbiosis—the theory of serial endosymbiosis—proposes that mitochondria and chloroplasts evolved from small prokaryotes that were engulfed by larger host cells. Eventually they became permanent functional parts of the cell.

■ **18. (E) is correct.** The key phrase in the question was *adaptations for terrestrial life.* Sporopollenin protects the male gamete from drying, as does the enclosure of the zygote in the female gametophyte.

■ **19. (A) is correct.** Bryophytes are one of the four main types of land plants. The most common bryophytes are the mosses, which do not have vascular tissue and are relatively simple.

■ **20. (D) is correct.** The angiosperms are the flowering plants. They have vascular tissue, and the flower is their reproductive structure. They also have seeds enclosed within their fruits.

■ **21. (B) is correct.** Pteridophytes are seedless plants with vascular tissue. One prominent member of this group is the ferns.

■ **22. (C) is correct.** Gymnosperms are seed plants that have "naked" seeds. These seeds are not enclosed in any specialized chambers. But these plants do possess vascular tissue. Pine trees are a common example of gymnosperms.

■ **23. (B) is correct.** The cells of the gametophyte are haploid (*n*) and have a single set of chromosomes. The fusion of sperm and egg cells during fertilization produces the zygote, which is diploid. The zygote divides mitotically to produce the sporophyte, which is also diploid.

24. (E) is correct. Photosynthesis in plants does not necessarily contribute to their ability to conserve water or to store water in their structure. The other plant structures listed are all evolutionary adaptations for terrestrial plants.

25. (E) is correct. Xylem and phloem are the defining trait of all vascular plants. Bryophytes (mosses and liverworts) lack vascular tissue.

26. (B) is correct. The sporophyte stage is the dominant one in the life cycle of the seedless vascular plants, which include the ferns. The sporophyte is the diploid stage, whereas gametophytes are tiny and exist only briefly.

27. (C) is correct. This plant is probably a monocot. Monocots have veins that run parallel, whereas dicot leaves have a netlike vein arrangement. Grasses are good examples of monocots.

28. (A) is correct. Hyphae are filaments made up of tube-like walls that surround cytoplasm and plasma membranes. They group together to form a woven mat called a mycelium.

29. (D) is correct. There are no fungi that are autotrophs. All fungi absorb nutrients and are consumers.

30. (B) is correct. Yeasts are unicellular fungi that live in moist environments or liquids. They reproduce by cell division or budding.

31. (C) is correct. Lichens are symbiotic associations of millions of photosynthetic organisms in a network of fungal hyphae. They are very hardy and frequently colonize newly broken rock faces.

32. (B) is correct. Mycorrhizae are mutualistic associations of fungi and plant roots. They exchange minerals extracted from the soil and nutrients produced by the plants.

33. (B) is correct. Mosses lack vascular tissue, ferns have vascular tissue and reproduce with spores, and gymnosperms reproduce with cones. Angiosperms (flowering plants) were the last to appear.

34. (E) is correct. The first four answer choices represent characteristics common to almost all animals; the last one does not. The characteristic of being diurnal (the opposite of nocturnal—being active during the day) is not a requirement for belonging to the kingdom Animalia.

35. (B) is correct. Sea anemones can best be described as having radial symmetry. Radial animals have a top and bottom but no head or rear end, or left and right side.

36. (B) is correct. "Protostome" means "first mouth." The protostomes have spiral, determinate cleavage, and the blastopore, the first opening into the embryo, develops into a mouth.

37. (A) is correct. The amniotic egg is composed of extraembryonic membranes that function in gas exchange, waste storage, and the delivery of nutrients to the embryo.

38. (B) is correct. Marsupials are predominantly located in Australia, where they have radiated to fill niches occupied by placental mammals elsewhere. Marsupials have a short period of pregnancy, and the embryo completes its development nursing within the pouch of the mother.

39. (C) is correct. Animals that have no cavity between their alimentary canal and their body wall are acoelomates. Platyhelminthes are acoelomates. Those that have a cavity not derived from the mesoderm are pseudocoelomates;

nematodes are pseudocoelomates. Those that have a true coelom derived from the mesoderm are coelomates and include echinoderms, annelids, and chordates.

■ **40. (C) is correct.** Birds and reptiles lay shelled eggs, and these must be fertilized internally before they are laid. Mammals also have internal fertilization. Amphibians and the fishes (Osteichthyes and Chondrichthyes) have external fertilization as their gametes are deposited in water.

Free-Response Question

(a) The old five-kingdom system contained the kingdoms Monera, Protista, Plantae, Fungi, and Animalia. However, scientists have determined that the Monera kingdom should be further separated into two groups. Thus, they split Monera into Archaea and Bacteria. They realized that the Archaea and Bacteria are as different from each other as they are different from us, the Eukarya. The new system is a three-domain system, with the domains Archaea, Bacteria, and Eukarya. Archaea and Bacteria are composed of prokaryotes, and Eukarya is made up of eukaryotes. From the old five-kingdom system, the Protista, Plantae, Fungi, and Animalia are all placed in the domain Eukarya.

(b) Three methods or types of evidence that scientists use to classify organisms and study their degree of evolutionary relatedness are fossil evidence, the structure and development of organisms, and molecular evidence.

Systematists can also study patterns of homologous structures to determine evolutionary relationships. Species that were derived from the same ancestor should have similarities, called homologies. Here, scientists would look for homologous structures—structures that are similar in different species—and use these to tie organisms together.

The third way that systematists can study evolutionary relationships is on a molecular level. Because DNA is heritable, related species should share common genes, and the more recently the species branched off from a common ancestor, the more similar their DNA should be. Studying the DNA of organisms makes it possible for systematists to determine the degree of evolutionary difference between two species that are nearly identical in appearance, and it also allows systematists to judge the relatedness of two species that they might not guess would be related at all, based on external appearance. It is a much more precise and quantitative method for appraising evolutionary relatedness.

This response uses the following key terms in context, showing the writer's knowledge of their meanings and relatedness:

Monera	Bacteria
Protista	Eukarya
Plantae	systematists
Fungi	homologous structures
Animalia	DNA
Archaea	

This response also accurately describes the rationale behind the process of the reorganizing of the five-kingdom system into the three-domain system. It clearly describes three methods for studying the evolutionary relatedness of species, thus answering the question in a complete and organized way. Always check your final response to make sure it addresses the original question.

Plant Form and Function

Chapter 35: Plant Structure, Growth, and Development

YOU MUST KNOW

- The function of xylem and phloem tissue.
- The specific functions of tracheids, vessels, sieve-tube elements, and companion cells.
- The correlation between primary growth and apical meristems versus secondary growth and lateral meristems.

Concept 35.1 The plant body has a hierarchy of organs, tissues, and cells

▌ Plants have a **root system** beneath the ground that is a multicellular organ which anchors the plant, absorbs water and minerals, and often stores sugars and starches. Additional structural characteristics of roots include the following:

■ *Fibrous roots* are made up of a mat of thin roots that are spread just below the soil's surface. *Taproots* are made up of one thick, vertical root with many lateral roots that come out from it.

■ At the tips of the roots vast numbers of tiny *root hairs* increase the surface area enormously, making efficient absorption of water and minerals possible. Plants may also have a symbiotic relationship with fungi at the tips of the roots, termed **mycorrhizae** ("fungus roots"). Mycorrhizae assist in the absorption process and are found in the vast majority of all plants.

▌ Plants have a **shoot system** above the ground that works as a multicellular organ consisting of stems and leaves.

■ **Stems** function primarily to display the leaves. Two types of buds help achieve this purpose:

1. A **terminal bud** is located at the top end of the stem where growth usually occurs. In apical dominance, the terminal bud prohibits the

growth of the axillary buds. This concentrates the growth of the plant upward, toward more light.

2. **Axillary buds** are located in the V formed between the leaf and the stem; these buds have the potential to form a branch (or lateral shoot).

- **Leaves** are the main photosynthetic organ in most plants.

- Plant organs—leaf, stem, and root—are composed of three tissue types:

    1. **Dermal tissue** is a single layer of closely packed cells that cover the entire plant and protect it against water loss (a waxy layer termed the **cuticle** in the leaves) and invasion by pathogens like viruses and bacteria.
    2. **Vascular tissue** is continuous through the plant and transports materials between the roots and shoots. Vascular tissue is made up of **xylem**, which transports water and minerals up from the roots, and **phloem**, which transports food from the leaves to the other parts of the plant.
    3. **Ground tissue** is anything that isn't dermal tissue or vascular tissue. Any ground tissue located inside the vascular tissue is *pith;* any ground tissue outside the vascular tissue is *cortex.*

- Plants have five major types of differentiated cells:

    1. **Parenchyma cells** are the most abundant cell type and are present throughout the plant. These cells perform most of the metabolism (including photosynthesis) in the plant.
    2. **Collenchyma cells** are grouped in cylinders and help support growing parts of the plant. The strings of celery are composed of vascular tissue surrounded and supported by collenchyma cells.
    3. **Sclerenchyma cells** exist in parts of the cell that are no longer growing. They have tough cell walls specialized for support.
    4. **Xylem cells** have two types of water conducting cells: tracheids and vessels that are dead at functional maturity.
        - Figure 8.1 shows several of the key features of tracheids, which are found in all vascular plants. **Tracheids** are long, thin cells with thick secondary cell walls strengthened with lignin. Water moves from cell to cell mainly through the pits, where the water does not have to cross the secondary cell walls.
        - **Vessels** are found primarily in angiosperms. They have both pits and perforated end walls for water movement. Note the difference in diameter between tracheids and pits in Figure 8.1.
    5. **Phloem cells** conduct sugar and other organic compounds. Phloem is composed of two types of cells, sieve tubes and companion cells, both alive at maturity.
        - Sieve tubes consist of chains of cells called **sieve-tube elements**. Figure 8.2 shows this relationship. Sieve-tube cells are highly modified for transport, lacking a nucleus, ribosomes, and a central vaculole.

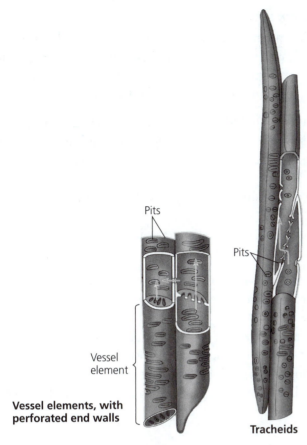

Pits

Vessel
element

**Vessel elements, with
perforated end walls**

**Figure 8.1 Tracheids are found in all vascular
plants while vessels are found primarily in
angiosperms. Both cell types conduct water**

Pits

Tracheids

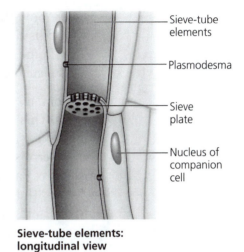

Sieve-tube
elements

Plasmodesma

Sieve
plate

Nucleus of
companion
cell

**Sieve-tube elements:
longitudinal view**

**Figure 8.2 Sieve-tube member cells
are the conducting cells of phloem.**

- **Companion cells** provide for the molecular needs of the sieve-tube elements. **Companion cells** are connected to the sieve-tube element cell by numerous plasmodesmata.

Concept 35.2 Meristems generate cells for new organs

▌ Based on their life cycle, flowering plants can be classified as **annuals** (life cycle completed in one year), **biennials** (life cycle completed in two years), or **perennials** (life cycle continues for many years).

▌ **Meristems** are perpetually embryonic tissues that are responsible for indeterminate growth (growth throughout the plant's life). Growth occurs only as a result of cell division in meristem tissue.

1. **Apical meristems** are located at the tips of roots and in buds of shoots. These are the sites of cell division, allowing the plant to grow in length. *Primary growth* occurs when the plant grows at the apical meristems (length).

2. **Lateral meristems** result in growth which thickens the shoots and roots. This is termed *secondary growth*.

Concept 35.3 Primary growth lengthens roots and shoots

▮ The **root cap** protects the delicate meristem of the root tip as it pushes through the soil. It also secretes a polysaccharide lubricant. The **root tip** contains three zones of cells in various stages of growth. In AP Lab 3, Mitosis and Meiosis, you looked at an onion root tip and should have observed the following:

1. The **zone of cell division** includes root apical meristem and its derivatives. New root cells are produced in this region, including the cells of the root cap. (This is where you would have seen many cells in various stages of the cell cycle.)
2. Above the zone of cell division is the **zone of elongation**, in which cells elongate significantly.
3. In the **zone of maturation**, the three systems in primary growth complete their differentiation and become functionally mature.

▮ At a shoot, the apical meristem is a dome of dividing cells at the tip of a terminal bud. This is primary growth and is accomplished by cell division and cell elongation.

▮ The epidermis of the underside of the leaf is interrupted by **stomata**, which are small pores flanked by *guard cells*, which open and close the stomata.

▮ In leaves, the ground tissue is sandwiched between the upper and lower epidermis, in the mesophyll. It is made up of parenchyma cells, the sites of photosynthesis.

Concept 35.4 Secondary growth adds girth to stems and roots in woody plants

▮ Two lateral meristems take part in plant growth. The **vascular cambium** produces secondary xylem (wood). The **cork cambium** produces a tough covering that replaces epidermis early in secondary growth.

▮ **Bark** is all the tissues outside the vascular cambium. Bark includes the phloem derived from the vascular cambium, the cork cambium, and the tissues derived from the cork cambium.

Concept 35.5 Growth, morphogenesis, and differentiation produce the plant body

▮ By increasing cell number, division in meristems increases the potential for growth. However, it is cell expansion that accounts for the actual increase in plant mass.

▮ **Morphogenesis**—the development of body form and organization—must occur in order for cells to be organized into multicellular arrangements such as tissues. The development of specific structures in specific locations is called **pattern formation**. Many developmental biologists postulate that pattern formation is determined by **positional information** in the form of signals that continuously indicate to each cell its location within a developing structure.

▮ One type of positional information is associated with **polarity**, the condition of having structural differences at opposite ends of an organism. Plants typically have an axis, with a root end and a shoot end.

Chapter 36: Resource Acquisition and Transport in Vascular Plants

YOU MUST KNOW

- The role of passive transport, active transport, and cotransport in plant transport.
- The role of diffusion, active transport, and bulk flow in the movement of water and nutrients in plants.
- How the transpiration cohesion-tension mechanism explains water movement in plants.
- How pressure flow explains translocation.

Concept 36.1 Transport occurs by short-distance diffusion or active transport and by long-distance bulk flow

▌ Transport begins with the movement of water and solutes across a cell membrane.

- Solutes diffuse down their electrochemical gradients (recall from Chapter 7 that electrochemical gradients are the combined effects of the concentration gradient of the solute and the voltage or charge differential across the membrane).
- If no energy is required to move a substance across the membrane, then the movement is termed *passive transport*. Diffusion is an example of passive tranport.
- If energy is required to move solutes across the membrane, it is termed *active transport*. As most solutes cannot move across the phospholipid barrier of the membrane, a **transport protein** is required. The most important transport protein in plants is the **proton pump**.
- A proton pump creates an electrochemical gradient by using the energy of ATP to pump hydrogen ions across the membrane. This potential energy can then be used in the process of **cotransport**—the coupling of the steep gradient of one solute (hydrogen in our example) with a solute like sucrose. The drop in potential energy experienced by the hydrogen ion pays for the transport of the sucrose.

▌ The uptake of water across cell membranes occurs through osmosis, the passive transport of water across a membrane.

- Water moves from areas of high water potential to low water potential. **Water potential** includes the combined effects of solute concentration and physical pressure.
- The water potential equation is $\psi = \psi_s + \psi_p$, where ψ is water potential, ψ_s is solute potential, and ψ_p is the pressure potential. The understanding of this formula is an objective from Laboratory 1 in the AP Lab book.

- By definition the ψ_s of pure water is 0. Adding solutes to pure water always lowers water potential. The solute potential of a solution is therefore always negative.
- Pressure potential is the physical pressure on a solution. An example of positive ψ_p occurs when the cell contents press the plasma membrane against the cell wall, a force termed *turgor pressure*. If the cell loses water, the pressure potential becomes more negative, resulting in wilting.

- **Aquaporins** are the transport proteins (channels) in the plant plasma membrane specifically designed for the passage of water.
- **Bulk flow** is the movement of water through the plant from regions of high pressure to regions of low pressure. Water and solutes move through xylem and phloem tissue by way of bulk flow.

Concept 36.2 Roots absorb water and minerals from the soil

- Most water absorption occurs near the root tips through the root hairs.
- Water and minerals from the soil enter the plant through the root epidermis, cross the cortex, pass into the vascular cylinder, and then flow up tracheids and vessels to the shoot system.
- The roots of many types of plants have symbiotic relationships with fungi. This type of relationship is called **mycorrhizae**, and leads to the absorption and transport of water and certain minerals into the plant.

Concept 36.3 Water and minerals are transported from roots to shoots

- Use Figure 8.3 to chart the movement of water and minerals out of the soil, across the root, and up through the xylem to the rest of the plant.

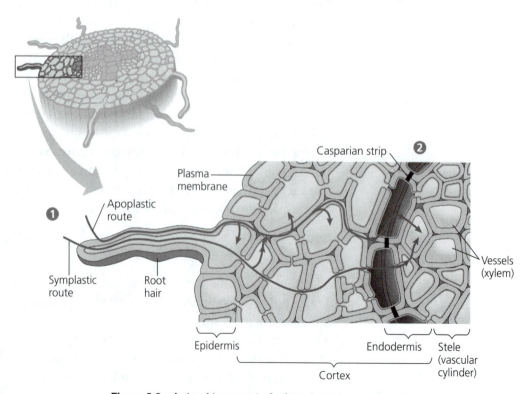

Figure 8.3   Lateral transport of minerals and water in roots

- At 1, water and minerals cross the walls of the root hairs and move between cells into the cortex of the root. This movement between cells is termed the *apoplastic* route. The *symplastic* route occurs only after the solution crosses a plasma membrane.
- Water can move inward by either route until reaching the innermost layer of the cortex, the **endodermis**, labeled number 2. The endodermal cells are sealed one to the other by the Casparian strip, a belt of waxy material that blocks the passage of water and dissolved materials. This is a critical control point for materials moving into the plant, because at the Casparian strip the soil solution must cross the plasma membrane. The plasma membrane determines what can cross into the xylem tissue and gain entrance to the rest of the plant.

- Once in the root xylem, water and minerals are transported long distances—to the rest of the plant—by bulk flow. The water and minerals, termed *xylem sap*, flow out of the root and up through the shoot, eventually exiting the plant primarily through the leaves.
- **Transpiration** is the loss of water vapor from the leaves and other parts of the plant that are in contact with air. It plays a key role in the movement of water from the roots.
- Two mechanisms influence how water is pulled up through the plant:

    1. **Root pressure** occurs when water diffusing in from the root cortex generates a positive pressure that forces fluid up through the xylem. Root pressure does not have the force to push water to the tops of trees.
    2. In the **transpiration-cohesion-tension mechanism**, water is lost through transpiration from the leaves of the plant due to the lower water potential of the air. The cohesion of water due to hydrogen bonding plus the adhesion of water to the plant cell walls enables the water to form a water column. Water is drawn up through the xylem as water evaporates from the leaves, each evaporating water molecule pulling on the one beneath it through the attraction of hydrogen bonds.

> *A Tip from the Readers:*
> *Hydrogen bonding plays a key role in cohesion-tension mechanisms. Be able to explain the importance of cohesion, adhesion (water hydrogen bonded to the xylem walls), and surface tension in this mechanism.*

Concept 36.4 Stomata help regulate the rate of transpiration

- Large surface area increases photosynthesis but also increases water loss by the plant through stomata. Guard cells open and close the stomata, controlling the amount of water lost by transpiration, but also the amount of carbon dioxide available from the atmosphere for photosynthesis.
- Guard cells control the size of the stomata opening by changing shape, widening or closing the gap between them. When the guard cells take up K^+ from the surrounding cells, this decreases water potential in the guard cells, causing

them to take up water. The guard cells then swell and buckle, increasing the size of the pore between them. When the guard cells lose K^+, the cells then lose water, become less bowed, and the pore closes.

▌ Guard cells are stimulated to open by the presence of light, loss of carbon dioxide in the leaf, and by normal circadian rhythms. Circadian rhythms are part of the plant's internal clock mechanism and cycle with intervals of about 24 hours. Even plants kept in the dark will open their stomata as dawn approaches.

Concept 36.5 Sugars are transported from leaves and other sources to sites of use or storage

▌ Phloem transports organic products of photosynthesis from the leaves throughout the plant, a process called **translocation**. The mechanism for translocation is **pressure flow**.

▌ Sieve tubes, a specialized cell type in phloem tissue, always carry sugars from a sugar source to a sugar sink. A **sugar source** is an organ that is a net producer of sugar, such as the leaves. A **sugar sink** is an organ that is a net consumer or storer of sugar, such as a fruit, or roots during the summer. Follow Figure 8.4 while noting the key steps.

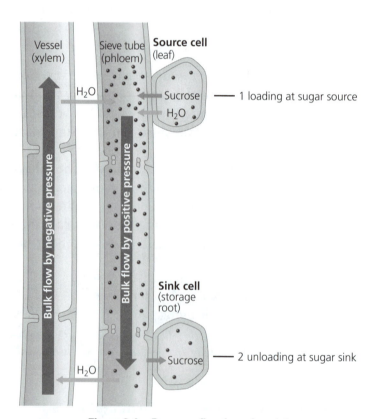

Figure 8.4   Pressure flow in a sieve tube

1. Sucrose is loaded into the sieve tubes at the sugar source. Proton pumps are used to create an electrochemical gradient that is utilized to load sucrose. This decreases water potential and causes the uptake of water, creating positive pressure.
2. The pressure is relieved at the sugar sink by the unloading of sucrose followed by the loss of water. In leaf-to-root translocation, xylem recycles the water back to the sugar source. Translocation via pressure flow is a second example of bulk flow.

Concept 36.6 *The symplasm is highly dynamic*

▌ The symplasm, or network of living phloem cells that connects all parts of the plant, is dynamic and interconnected.

  ■ Plasmodesmata allow for the movement of informational molecules like RNAs and proteins that coordinate development between cells.
  ■ In some plants the phloem carries out rapid, long-distance electrical signaling. This may lead to changes in gene transcription, respiration, photosynthesis, and other cellular functions in widely spaced organs. This is a nerve-like function, allowing for swift communication.

Chapter 37: Soil and Plant Nutrition

YOU MUST KNOW

- The difference between macronutrients and micronutrients.
- The importance of mutualistic relationships between plant roots and the bacteria and fungi that grow in the rhizosphere.
- Examples of nonmutualistic nutritional adaptations in plants.

Concept 37.2 *Plants require essential elements to complete their life cycles*

▌ An essential element is required for a plant to complete its life cycle and produce another generation.

  ■ Essential nutrients required in relatively large amounts are termed **macronutrients**. Nine elements are macronutrients: carbon, hydrogen, nitrogen, oxygen, phosphorus, sulfur, (note CHNOPS) plus potassium, calcium, and magnesium. Of all the macronutrients, the one that contributes the most to plant growth is nitrogen.
  ■ **Micronutrients** are elements needed in minute quantities. Plants need at least eight micronutrients, which function in plants mainly as cofactors—nonprotein helpers in enzymatic reactions. A few examples include iron, manganese, zinc, and copper. Although these elements may be needed in very small amounts, a deficiency can kill or weaken the plant.

■ A unique ecosystem, the **rhizosphere**, is the layer of soil that is bound to the plants' roots and is rich in microbial activity.

   ■ The rhizosphere is characterized by a mutually beneficial (mutualistic) relationship between the plant's roots and the rhizobacteria that live here. The plants provide nutrients in the form of sugars and amino acids for the rhizobacteria. The rhizobacteria may produce hormones that stimulate plant root growth, antibiotics that protect roots from disease, and make nutrients more available to the plant's roots.

■ A second mutualistic relationship is between nitrogen-fixing bacteria and plants. Nitrogen-fixing bacteria from the genus *Rhizobium* live in the root nodules of the legume plant family, including plants like peas, soybeans, alfalfa, peanuts, and clover. *Rhizobium* bacteria can fix atmospheric nitrogen into a form that can be used by plants. The plant provides food into the root nodule where the bacteria live, hence the designation as a mutualistic relationship.

■ Bacteria also play a critical role in the nitrogen cycle, discussed in Chapter 55. Ammonia is recycled in the soil by bacteria into forms that can be absorbed and used by plants.

■ Mycorrhizae are another example of mutualistic relationships with roots, this time between the roots and fungi in the soil. In mycorrhizae, the fungus benefits from a steady supply of sugar donated by the host plant. In return, the fungus increases the surface area for water uptake, selectively absorbs minerals that are taken up by the plant, and secretes substances that stimulate root growth and antibiotics that protect the plant from invading bacteria.

■ Plants also form symbiotic relationships that are not mutualistic:

   ■ **Parasitic plants**, such as mistletoe, are not photosynthetic and rely on other plants for their nutrients. They tap into the host plant's vascular system.
   ■ **Epiphytes** are not parasitic but just grow on the surfaces of other plants instead of the soil. Many orchids grow as epiphytes.
   ■ **Carnivorous plants** are photosynthetic, but they get some nitrogen and other minerals by digesting small animals. They are commonly found in nitrogen-poor soil, such as in bogs.

Chapter 38: Angiosperm Reproduction and Biotechnology

YOU MUST KNOW

- The process of double fertilization, a unique feature of angiosperms.
- The relationship between seed and fruit.
- The structure and functions of all parts of the flower.

Concept 38.1 Flowers, double fertilization, and fruits are unique features of the angiosperm life cycle

■ Review the concepts covered in this guide for 30.1 and 30.3. This will cover basic flower parts and the significant trends in evolution for angiosperms. Also review Figures 7.6 and 7.7.

■ The angiosperm life cycle has three unique features, all of which start with the letter **F**, a good memory aid—Flowers, Fruits, and double Fertilization.

■ In the pollen sacs (**microsporangia**) of an anther, diploid cells undergo meiosis to produce haploid microspores. Each microspore develops into a male gametophyte, also called a pollen grain. A pollen grain has two haploid nuclei:

■ The *tube nucleus* will eventually produce the **pollen tube**, a long cellular extension that delivers sperm to the female gametophyte, which houses the egg.

■ The *generative nucleus* usually divides to yield two sperm cells, which remain inside the pollen tube (see Figure 7.7).

■ In the ovary, ovules form with a diploid cell that undergoes meiosis to produce four haploid megaspores. In many angiosperms, only one megaspore survives and divides by mitosis three times to form eight haploid nuclei. Three of the haploid nuclei are particularly important:

■ One haploid nucleus is the egg. It will combine with a sperm nucleus to form the zygote.

■ Two other haploid nuclei are called the polar nuclei and will fuse with a sperm nucleus to make the 3n endosperm.

■ **Pollination** is the transfer of pollen from an anther to a stigma. If pollination is successful, a pollen grain produces a pollen tube, which grows down into the ovary. In Figure 8.5 pollination has occurred, the pollen tube has formed, and the unique features of angiosperm fertilization are visible and explained next.

■ When the pollen tube reaches the ovule, two fertilization events occur:

1. One sperm fertilizes the egg, forming the zygote. The zygote will develop into the embryo and eventually the new sporophyte plant.
2. The other sperm combines with *both* polar nuclei, forming a triploid (3n) nucleus. This unique 3n tissue will give rise to the **endosperm**, a food-storing tissue in the seed. The union of two sperm cells forming both zygote and endosperm is unique to angiosperms and known as **double fertilization**.

■ After double fertilization, the ovule develops into a seed, and the ovary develops into the fruit, which encloses the seed. The **fruit** protects the enclosed seeds and aids in their dispersal by wind or animals.

■ The food reserves of the endosperm may be completely exported to structures termed *cotyledons* (the number of which divide the angiosperms into monocotyledons and dicotyledons).

■ The **seed coat** protects the embryo and its food supply. A **radicle** is the embryonic root. The portion of the embryonic axis above where the **cotyledons** are attached is the **epicotyl** (from the Greek *epi*—on or over). It consists of the shoot tip with a pair of miniature leaves.

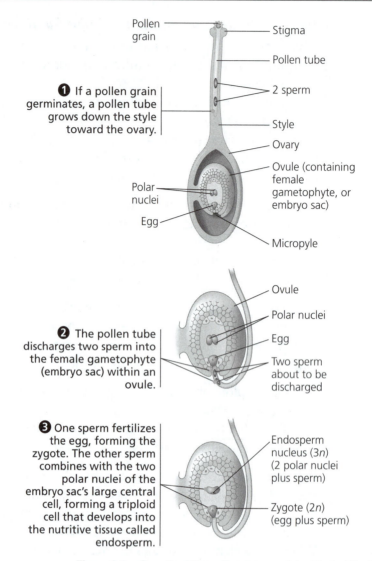

Figure 8.5 Growth of the pollen tube and double fertilization

As the seed matures, it enters dormancy, in which it has a low metabolic rate and its growth and development are suspended. The seed resumes growth when there are suitable environmental conditions for germination. (If you did AP Lab 5 on Cell Respiration, you will recall that the dry peas had a very low rate of respiration because they were dormant, not dead.)

> **A Tip from the Readers:**
> Review the evolution of plants at this point. You will recall that the bryophytes have a dominant gametophyte generation and a parasitic sporophyte. With the angiosperms, the dominant generation is the sporophyte, and the male gametophyte is now reduced to the pollen grain, whereas the female gametophyte is only seven cells with eight nuclei.

Concept 38.2 Plants reproduce sexually, asexually, or both

▌ Asexual reproduction, or **vegetative reproduction,** produces clones. Fragmentation is an example, in which pieces of the parent plant break off to form new individuals that are exact genetic replicas of the parent.

▌ Agriculture uses several techniques of artificial vegetative reproduction such as grafting, growing clones from cuttings, and test-tube cloning.

▌ While some flowers self-fertilize, others have methods to prevent self-fertilization and maximize genetic variation. One of these is self-incompatibility, in which a plant rejects its own pollen or that of a closely related plant, thus insuring cross-pollination.

Concept 38.3 Humans modify crops by breeding and genetic engineering

▌ Humans have intervened in the reproduction and genetic makeup of plants for thousands of years through **artificial selection.**

▌ **Genetically modified organisms** are engineered to express a gene from another species. Examples are Golden Rice, engineered to include large amounts of vitamin A; and *Bt* corn, engineered to contain a toxin that kills specific crop pests. There is some debate over the creation of these crops due to fear of human allergies and possible effects on nontarget organisms, among other concerns.

▌ The conversion of plant material to sugars which can be fermented to form alcohols and distilled to yield **biofuels** is currently under study.

Chapter 39: Plant Responses to Internal and External Signals

YOU MUST KNOW

- The three steps to a signal transduction pathway.
- The role of auxins in plants.
- The survival benefits of phototropism and gravitropism.
- How photoperiodism determines when flowering occurs.

Concept 39.1 Signal transduction pathways link signal reception to response

A Tip from the Readers:
Essays on the AP Biology Exam often cover ideas in different units in the textbook. This concept brings together the general ideas on cell communication from Chapter 11 with the specific results of cell communication with plants. This would be an excellent place for an essay question.

■ As a review from Chapter 11, signal transduction pathways involve three steps:

1. **Reception:** Cell signals are detected by receptors that undergo changes in shape in response to a specific stimulus. Two common plasma membrane receptors are G-protein-coupled receptors and tyrosine kinase receptors.
2. **Transduction:** Transduction is a multistep pathway that amplifies the signal. This allows a small number of signal molecules to produce a large cellular response.
3. **Response:** Cellular response is primarily accomplished by two mechanisms: (1) increasing or decreasing mRNA production, or (2) activating existing enzyme molecules.

Concept 39.2 Plant hormones help coordinate growth, development, and responses to stimuli

■ **Hormones** are defined as chemical messengers that coordinate the different parts of a multicellular organism. They are produced by one part of the body and transported to another.

■ A **tropism** is a plant growth response from hormones that results in the plant growing either toward or away from a stimulus.

■ **Phototropism** is the growth of a shoot in a certain direction in response to light. **Positive phototropism** is the growth of a plant toward light; **negative phototropism** is growth of a plant away from light.

■ Following is a survey of the most important actions of hormones.

STUDY TIP: Pay particular attention to the auxin hormone group.

1. The natural auxin in plants is *indoleacetic acid*, usually abbreviated as *IAA*. Auxins have many functions and play key roles in phototropisms and gravitropisms.
   • **Auxins** stimulate elongation of cells within young developing shoots. Auxins produced in the apical meristems activate proton pumps in the plasma membrane, which result in a lower pH (acidification of the cell wall). This weakens the cell wall, allowing turgor pressure to expand the cell wall, resulting in cell elongation.
   • Synthetic auxins are often used as herbicides. Monocots, like grasses, can quickly inactivate synthetic auxins, but eudicots cannot. Thus, the high concentrations of auxins kill broadleaf weeds (dicots) while grasses (monocots like turf grass or corn) are not harmed.
2. **Cytokinins** play an essential role in cell division and differentiation. These hormones stimulate cytokinesis or cell division. When the proper ratio of cytokinins to auxins exists, cytokinins stimulate cell division and the pathways to cell differentiation.
3. **Gibberellins** work in concert with auxins to stimulate stem elongation. Gibberellins help loosen cell walls, allowing expansion of cells and

therefore of stems. Many dwarf varieties of plants do not produce working gibberellins—including Mendel's dwarf pea variety. Gibberellins are also used to signal the seed to break dormancy and germinate.

4. In general, **abscisic acid** slows growth, often acting antagonistically to the other hormones mentioned. For example, abscisic acid promotes seed dormancy, thus keeping seeds from germinating too quickly. When leaves are under water stress, it is abscisic acid that signals the stomata to close, saving water.

5. **Ethylene** is unusual as a hormone because it is a gas. Ethylene plays a critical role in programmed cell death, or apoptosis. The shedding of leaves and the death of an annual after flowering are examples. Ethylene also promotes the ripening of fruits. Ethylene triggers ripening, and ripening triggers more ethylene. This rare positive feedback loop can rapidly promote the ripening of fruit. Because of ethylene, one rotten apple *can* spoil the bunch!

Concept 39.3 *Responses to light are critical for plant success*

▋ Plants can detect not only the presence of light, but also its direction, intensity, and wavelength. Action spectra (introduced in Chapter 10 on photosynthesis) reveal that red and blue light are the most important colors in plant responses to light.

▋ Blue-light photoreceptors initiate a number of plant responses to light including phototropisms and the light-induced opening of stomata.

▋ Light receptors termed **phytochromes** absorb mostly red light.

  ■ Phytochromes exist in two isomer forms, P_r and P_{fr}, that can switch back and forth depending on the wavelength of light in greatest supply. P_{fr} is the form of phytochrome that triggers many of a plant's developmental responses to light. The plant produces phytochrome in the P_r form, but upon illumination, the P_{fr} level increases by rapid conversion from P_r. The relative amounts of the two pigments provide a baseline for measuring the amount of sunlight in a day.

  ■ **Circadian rhythms** are physiological cycles that have a frequency of about 24 hours and that are not paced by a known environmental variable. In plants, the surge of P_{fr} at dawn resets the biological clock. The combination of a phytochrome system and a biological clock allow the plant to accurately assess the amount of daylight or darkness and hence the time of year.

▋ A physiological response to a photoperiod (the relative lengths of night and day), such as flowering, is called **photoperiodism**. Photoperiodism controls when plants will flower. When this was first discovered, scientists thought the critical factor was the length of the day. It was later determined that the length of the night is the actual critical factor. The old terminology, however, is still used, so take note of that in the following categories:

- **Short-day plants** require a period of continuous darkness longer than a critical period in order to flower. These plants flower in early spring or fall. Short-day plants are actually long-night plants; that is, what the plant measures is the length of the night.
- **Long-day plants** flower only if a period of continuous darkness was shorter than a critical period. They often flower in the late spring or early summer. Long-day plants are actually short-night plants.
- **Day-neutral plants** can flower in days of any length.

Concept 39.4 Plants respond to a wide variety of stimuli other than light

- What environmental cue causes the shoot of a young seedling to grow up and the root to grow down? **Gravitropism** is a plant's response to gravity. Roots show **positive gravitropism** and grow toward the source of gravity, whereas shoots show **negative gravitropism** and grow away from gravity.

  - The hormone **auxin** plays a key role in gravitropism. In the young root gravity causes a high concentration of auxin on the root's lower side. *High* concentrations of auxins inhibit cell elongation, causing the lower side to grow more slowly, whereas more rapid elongation of cells on the upper side causes the root to curve as it grows.

- **Thigmotropism** is directional growth in a plant as a response to a touch. Vines display thigmotropism when their tendrils coil around supports.
- Plants have various responses to stresses. In times of **drought**, the guard cells lose turgor. This causes the stomata to close; young leaves will stop growing, and they will roll into a shape that slows transpiration rates. Also, deep roots continue to grow, while those near the surface (where there isn't much water) do not grow very quickly.

Concept 39.5 Plants respond to attacks by herbivores and pathogens

- Some physical defenses plants have against predators (herbivores) are thorns, chemicals such as distasteful or poisonous compounds, and airborne attractants that attract other animals to kill the herbivores.
- The first line of defense against viruses for a plant (as for humans) is the epidermal layer.
- Plants are capable of recognizing plant pathogens and dealing with them in complex biochemical ways.

For Additional Review

Describe the plant structures that make plants ideally suited for trapping and processing the sun's energy, and for the process of carbon fixation.

Multiple-Choice Questions

1. Which of the following is primarily responsible for fruit ripening in plants?
   (A) ethylene
   (B) auxin
   (C) gibberellins
   (D) abscisic acid
   (E) brassinosteroids

2. Which of the following processes is responsible for the bending of the stem of a plant toward a light source?
   (A) The amount of chlorophyll produced on the side facing the light increases.
   (B) The rate of cell division on the side facing the light increases.
   (C) The rate of cell division on the side away from the light increases.
   (D) The cells on the side of the stem facing the light elongate.
   (E) The cells on the side of the stem away from the light elongate.

3. The driving force for the movement of materials in the xylem of plants is
   (A) gravity.
   (B) root pressure.
   (C) transpiration.
   (D) the difference in osmotic pressure between the source and the sink.
   (E) osmosis.

4. Hydrogen bonding plays a particularly important role in which plant process?
   (A) Pressure-flow hypothesis
   (B) Mutualistic relationships
   (C) Attraction of the sperm and egg
   (D) The transpiration-cohesion-tension mechanism
   (E) Attraction of pollen tube to female gametophyte

5. Which statement below describing alternation of generations in angiosperms is true?
   (A) The gametophyte stage is dominant.
   (B) The sporophyte stage deals only with reproduction.
   (C) The sporophyte is the form of the plant that is independent and conspicuous.
   (D) The gametophyte is the form of the plant that is independent and conspicuous.
   (E) The sporophyte has bright colors to attract pollinators.

6. In plants, translocation occurs as a result of
   (A) a difference in water potential between a sugar source and a sugar sink.
   (B) transpiration.
   (C) cohesion-adhesion.
   (D) active transport by sieve-tube members.
   (E) active transport by tracheid and vessel elements.

Directions: The group of questions below consists of five lettered choices followed by a list of numbered phrases or sentences. For each numbered phrase or sentence, select the one choice that is most closely related to it. Each choice may be used once, more than once, or not at all.

Questions 7–11
   (A) Abscisic acid
   (B) Auxin
   (C) Cytokinins
   (D) Ethylene
   (E) Gibberellins

7. Act with auxins to promote cell division and differentiation

8. Primary hormone involved with phototropism and gravitropism

9. This hormone is lacking in dwarf varieties of plants

10. Inhibits growth; closes stomata during drought

11. This hormone is used at high concentrations as an herbicide

12. Which of the following is the correct name for a system of roots that grows as one large vertical root with smaller lateral offshoots?
(A) fibrous root
(B) taproot
(C) secondary root
(D) prop root
(E) mycorrhizae

13. The three types of plant tissue, in order from the outside of a root to the inside, are
(A) vascular, ground, dermal.
(B) vascular, dermal, ground.
(C) ground, vascular, dermal.
(D) ground, dermal, vascular.
(E) dermal, ground, vascular.

14. A plant whose life span occurs over the course of two years is known as a(n)
(A) annual.
(B) diannual.
(C) biennial.
(D) perennial.
(E) seasonal.

15. The region of the plant in which the parenchyma cells are located that are involved in photosynthesis is called
(A) vascular cylinder.
(B) mesophyll.
(C) epidermis.
(D) xylem.
(E) phloem.

16. In a mesophyll cell of a leaf, the synthesis of ATP takes place in the mitochondria and which of the following other cell organelles?
(A) chloroplasts
(B) Golgi apparatus
(C) nucleus
(D) ribosomes
(E) lysosomes

17. All of the following enhance the uptake of water by a plant's roots EXCEPT
(A) root hairs.
(B) the large surface area of cortical cells.
(C) mycorrhizae.
(D) the attraction of water and dissolved minerals to root hairs.
(E) gravitational force.

18. The waxy barrier found in the endodermal wall of a root which prevents the passage of unwanted materials into the vascular tissue is called the
(A) pericycle.
(B) ghostly strip.
(C) Casparian strip.
(D) cortex.
(E) epidermis.

19. All of the following contribute to the closing of stomata during the day EXCEPT
(A) water deficiency.
(B) wilting.
(C) high temperatures.
(D) excessive rainfall.
(E) excessive transpiration.

20. Which of the following constitute plant macronutrients?
(A) carbon, oxygen, nitrogen, and hydrogen
(B) carbon, boron, nitrogen, and chlorine
(C) phosphorus, oxygen, nitrogen, and iron
(D) potassium, oxygen, hydrogen, and zinc
(E) carbon, oxygen, nitrogen, and copper

21. Which term describes the symbiotic relationship between the roots of many species of plants and fungi?
(A) *Rhizobium* association
(B) humus and earthworms
(C) pollinators and flowers
(D) apoplast and symplast
(E) mycorrhizae

22. In double fertilization, how is the endosperm formed?
    (A) from the endodermis
    (B) from the fertilization of the egg
    (C) from the zygote during development
    (D) from the fusing of a sperm with two polar bodies
    (E) the fusing of two sperm and an egg

23. The angiosperm life cycle has three unique features:
    (A) fertilization, development, and meiosis
    (B) flowers, fruit, and double fertilization
    (C) cell growth, elongation, and differentiation
    (D) spores, cones, and microspores
    (E) alternation of generations, sporophytes, and gametophytes

24. Which vascular tissue in plants is responsible for carrying sugars down from the leaves to the rest of the plant?
    (A) xylem
    (B) phloem
    (C) dermal tissue
    (D) tracheids
    (E) vessel elements

25. In order to flower, short-day plants require a period of
    (A) light greater than a critical period.
    (B) darkness greater than a critical period.
    (C) light less than a critical period.
    (D) darkness less than a critical period.
    (E) equal amounts of light and darkness.

Free-Response Question

1. *Describe the following processes in the plant life cycle and why they are important for the plant to complete, listing all of the plant hormones involved and describing their function:*

   (**a**) elongation of the plant shoot.
   (**b**) the process by which seeds orient themselves.
   (**c**) photoperiodism.

ANSWERS AND EXPLANATIONS

Multiple-Choice Questions

1. (A) is correct. Ethylene is a plant hormone that causes fruit to ripen. It also causes apoptosis, or programmed cell death in plant cells; and it causes the loss of leaves in autumn.

2. (E) is correct. When the light source on a plant is uneven, the plant will grow toward the light source. In some plants this is the result of auxin moving from the apex down to the cells that are less exposed to light, and causing them to elongate faster than the cells on the side that is illuminated.

3. (C) is correct. The driving force behind the movement of sap in xylem (in the direction from the roots to the leaves) is the transpiration of water through the stomata on the leaves. The mechanism responsible for movement up through the xylem is the transpiration-cohesion-tension mechanism, and it occurs through bulk flow, in which fluid moves because of a pressure difference at opposite ends of a tube. The pressure is created by transpiration from the leaves, and contributing to the movement of water and minerals up the plant are gradients of water potential from cell to cell within the plant.

4. (D) is correct. In the transpiration-cohesion-tension mechanism, water is lost through transpiration from the leaves of the plant due to the lower water potential of the air. The cohesion of water due to hydrogen bonding plus the adhesion of water to the plant cell walls by hydrogen bonding enables the water to form a water column. Water is drawn up through the xylem as water evaporates from the leaves, each evaporating water molecule pulling on the one beneath it through the attraction of hydrogen bonds.

5. (C) is correct. In vascular plants the dominant form of the plant is the sporophyte—it is the full-grown plant that we see growing in a field. It has a reproductive structure called a flower, which creates gametes. The fusion of gametes results in fertilization and the possible development of a sporophyte embryo, which may develop into a full-grown mature sporophyte.

6. (A) is correct. In plants, phloem is responsible for carrying sugar made in the leaves to other locations that are incapable of photosynthesis. This process is called translocation. In angiosperms phloem is made up of sieve-tube members that are arranged end to end in long sieve tubes. In phloem, sugar travels from a sugar source—any site in the plant involved in photosynthesis, for example—to a sugar sink, which is any site in the plant not engaged in photosynthesis.

7. (C) is correct. Cytokinins are plant hormones that are involved in the control of cell division and differentiation. They also influence apical dominance and have some anti-aging effects on plant tissues. Cytokinins often act in concert with auxin.

8. (B) is correct. Auxins are the primary hormone involved in tropisms. A tropism is a plant growth response that results in the plant growing either toward or away from a stimulus. Auxins allow for increased cell elongation by weakening cell walls and allowing cell expansion. When auxins act on only one area, different growth rates result, which causes the plant to alter its growth form, either toward or away from the stimulus.

9. (E) is correct. Gibberellins are plant hormones that are responsible for many different effects in plants, but three main ones are the elongation of the plant stem (they stimulate both cell division and cell elongation); fruit growth; and seed germination (the process by which a seed breaks dormancy and begins to grow). A lack of gibberellins—a homozygous recessive situation—means no stem elongation and, consequently, dwarf plants.

10. (A) is correct. Abscisic acid is a plant hormone that prevents the seed from immediately germinating—it is responsible for seed dormancy. This hormone is also responsible for closing the stomata on plant leaves in times of drought; when plants start to wilt, abscisic acid causes changes in guard cells, closing the stomata and preventing further water loss through transpiration.

11. (B) is correct. Auxins promote cell elongation and growth at the proper concentrations, but at high concentrations auxins stop growth and may kill eudicots. Monocots can break down the excess auxins in the herbicide and generally do not suffer any long-term consequences.

12. (B) is correct. There are two main root systems in plants: the taproot system and the fibrous root system. Monocots, including grasses, usually have the fibrous root system, which firmly anchors them into the ground, whereas many eudicots have a taproot system, with one long, thick root extending down and smaller lateral branch roots shooting from it. This question is actually covered in Chapter 30. Although the text separates the evolution of plants from plant form and function, the AP exam may combine questions from the two areas.

■ **13. (E) is correct.** The three types of tissue that make up plant organs, from the outermost layer to the innermost layer, are dermal, ground, and vascular. The dermal layer is a single layer of very tightly packed cells that serves to protect the plant from harm. The vascular tissue consists of xylem and phloem, and it is responsible for transporting water, minerals, and food throughout the plant. The ground tissue has various functions but in the root serves primarily to store starch.

■ **14. (C) is correct.** A plant whose life cycle spans two years is known as a biennial. One whose life cycle spans one year is known as an annual, and plants that live for many years are known as perennials.

■ **15. (B) is correct.** The name of the region of the plant leaf in which parenchyma cells are situated—and the site of photosynthesis in plants—is called mesophyll. The mesophyll lies between the upper and lower epidermis of the leaf and consists mainly of parenchyma cells. Eudicots have two regions of mesophyll: spongy mesophyll and palisade mesophyll.

■ **16. (A) is correct.** The production of ATP in plant cells occurs in the mitochondria, as it does in the cells of animals, but it also occurs in the chloroplasts during photosynthesis. The light reactions of photosynthesis convert solar energy to the chemical energy of ATP and NADPH, and these light reactions take place in the chloroplasts, in the mesophyll cells of the plant leaf.

■ **17. (E) is correct.** All of the factors listed aid in the uptake of water and minerals by the roots of a plant except the last choice, gravity. Water and minerals flow from the soil into the epidermis of the plant, then through the root cortex and into the xylem of the plant. The xylem then transports water and minerals throughout the plant body.

■ **18. (C) is correct.** The Casparian strip is a belt made of a waxy material that runs through all of the endodermal cells, creating a ring that protects the vascular tissue from unwanted materials. The water and mineral solution must pass through the endodermis to enter the xylem in the stele of the root. The Casparian strip forces the solution through the plasma membrane of an endodermal cell, allowing the screening of the solution.

■ **19. (D) is correct.** All of the answers listed, with the exception of excessive rainfall, are factors that would cause the stomata of a leaf to close during the day. All of the factors with the exception of answer *D* could lead to dehydration.

■ **20. (A) is correct.** The macronutrients (elements required in large amounts) in plants are carbon, oxygen, hydrogen, nitrogen, sulfur, phosphorus, potassium, calcium, and magnesium. The micronutrients (elements needed only in trace amounts) are chlorine, iron, boron, manganese, zinc, copper, molybdenum, and nickel.

■ **21. (E) is correct.** Mycorrhizae are mutualistic symbiotic associations of roots and fungi. The fungus benefits from having a "home" and a steady sugar supply from the plant. The plant benefits because the fungus increases the surface area for water uptake and supplies the plant with certain minerals.

■ **22. (D) is correct.** In double fertilization the second sperm combines with *both* polar nuclei, forming a triploid (3n) nucleus. This unique 3n tissue will give rise to the endosperm, a food-storing tissue in the seed. The first sperm cell combines with the egg to form the zygote, hence the term *double fertilization.*

■ **23. (B) is correct.** The angiosperm life cycle has three unique features, all of which start with the letter **F**, a good memory aid—**F**lowers, **F**ruits, and double **F**ertilization. The other choices are not uniquely common to angiosperms.

■ **24. (B) is correct.** The phloem transports the food made in mature leaves to the roots and other nonphotosynthetic parts of the plant, such as roots. The xylem is responsible for carrying water and minerals up through the plant from the roots. Those are the two types of vascular tissue in plants.

■ **25. (B) is correct.** Short-day plants require a period of continuous darkness longer than a critical period in order to flower. These plants flower in early spring or fall. Short-day plants are actually long-night plants; that is, what the plant measures is the length of the night.

Free-Response Question

(a) The elongation of the stem in plants is an important process in the plant life cycle because it enables the plant to reach its full size and complete development of the sporophyte stage. The sporophyte needs to complete development in order to produce reproductive structures. Elongation of the plant shoot also allows for efficient photosynthesis.

The apical meristems, which are located at the tips of roots and in the buds of shoots, provide additional cells that enable the plant to grow in length. Both auxins and gibberellins are hormones involved in stem elongation. Auxins and gibberellins work together to stimulate growth via cell wall elongation and subsequent stem elongation.

(b) It is important for the seed to orient itself so that its roots reach into the soil and extract water and minerals, while its leaves point up to the sun so that they can trap light energy (to use in photosynthesis) and gas exchange can occur through transpiration. *Gravitropism* is the term used to describe plant growth in response to the force of gravity. Roots show positive gravitropism—they grow into the soil, toward the source of gravity—whereas shoots show negative gravitropism—that is, they grow away from the source of gravity, toward the sun. The major hormone involved in gravitropism is auxin. It has not yet been determined exactly how auxin influences the way roots and shoots grow, but it is theorized that certain dense molecules settle with gravity to one end of the plant root, and auxin accumulates as a result of this. The accumulation of high concentrations of auxin prevents cell elongation on the underside of the root, while cells on the upper side elongate so that the root curves down, into the soil.

(c) Photoperiodism is defined as any physiological response to a photoperiod (meaning a specific length of daylight or darkness). One example of a photoperiodism is flowering. It was originally thought that plant flowering depended on the length of the daylight, but then scientists concluded that it is actually night length that determines when a plant will flower. Plants monitor the length of the night by the molecular switching of two forms of the phytochrome pigment. The accumulation of specific phytochrome isomers combined with the biological clock of plants allows for the specific monitoring of the length of the night. Long-day plants and short-day plants actually monitor the length of the night, not the day. Short-day plants must have a period of darkness longer than a specific critical period to flower, whereas long-day plants must have a period of darkness shorter than a critical period.

This response uses the following key terms in context, showing the writer's knowledge of their meanings and relatedness:

sporophyte
flower
apical meristem
auxins
gibberellins

gravitropism
photoperiodism
phytochromes
short-day and long-day plants

It also shows an understanding of the following processes: plant growth via stem elongation, gravitropism, and photoperiodism.

Animal Form and Function

Chapter 40: Basic Principles of Animal Form and Function

YOU MUST KNOW

- The four types of tissues and their general functions.
- The importance of homeostasis and examples.
- How feedback systems control homeostasis, and one example of positive feedback and one example of negative feedback.

Concept 40.1 Animal form and function are correlated at all levels of organization

▌ **Tissues** are groups of cells that have a common structure and function. Tissues are further organized into functional units called **organs**. Groups of organs that work together make up **organ systems** (for example, the digestive, circulatory, and excretory systems).

▌ The four types of tissue are listed below:

1. **Epithelial tissue** occurs in sheets of tightly packed cells, covers the body, lines the organs, and acts as a protective barrier. One side of the epithelium is always bound to an underlying supportive surface called the basement membrane. The outside surface is facing either air or a fluid environment.
2. **Connective tissue** mainly supports and binds other tissues. It consists of scattered cells within an extracellular matrix. Some connective tissues are cartilage, tendons, ligaments, bone, and blood.
3. **Muscle tissue** is responsible for nearly all types of body movement. Muscle filaments are made of the proteins actin and myosin. Muscle fibers contract when they are stimulated by a nerve impulse. Muscle tissue is the most abundant tissue in most animals. There are three types of muscle: *skeletal muscle, cardiac muscle,* and *smooth muscle.*

4. The functional unit of **nervous tissue** is the nerve cell, or *neuron*. This tissue senses stimuli and transmits signals from one part of the body to another part of the body, including to other neurons, glands, muscles, and the brain.

■ For animal survival tissues, organs, and organ systems must act in a coordinated manner. Two major organ systems specialize in control and coordination:

1. In the **endocrine system**, chemical signals called **hormones** are released into the bloodstream and are broadcast throughout the body. Different hormones cause specific effects, but only in cells with specific receptors for the released hormone.
2. In the **nervous system, neurons** transmit information between specific locations. Only three types of cells receive nerve impulses: neurons, muscle cells, and endocrine cells.

Concept 40.2 *Feedback control loops maintain the internal environment in many animals*

■ In **homeostasis** animals maintain a relatively constant internal environment, even when the external environment changes significantly.

> **Tip from the Readers:**
> Homeostasis is a key concept in animal physiology. Essay questions on homeostasis are common, so have an example of homeostasis that you can write about.

■ Homeostatic control systems function by having a **set point** (like a body temperature to maintain), sensors to detect any stimulus above or below the set point, and a physiological response that helps return the body to its set point.

■ In **negative feedback systems**, the animal responds to the stimulus in a way that reduces the stimulus. For example, in response to exercise, the body temperature rises, which initiates sweating to cool the body.

■ In **positive feedback systems**, a change in some variable triggers mechanisms that amplify rather than reverse the change. For example, during childbirth, pressure of the baby's head against receptors near the opening of the uterus stimulates greater uterine contractions, which cause greater pressure against the uterine opening, which heightens the contractions, and so forth.

Concept 40.3 *Homeostatic processes for thermoregulation involve form, function, and behavior*

■ **Thermoregulation** refers to how animals maintain their internal temperature within a tolerable range.

■ **Endotherms** (such as mammals and birds) are warmed mostly by heat generated by metabolism. **Ectotherms** (such as most invertebrates, fishes, amphibians, and reptiles) generate relatively little metabolic heat, gaining most of their heat from external sources.

■ In many birds and mammals, reduction of heat loss relies on **countercurrent exchange**. Heat transfer involves antiparallel arrangement of blood vessels such that warm blood from the core of the animal, en route to the extremities, transfers heat to colder blood returning from the extremities. Heat that would have been lost to the environment is conserved in the blood returning to the core of the animal.

Chapter 41: Animal Nutrition

YOU MUST KNOW

- The major compartments of the alimentary canal—oral cavity, pharynx, esophagus, stomach, small intestines, and large intestines—and their contributions to animal nutrition.
- The major digestive glands—salivary glands, pancreas, liver, and gallbladder—and their contributions to animal nutrition.
- The general scheme of chemical digestion of carbohydrates, fats, proteins, and nucleic acids.

Concept 41.1 An animal's diet must supply chemical energy, organic molecules, and essential nutrients

■ The **essential nutrients** required by an animal include both minerals and pre-assembled organic molecules that the animal cannot produce from raw materials. There are four classes of essential nutrients:

1. About half of the 20 amino acids required by animals are **essential amino acids** and must be obtained from food.
2. There are also **essential fatty acids** that animals cannot make and must ingest. For example, humans require linoleic acid to make some phospholipids found in the cell membrane.
3. **Vitamins**, such as the B vitamins and vitamin E, are organic molecules that are required in the diet in small amounts.
4. **Minerals**, such as calcium and phosphorus, are simple inorganic nutrients that are also required in the diet in small amounts.

STUDY TIP: As you work with animal nutrition, use Figure 9.1 to identify each part of the complex digestive system. This will make the structures and enzymatic conversions more concrete and easier to remember.

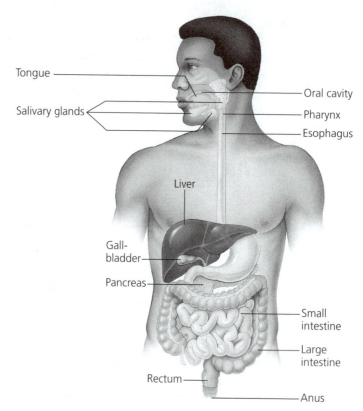

Figure 9.1 The human digestive system

Concept 41.2 The main stages of food processing are ingestion, digestion, absorption, and elimination

▌ The four main stages of food processing are explained below:

1. **Ingestion** is the act of taking in food.
2. **Digestion** is the breakdown of food into small molecules capable of being absorbed by the cells of the body. Enzymatic hydrolysis, the breaking of bonds with the addition of a water molecule, is the reaction type by which macromolecules are digested.
3. **Absorption** is the stage in food processing when the body's cells take up small molecules such as amino acids and simple sugars from the digestive tract.
4. **Elimination** is the passing of undigested material from the digestive tract.

▌ **Intracellular digestion** occurs within a cell enclosed by a protective membrane. Sponges digest their food this way.

▌ **Extracellular digestion** is carried out by most animals; in this type of digestion, food is broken down outside cells. This process allows the animal to devour much larger sources of food than can be handled using only intracellular digestion.

▌ In many simple animals, digestion takes place in a **gastrovascular cavity**. These animals have a single opening through which food enters and waste is eliminated. Cnidarians, including hydras, and flatworms like planarians are examples of animals with a gastrovascular cavity.

■ More complex animals have **complete digestive tracts (alimentary canals)**, which are one-way digestive tubes that begin at the mouth and terminate at the anus.

Concept 41.3 Organs specialized for successive stages of food processing form the mammalian digestive system

■ The movement of food through the digestive system is controlled by **peristalsis**, the rhythmic waves of contraction by smooth muscle in the walls of the alimentary canal, and by **sphincters**—muscular, ringlike valves that regulate the passage of material between digestive compartments.

■ When food is in the mouth, or **oral cavity**, a nervous reflex occurs that causes saliva to be secreted into the mouth. Saliva lubricates the food to facilitate swallowing. It also starts chemical digestion, because saliva contains the enzyme **amylase**, which hydrolyzes starch and glycogen into smaller polysaccharides and the disaccharide maltose.

■ During chewing, food is shaped into a ball called a **bolus**. During swallowing, the bolus enters the **pharynx**—a junction that opens to the esophagus and the trachea (generally referred to as the throat). During swallowing, the **epiglottis**, a flap made of cartilage, covers the trachea. This diverts the food down the esophagus and out of the airway.

■ The **esophagus** moves food from the pharynx down to the stomach through peristalsis.

■ The stomach's functions include storing food and secreting the digestive fluid termed gastric juice. Two components of **gastric juice** carry out chemical digestion:

1. **Hydrochloric acid**, with a pH of about 2, breaks down the extracellular matrix of meat and plant materials, and it also kills most of the bacteria ingested with food.
2. **Pepsin** is an enzyme in gastric juice that begins to hydrolyze proteins into smaller polypeptides. Pepsin is secreted in an inactive form called *pepsinogen*, which is activated by hydrochloric acid in the stomach. The inactive form protects the cells that produce the protein-digesting enzyme from self-digestion. For further protection, a thick **mucus** is produced by the lining of the stomach.

■ The result of digestion in the stomach is a substance called **acid chyme**. The acid chyme is shunted from the stomach into the small intestine via the **pyloric sphincter**.

■ The first section of the **small intestine** is known as the duodenum. The **duodenum** is the major site of chemical digestion. In the duodenum, acid chyme mixes with secretions from the pancreas and the liver. The pancreas releases a **bicarbonate fluid**, which acts as a buffer against acidic contents from the stomach. **Bile** is made in the liver and stored in the gall bladder. Bile emulsifies fats; that is, bile coats fat droplets turning large fat droplets into small fat droplets, which are easier to digest.

■ Chemical digestion in the duodenum can be summarized as follows:

1. **Carbohydrates:** The breakdown of starch and glycogen began with salivary amylase in the mouth. In the small intestine, pancreatic **amylases** break starch, glycogen, and small polysaccharides into the disaccharide maltose. The breakdown of maltose and other disaccharides (sucrose and lactose) into their monomers occurs at the wall of the duodenal epithelium.

2. **Proteins:** Pepsin begins the breakdown of proteins in the stomach. In the duodenum, **trypsin** and **chymotrypsin** break polypeptides into smaller chains. **Dipeptidases, carboxypeptidase**, and **aminopeptidase** break apart polypeptides into amino acids.

3. **Nucleic acids:** The breakdown of nucleic acids starts with the hydrolysis of DNA and RNA to their respective nucleotides. The nucleotides are then broken down to nitrogenous bases, sugars, and phosphate groups. Most of the enzymes responsible for nucleic acid digestion enter the duodenum from the pancreas.

4. **Fats:** Digestion of fats starts in the small intestine. Bile coats the fat droplets and keeps them from clumping (emulsifying). The enzyme **lipase**, which is produced in the pancreas, hydrolyzes the small fat droplets.

■ The epithelial lining of the small intestine has folds called **villi**, which in turn bear projections called **microvilli**—both of which radically increase the surface area available for absorption.

■ In each villus are capillaries for the absorption of monomers and a lymph vessel, termed a **lacteal**, which absorbs small fatty acids. Passive-facilitated diffusion and active transport are used to move monomers across the intestinal membrane and into blood vessels.

■ The capillaries and veins that drain the nutrients away from the villi form the **hepatic portal vessel**, a blood vessel that goes to the liver. The liver then regulates the distribution of nutrients to the body.

■ Hormones are chemical messengers that travel to target tissues through the blood stream. Hormones help to coordinate the digestive process as follows:

1. **Gastrin** is produced by the stomach and increases the production of gastric juices.

2. **Enterogastrone**, produced by the duodenum in the presence of fats, slows peristalsis. This allows more time for fat digestion.

3. **Secretin** and **cholecystokinin (CCK)**, secreted by the walls of the duodenum, increase the flow of digestive juices from the pancreas and gall bladder.

■ The **large intestine**, also called the **colon**, is connected to the small intestine by a sphincter. The point of the connection is the site of the **cecum**, a small pouch with an extension called the **appendix**.

■ The main functions of the large intestine are to compact waste and recover water. The large intestine includes a rich flora of mostly harmless bacteria, including *Escherichia coli*. The presence of *E. coli* in lakes and streams is a useful indicator of contamination by untreated sewage.

▌ At the end of the colon is the **rectum**, where feces are stored until they are eliminated.

Concept 41.5 Evolutionary adaptations of vertebrate digestive systems correlate with diet

▌ A mammal's **dentition** is generally correlated with its diet. In particular, mammals have specialized dentition that best enables them to ingest their food.

▌ Herbivores generally have longer alimentary canals than carnivores, reflecting the longer time needed to digest vegetation. Much of the chemical energy in herbivore diets comes from the cellulose of plant cell walls. Many vertebrates (as well as termites) house large populations of symbiotic bacteria and protists whose enzymes actually digest the cellulose.

Chapter 42: Circulation and Gas Exchange

YOU MUST KNOW

- The circulatory vessels, heart chambers, and route of mammalian circulation.
- How RBCs demonstrate the relationship of structure to function.
- The general characteristics of a respiratory surface.
- The pathway a molecule of oxygen takes from the air until it is picked up by the hemoglobin of a red blood cell.

Concept 42.1 Circulatory systems enable exchange at a distance

▌ Exchange of gases, nutrients, and wastes occurs at the cellular level. Since diffusion is rapid only across small distances, natural selection has led to two general solutions.

1. Body size and shape that keep many or all cells in direct contact with the environment, such as with sponges. Cnidarians and flatworms possess a **gastrovascular cavity** that serves both in digestion and distribution of substances throughout the body.
2. Other animals have a **circulatory system** that moves fluid to the tissues and cells for exchange.

▌ A circulatory system has three components: **blood** (a circulatory fluid), **vessels** (tubes through which blood moves), and a **heart** (a structure that pumps the blood).

▌ In **open circulatory systems**, blood bathes the organs directly. The blood and lymph combined are called **hemolymph**, and a heart pumps hemolymph into cavities called **sinuses**. This is seen in arthropods and most mollusks.

- In **closed circulatory systems**, blood is contained within vessels and pumped around the body; the blood is separate from the interstitial fluid. This is seen in annelids, cephalopods and all vertebrates.
- There are three main types of blood vessels:

  1. **Arteries** carry blood *away* from the heart and branch into smaller **arterioles**. Their walls are relatively thick and include a significant amount of smooth muscle. The *pulse* is felt in an artery.
  2. **Capillaries** are microscopic vessels that are composed of only a single layer of cells, the *endothelium*, on a basement membrane. All diffusion occurs here.
  3. **Veins** carry the blood back to the heart. They have valves to prevent backflow.

- **Atria** are heart chambers that receive blood and convey it to **ventricles**, which pump blood.
- There are several variations on the circulatory system theme. Study Figure 9.2.

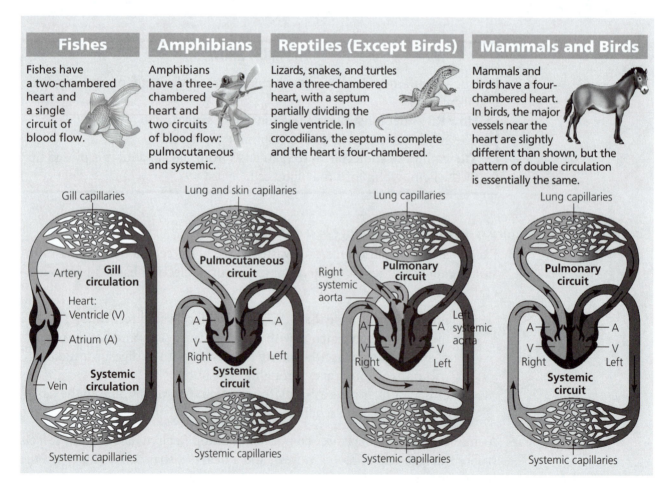

Figure 9.2

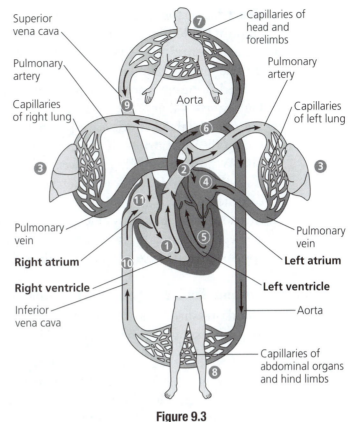

Figure 9.3

Concept 42.2 *Coordinated cycles of heart contraction drive double circulation in mammals.*

▌ Study Figure 9.3. Be able to trace these steps!

1. Blood is pumped from the **right ventricle**.
2. It enters the **pulmonary arteries** and is carried to the lungs.
3. The blood flows through **capillary beds** in the lungs and picks up oxygen and releases CO_2.
4. The blood returns to the **left atrium** of the heart via **pulmonary veins**.
5. Then it continues to the **left ventricle**.
6. It leaves the heart via the **aorta**, which branches off and sends blood through arteries throughout the body.
7. The blood enters **capillary beds** in the neck, head, and arms, and in the abdomen and legs, giving up oxygen and picking up CO_2.
8. The capillaries form **venules**, and blood from the neck, head, and arms travels back to veins and back to the **right atrium** via the anterior vena cava.
9. Blood from the legs and trunk travels through the **posterior vena cava** back to the right atrium.
10. Blood is pumped into the right ventricle and the cycle begins again.

▌ The **cardiac cycle** is the complete cycle of contraction and relaxation of the heart. The contraction phase is called **systole**, and the relaxation phase is called **diastole**.

- **Heart rate** is the rate of contraction per minute, and the **stroke volume** is the amount of blood pumped by the left ventricle during each contraction.
- An **atrioventricular (AV) valve** between each atrium and ventricle prevents the backflow of blood into the atria; there are also two **semilunar valves**—one located at the entrance to the pulmonary artery and the second at the entrance to the aorta—that prevent backflow of blood into the ventricles.
- The **sinoatrial (SA) node** is the pacemaker of the heart. It is located in the upper wall of the right atrium. It sets the rate at which cardiac muscle cells contract.
- The **AV node**, located in the lower wall of the right atrium, delays the impulses from the SA node to allow the atria to completely empty before the ventricles contract.
- Heart rate is regulated by at least three factors. The sympathetic nerves accelerate heart rate, and the parasympathetic nerves slow it down. Hormones such as epinephrine increase heart rate, as does an increase in body temperature.

Concept 42.3 Blood pressure and flow reflect the structure and arrangement of blood vessels

- **Blood pressure** is measured with a *sphygmomanometer*. (In AP Lab 10, you will assess fitness by measuring pulse and blood pressure under different conditions).

  - For a healthy 20-year old at rest, typical blood pressure is 120/70. The first number is the **systolic pressure** (when the heart contracts); the second is the **diastolic pressure** (when the heart is relaxed).
  - Short-term regulation of blood pressure occurs when the smooth muscle of the arterioles contracts or relaxes, due to changes in activity or hormonal signals.
  - Long-term regulation of blood pressure is accomplished by changes in blood volume due to the rennin-angiotensin-aldosterone system. (See Chapter 44).

- Blood moves through the veins and venules as rhythmic contractions of smooth muscles in the walls propel it, and it is squeezed by contraction of skeletal muscles during exercise. Valves prevent backflow.
- **The lymphatic system** is responsible for returning lost fluid and proteins to the blood in the form of **lymph**. Along a lymph vessel are **lymph nodes** that filter lymph and attack viruses and bacteria, playing an important role in immunity.

Concept 42.4 Blood components mediate exchange, transport, and defense

- **Plasma** is mostly water, but it also contains ions, electrolytes, and plasma proteins. It transports nutrients, metabolic wastes, gases, and hormones. In addition, blood plasma carries

  1. **Red blood cells (erythrocytes or RBCs)**, which transport oxygen via *hemoglobin* (an iron-containing protein).
  2. **White blood cells (leukocytes or WBCs)**, which are part of the immune system.
  3. **Platelets**, which are fragments of cells responsible for blood clotting.

- Red blood cells are *biconcave disks*. This shape increases surface area to enhance oxygen transport. Each RBC contains about 250 million molecules of hemoglobin, and each hemoglobin molecule can bind up to four molecules of oxygen.
- RBCs lack nuclei, which increases space for hemoglobin.
- RBCs lack mitochondria, so the oxygen they carry is not consumed.
- Blood contains a soluble plasma protein called **fibrinogen**, which forms clots when it is converted to its active form, **fibrin**.
- RBCs, WBCs, and platelets all develop from **stem cells** found in red marrow of flat bones such as the sternum, ribs, and pelvis.

Concept 42.5 Gas exchange occurs across specialized respiratory surfaces

- **Gas exchange**, or **respiration**, is the uptake of molecular oxygen (O_2) from the environment and the discharge of carbon dioxide (CO_2) to the environment.
- At sea level, the atmosphere exerts a downward force of 760mm Hg. Each gas exerts a partial pressure based on the fraction it comprises of the mixture. For example, O_2 is 21% of the mixture and so exerts a partial pressure of $(0.21)(760 \text{ mm Hg}) = 160$mm Hg.
- The diffusion of a gas depends on **partial pressure**. Gases always diffuse from regions of higher partial pressure to regions of lower partial pressure.
- The **respiratory medium** is the source of the O_2. It is air for terrestrial animals and water for most aquatic animals.

  - Recall AP Lab 12, Dissolved Oxygen and Aquatic Primary Productivity. Water holds less oxygen than an equivalent volume of air and is more dense and viscous. Therefore, aquatic animals expend considerable energy to carry out gas exchange. The amount of dissolved O_2 in water decreases as temperature or salinity increase.

- The **respiratory surface** is the part of an animal's body where gases are exchanged with the surrounding environment. It can be the body wall, the skin, gills, tracheae, or lungs.

  - General characteristics of respiratory surfaces include

    1. Must be moist
    2. Favorable surface area/volume ratio. Respiratory surfaces are often extensively folded or branched. (Think structure/function.)
    3. Closely associated with the vascular system of larger animals

- **Gills** are respiratory organs in aquatic animals. Water flows through them, and blood flowing through capillaries within the wall of the gill picks up oxygen from the water. Blood flows in a direction opposite to the flow of water. This is called **countercurrent exchange**, and it maximizes the absorption of oxygen.
- Countercurrent exchange mechanisms allow for more diffusion to occur than would otherwise be possible. To quickly review, you will see this mechanism again in the flow of filtrate in loops of Henle in the kidney and in the capillary beds of the feet of many aquatic birds (temperature regulation, Chapter 40).
- Insects have **tracheal systems**, which are made up of air tubes that branch through the body and open to the outside. They extend to almost all cells, and gas exchange occurs directly across the epithelial membrane inside the tracheal walls.
- **Lungs** are infoldings of the body surface and found in most terrestrial vertebrates.
- The **larynx** (voice box) is the upper part of the respiratory tract. It is a tube with cartilage-reinforced walls that leads to the trachea (windpipe).
- The **trachea** divides into two bronchi, each of which leads to a lung. *Cilia* and *mucus-producing cells* line the trachea, and their action keeps particulate matter from reaching the lungs. The trachea has *C-shaped cartilage rings*, which keep it from collapsing, much like the reinforcements in a vacuum cleaner's hose.
- In the lungs, the **bronchi** branch into **bronchioles**.
- The **alveoli** are air sacs clustered at the ends of bronchioles. They are thin, moist, have a large surface area (in a human, the size of a tennis court if spread out!), and are associated with capillary beds. Here, O_2 diffuses into the blood by passing through the simple squamous membrane of an alveolus and through the simple squamous membrane of a capillary, into the blood, and attaches to a hemoglobin molecule in a red blood cell.

> *STUDY TIP:* Refer again to the characteristics of a respiratory surface and note how the mammalian lung fulfills the requirements.

Concept 42.6 Breathing ventilates the lungs

- **Breathing** is the inhalation and exhalation of air that ventilates lungs.
- In mammals, breathing involves movement of the **diaphragm**—a dome-shaped muscle separating the thoracic cavity from the abdominal cavity. Lung volume increases when the rib muscles and diaphragm contract, pressure within the lungs decreases, and air flows into the lungs. During exhalation, lung volume is decreased as the diaphragm relaxes and moves up, and pressure within the lungs increases.
- Breathing is under control of regions located in the medulla and the pons. It is influenced by pH, which is an indirect indication of blood CO_2 levels.
- Increased metabolic activity lowers pH by increasing the concentration of CO_2 in the blood. CO_2 reacts with water to form carbonic acid, which dissociates into a bicarbonate ion and a hydrogen ion.

Concept 42.7 Adaptations for gas exchange include pigments that bind and transport gases

▌ **Hemoglobin** is the respiratory pigment found in almost all vertebrates. It consists of four subunits, each of which has a heme group with an embedded iron atom. The iron atom binds O_2, so each hemoglobin can carry four oxygen molecules.

▌ A lowering of the pH in blood lowers the affinity of hemoglobin for oxygen, and oxygen dissociates. This is called the **Bohr shift**.

▌ CO_2 is most commonly carried in the blood in the form of bicarbonate ions (70%). Less commonly, it is transported via hemoglobin (23%) and in solution in the blood plasma (7%)

▌ **Carbonic anhydrase** is an enzyme found in RBCs that catalyzes the formation of carbonic acid, which dissociates into a bicarbonate ion and a hydrogen ion. As blood pH drops, the rate and depth of respiration will increase.

> **STUDY TIP:** Know how oxygen and carbon dioxide are transported in the blood!

Chapter 43: The Immune System

> ## YOU MUST KNOW
>
> - Several elements of an innate immune response.
> - The differences between B and T cells relative to their activation and actions.
> - How antigens are recognized by immune system cells.
> - The differences in humoral and cell-mediated immunity.
> - Why Helper T cells are central to immune responses.

Concept 43.1 In innate immunity, recognition and response rely on shared pathogen traits

▌ **Innate immune** responses include barrier defenses as well as defenses to combat pathogens that enter the body.

1. **Barrier defenses** include skin and the mucous membranes that cover the surface and line the openings of the animal body. These provide a physical barrier and also produce secretions that result in a skin pH from 3 to 5, and the antimicrobial **lysozyme** found in saliva, mucous secretions, and tears.

2. **Cellular innate defenses** combat pathogens that get through the skin—for example, in a cut. They include phagocytic white blood cells and antimicrobial proteins.

Phagocytic White Blood Cells

- **Neutrophils** are white blood cells that ingest and destroy microbes in a process called **phagocytosis**.
- **Monocytes** are another type of phagocytic leukocyte. They migrate into tissues and develop into **macrophages**, which are giant phagocytic cells.
- **Eosinophils** are leukocytes that defend against parasitic invaders such as worms by positioning themselves near the parasite's wall and discharging hydrolytic enzymes.

Antimicrobial Proteins

- **Interferon** proteins provide innate defense against viral infections. They cause cells adjacent to infected cells to produce substances to inhibit viral replication.
- The **complement system** consists of roughly 30 proteins with a variety of functions. One function is to lyse invading cells.

3. A local **inflammatory response** is triggered by damage to tissue by physical injury or the entry of pathogens. It leads to release of numerous chemical signals. For example, **histamines** are released by basophils and mast cells (two types of leukocytes) in response to injury. Histamines trigger the dilation and permeability of nearby capillaries. This aids in delivering clotting agents and phagocytic cells to the injured area. Systemic inflammatory responses include fever and septic shock.
4. **Natural killer (NK) cells** help recognize and remove diseased cells.

Concept 43.2 *In acquired immunity, lymphocyte receptors provide pathogen-specific recognition*

- Vertebrates have two types of lymphocytes: **B lymphocytes (B cells)**, which proliferate in the bone marrow, and **T lymphocytes (T cells)**, which mature in the thymus. They circulate through the blood and lymph, and both recognize particular microbes. All blood cells proliferate from stem cells in the bone marrow.
- **Antigens** are foreign molecules that elicit a response by lymphocytes. B and T cells recognize them by specific receptors imbedded in their plasma membranes.
- **Antibodies** are soluble proteins secreted by B cells during an immune response.
- **B- or T-cell activation** occurs when an antigen binds to a B or T cell. B-cell activation is enhanced by cytokines. The lymphocyte forms two clones of cells in a process called **clonal selection**. This results in thousands of cells, all specific to this antigen.

  1. **Effector cells** combat the antigen.
  2. **Memory cells**, which are long-lived, bear receptors for the same antigen, thus allowing them to quickly mount an immune response in subsequent infections.

- B-cell receptors bind intact antigens.
- T-cell receptors bind antigens that are displayed by **antigen-presenting cells (APCs)** on their **MHCs**.

■ **MHCs (major histocompatibility complex molecules)** are proteins that are the product of a group of genes. (Individuals differ in their MHCs. This is a major component of "self.")

■ There are two types of MHCs:

1. **Class I MHCs** are found on almost all cells of the body, except RBCs.
2. **Class II MHCs** are made by dendritic cells, macrophages, and B cells.

■ The specificity of B and T cells is a result of the shuffling and recombination of several gene segments and results in more than 1 million *different* B cells and 10 million different T cells.

■ Each B or T cell responds to only one antigen.

■ A **primary immune response** occurs when the body is first exposed to an antigen and a lymphocyte is activated.

■ A **secondary immune response** occurs when the same antigen is encountered at a later time. It is faster and of greater magnitude. (Refer to Figure 9.4.)

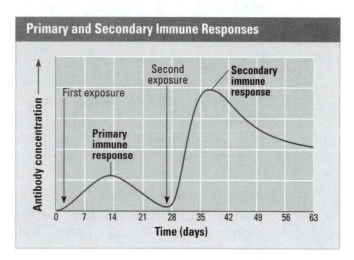

Figure 9.4

Concept 43.3 *Acquired immunity defends against infection of body cells and fluids*

■ Acquired immunity has two branches:

1. **Humoral immune response** involves the activation and clonal selection of effector B cells, which produce antibodies that circulate in the blood.
2. **Cell-mediated immune response** involves the activation and clonal selection of cytotoxic T cells, which identify and destroy infected cells.

■ **Helper T cells** aid both responses. When activated by interaction with the class II MHC molecule of an APC, they secrete *cytokines* that stimulate and activate both B cells and cytotoxic T cells. Refer to Figure 9.5 to see the central role of helper T cells. The helper T cell is bound to the class II MHC by a **CD4** protein.

■ **Cytotoxic T cells** bind to class I MHC molecules, displaying antigenic fragments on the surface of infected body cells. The cytotoxic T cell is bound to the infected cell's class I MHC by its **CD8** protein. Cytotoxic T cells destroy infected body cells.

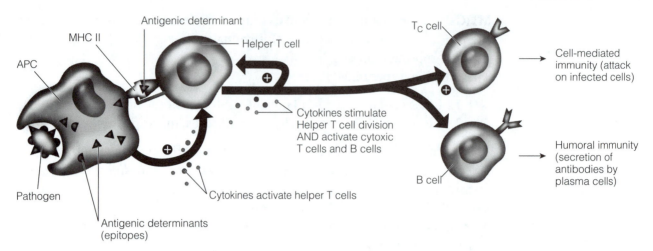

Figure 9.5

> ***ORGANIZE YOUR THOUGHTS***
>
> Confused by all the different cell names?
>
> 1. **B cells** make antibodies, which provide humoral immunity. This helps fight pathogens that are circulating in the blood.
> 2. **Cytotoxic T cells** destroy body cells that are infected by a pathogen or cancer cells.
> 3. **Helper T cells** activate both B and T cells.

▌ Recall that activated B cells produce memory cells as well as plasma cells. The plasma cells secrete antibodies in prodigious numbers. These will circulate in the blood, and bind and destroy the antigen.

▌ Modes of antibody action include

1. **Neutralization:** Antibodies bind the pathogen's surface proteins, which prevents it from entering and infecting cells.
2. **Opsonization:** Results in increased phagocytosis of the antigen.
3. **Lysis:** Caused by activation of the complement system.

▌ **Active immunity** develops naturally in response to an infection; it also develops artificially by immunization (vaccination). In immunization, a nonpathogenic form of a microbe or part of a microbe elicits an immune response resulting in immunological memory for that microbe.

▌ **Passive immunity** occurs when an individual receives antibodies, such as those passed to the fetus across the placenta and to infants via milk.

▌ Certain antigens on red blood cells determine whether a person has **type A, B, AB**, or **O blood**. Because antibodies to nonself blood antigens already exist in the body, transfusion with incompatible blood leads to destruction of the transfused cells and a life-threatening situation for the patient.

- **MHC molecules** are responsible for stimulating the rejection of tissue grafts and organ transplants. The chances of successful transplantation are increased if the donor's tissue-bearing MHC molecules closely match the recipient's. The recipient also must take immunosuppressant drugs.

Concept 43.4 Disruptions in immune system function can elicit or exacerbate disease

- In localized **allergies** such as hay fever, IgE antibodies produced after first exposure to an allergen attach to receptors on mast cells. The next time the same allergen enters the body, it bonds to mast cell–associated IgE molecules, inducing the cell to release histamine and other mediators that cause vascular changes and typical symptoms.
- **Lupus, rheumatoid arthritis, and multiple sclerosis** are examples of autoimmune diseases. In each case, the immune system turns against particular molecules of the body, generating antibodies that attack and damage the body's own healthy cells.
- **HIV** infects helper T cells. Refer again to Figure 9.5. Can you see why people with AIDS are immune suppressed? Note what cell is central in both humoral and cell-mediated immunity.

Chapter 44: Osmoregulation and Excretion

YOU MUST KNOW

- Three categories of nitrogenous waste, which animal groups produce each, and why.
- The components of a nephron, and what occurs in each region.
- How hormones affect water balance by acting on the nephron.

Concept 44.1 Osmoregulation balances the uptake and loss of water and solutes

- **Osmoregulation** is the process by which animals control solute concentrations and balance water gain and loss.
- Most metabolic wastes must be excreted from the body. One of the most important types is **nitrogenous wastes** from the breakdown of proteins and nucleic acids.
- **Excretion** includes the removal of nitrogenous wastes from the body.
- **Transport epithelia** regulate water balance and waste disposal. They consist of layers of specialized cells with extensive surface area, typically arranged into complex tubular networks, connected to the outside.

Concept 44.2 An animal's nitrogenous wastes reflect its phylogeny and habitat

▌ There are three types of nitrogenous wastes:

1. **Ammonia** is very water soluble and very toxic. It is generally produced only by aquatic animals where water loss is not a competing problem.
2. **Urea** is produced in the liver of most vertebrates, where ammonia is combined with carbon dioxide in an energetically expensive process. However, because it is less toxic, water is conserved.
3. **Uric acid** is more energetically expensive to produce than urea, but it is insoluble in water and can be excreted as a paste or crystals. It is produced by both birds and reptiles because it can be stored within the shelled eggs of these groups and then left as a harmless solid when the young hatch.

Concept 44.3 Diverse excretory systems are variations on a tubular theme

▌ **Survey of Excretory Systems:** Recognize the following as excretory systems, and be able to identify them by phyla or group:

- **Protonephridia/Flame-bulb system:** Platyhelminthes (Planaria)
- **Metanephridia:** Annelida (Earthworms)
- **Malpighian tubules:** Insects and terrestrial arthropods
- **Kidneys:** Vertebrates

▌ Most excretory systems produce urine in a four-step process. Focus on the location of each of these processes in the mammalian kidney, which is introduced in the next two concepts and shown in Figure 9.6.

▌ Study the mammalian kidney, seen in Figure 9.7.

▌ Mammals have two kidneys, and each is supplied with a **renal artery** and a **renal vein**.

▌ **Urine** leaves the kidneys through the **ureters**, which drain into the urinary bladder. Urine is expelled from the body through the **urethra**.

▌ **Nephrons** are the functional units of the kidney. Each kidney has approximately one million! They are made up of a single long tubule and the **glomerulus**, a ball of capillaries. At one end of the tubule is the **Bowman's capsule**, a C-shaped structure that surrounds the glomerulus.

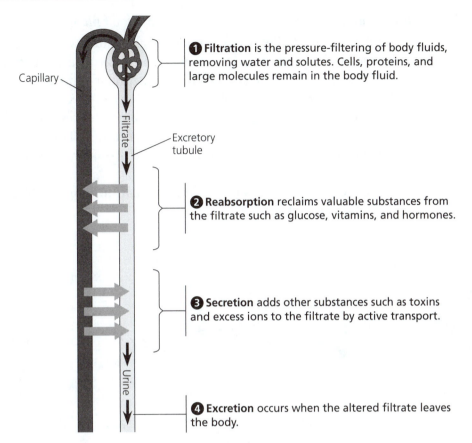

❶ Filtration is the pressure-filtering of body fluids, removing water and solutes. Cells, proteins, and large molecules remain in the body fluid.

Capillary

Filtrate

Excretory tubule

❷ Reabsorption reclaims valuable substances from the filtrate such as glucose, vitamins, and hormones.

❸ Secretion adds other substances such as toxins and excess ions to the filtrate by active transport.

Urine

❹ Excretion occurs when the altered filtrate leaves the body.

Figure 9.6

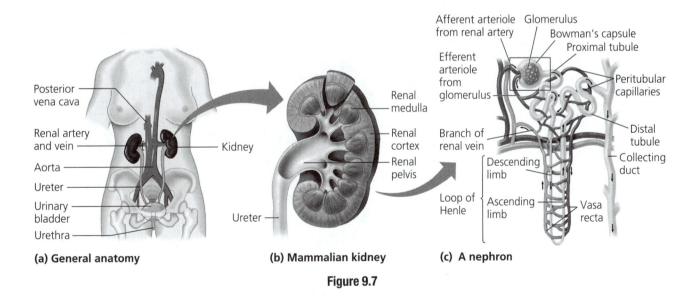

Posterior vena cava

Renal artery and vein

Aorta

Ureter

Urinary bladder

Urethra

Kidney

(a) General anatomy

Renal medulla

Renal cortex

Renal pelvis

Ureter

(b) Mammalian kidney

Afferent arteriole from renal artery

Glomerulus

Bowman's capsule

Proximal tubule

Efferent arteriole from glomerulus

Branch of renal vein

Descending limb

Loop of Henle

Ascending limb

Peritubular capillaries

Distal tubule

Collecting duct

Vasa recta

(c) A nephron

Figure 9.7

Concept 44.4 The nephron is organized for stepwise processing of blood filtrate

▌ Capillaries called **afferent arterioles** are associated with the nephrons, and as they leave the glomerulus, the capillaries converge into an **efferent arteriole**. This vessel subdivides again to form **peritubular capillaries**, which surround the proximal and distal tubules.

▌ The filtrate flows through the **proximal tubule**, the descending **loop of Henle**, the ascending loop of Henle, and the **distal tubule**. The distal tubule empties into a **collecting duct**, which receives wastes from many nephrons. The filtrate empties into the renal pelvis.

▌ There are five main steps in the **transformation of blood filtrate to urine**, as shown in Figure 9.8

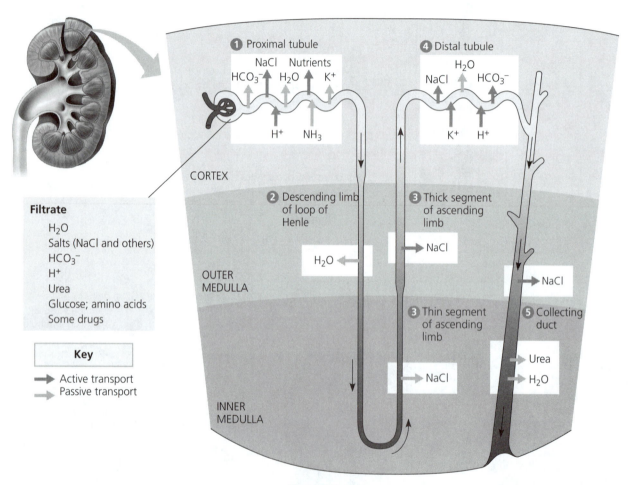

Figure 9.8

1. *In the proximal tubule*, secretion and reabsorption change the volume and composition of the filtrate. The pH of body fluids is controlled, and bicarbonate is absorbed, as are NaCl and water.
2. *In the descending loop of Henle*, reabsorption of water continues.
3. *In the ascending loop of Henle*, the filtrate loses salt without giving up water and becomes more dilute.
4. In the *distal tubule*, K^+ and NaCl levels are regulated, as is filtrate pH.

5. *The collecting duct* carries the filtrate through the medulla to the renal pelvis, and the filtrate becomes more concentrated by the movement of salt.

▌ Let's go back to the four steps of urine formation introduced earlier. Make specific note of where each occurs in the mammalian kidney.

- **Filtration** occurs in the glomerulus, when filtrate is forced into the Bowman's capsule. Blood cells and proteins do *not* enter the filtrate.
- **Reabsorption** occurs in the proximal and distal tubules, as well as the loop of Henle.
- **Secretion** occurs in the proximal tubule.
- **Excretion** occurs when the filtrate leaves the body as the collecting tubules carry urine toward the ureters, then to the bladder, then out the urethra.

▌ Note that the flow of filtrate in the loop of Henle is an example of a **countercurrent** system, and allows the kidney to form a concentrated urine with minimal water loss.

Concept 44.5 *Hormonal circuits link kidney function, water balance, and blood pressure*

▌ **Antidiuretic hormone (ADH)** is an important hormone in the regulation of water balance. It is produced in the hypothalamus and stored in and released from the pituitary gland. It makes the collecting ducts more permeable to water, so more water leaves the filtrate, resulting in more concentrated urine and reduced loss of water from the body.

▌ Also involved in regulation of water balance are **renin**, **angiotensin**, and **aldosterone**. Let's look at what happens if blood pressure or blood volume drop.

1. **Renin**, an enzyme, is released in the kidney. Renin activates angiotensin II.
2. **Angiotensin II** acts as a hormone and causes arterioles to constrict, which will raise blood pressure. It will also cause the adrenal glands to release the hormone **aldosterone**.
3. **Aldosterone** causes the kidney to reabsorb more Na+, which increases retention of water and blood volume and pressure.

Chapter 45: Hormones and the Endocrine System

YOU MUST KNOW
- Two ways hormones affect target organs.
- The secretion, target, action, and regulation of at least three hormones.
- An illustration of both positive and negative feedback in the regulation of homeostasis by hormones.

Concept 45.1 **The endocrine system and the nervous system act individually and together in regulating an animal's physiology**

▌ The **endocrine system** of an animal is the sum of all its hormone-secreting cells and tissues.

▌ **Endocrine glands** are *ductless* and secrete hormones directly into body fluids.

▌ **Hormones** are chemical signals that cause a response in *target cells*.

▌ **Positive and negative feedback** regulates most endocrine secretion.

Concept 45.2 **Hormones and other chemical signals bind to target cell receptors, initiating pathways that culminate in specific cell responses**

▌ Study two mechanisms of hormone action. Recall that these mechanisms were covered in some detail in Chapter 11, Cell Signaling.

1. **Cell-surface receptors** bind the hormone, and a *signal transduction pathway* is triggered. A signal transduction pathway consists of a series of molecular events that initiate a response to the signal. *Example:* The binding of epinephrine to liver cells causes a cascade that leads to the conversion of glycogen to glucose.

2. **Intracellular receptors** are bound by hormones that are lipid-soluble. The receptor then acts as a transcription factor, causing a change in gene expression. *Example:* Testosterone and estrogen enter the nuclei of target cells, bind the DNA, and stimulate transcription of certain genes.

▌ Hormones in the body can affect one tissue, a few tissues, or most of the tissues in the body (as with the sex hormones), or they may affect other endocrine glands (these last are referred to as **tropic hormones**).

Concept 45.3 **The hypothalamus and pituitary integrate many functions of the vertebrate endocrine system**

▌ The **hypothalamus** receives information from nerves throughout the body and from other parts of the brain and then initiates endocrine signals in response.

▌ The **posterior pituitary** is an extension of the hypothalamus that stores and secretes two hormones:

1. **Oxytocin** causes contraction of the uterine muscles in childbirth and ejection of milk in nursing

2. **Antidiuretic hormone (ADH)** makes the collecting tubules of the kidney more permeable to water, increasing water retention.

▌ The **anterior pituitary** consists of endocrine cells that synthesize and secrete several hormones. Some of these are *tropic hormones*, which means they stimulate the activity of other endocrine tissues (FSH, LH, TSH, and ACTH).

1. **Follicle-stimulating hormone (FSH)** stimulates development of the ovarian follicles in females and promotes spermatogenesis in males by acting on the cells in the seminiferous tubules.

2. **Luteinizing hormone** triggers ovulation in females and stimulates the production of testosterone by the interstitial cells of the testes.

Other endocrine glands, their secretion, targets, action, and regulation will be covered in Table 9.1.

Table 9.1 Major Human Endocrine Glands and Some of Their Hormones

Gland	Hormone	Representative Actions	Regulated By
Hypothalamus	Hormones released from the posterior pituitary and hormones that regulate the anterior pituitary (see below)		
Posterior pituitary gland (releases neurohormones made in hypothalamus)	Oxytocin	Stimulates contraction of uterus and mammary gland cells	Nervous system
	Antidiuretic hormone (ADH)	Promotes retention of water by kidneys	Water/salt balance
Anterior pituitary gland	Growth hormone (GH)	Stimulates growth (especially bones) and metabolic functions	Hypothalamic hormones
	Prolactin (PRL)	Stimulates milk production and secretion	Hypothalamic hormones
	Follicle-stimulating hormone (FSH)	Stimulates production of ova and sperm	Hypothalamic hormones
	Luteinizing hormone (LH)	Stimulates ovaries and testes	Hypothalamic hormones
	Thyroid-stimulating hormone (TSH)	Stimulates thyroid gland	Hypothalamic hormones
	Adrenocorticotropic hormone (ACTH)	Stimulates adrenal cortex to secrete glucocorticoids	Hypothalamic hormones
Thyroid gland	Triiodothyronine (T_3) and thyroxine (T_4)	Stimulate and maintain metabolic processes	TSH
	Calcitonin	Lowers blood calcium level	Calcium in blood
Parathyroid glands	Parathyroid hormone (PTH)	Raises blood calcium level	Calcium in blood
Pancreas	Insulin	Lowers blood glucose level	Glucose in blood
	Glucagon	Raises blood glucose level	Glucose in blood
Adrenal glands Adrenal medulla	Epinephrine and norepinephrine	Raise blood glucose level; increase metabolic activities; constrict certain blood vessels	Nervous system
Adrenal cortex	Glucocorticoids	Raise blood glucose level	ACTH
	Mineralocorticoids	Promote reabsorption of Na^+ and excretion of K^+ in kidneys	K^+ in blood; angiotensin II
Gonads Testes	Androgens	Support sperm formation; promote development and maintenance of male secondary sex characteristics	FSH and LH
Ovaries	Estrogens	Stimulate uterine lining growth; promote development and maintenance of female secondary sex characteristics	FSH and LH
	Progestins	Promote uterine lining growth	FSH and LH
Pineal gland	Melatonin	Involved in biological rhythms	Light/dark cycles

Concept 45.4 Nonpituitary hormones help regulate metabolism, homeostasis, development, and behavior

▌ The maintenance of blood calcium level is one example of how homeostasis is maintained by *negative feedback*. Remember, in negative feedback, *more gets you less.*

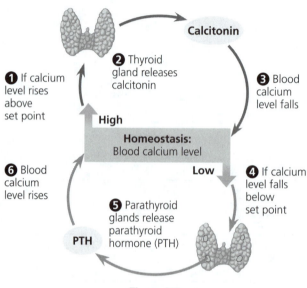

Figure 9.9

▌ While we are on regulation by feedback loops, let's jump ahead to Chapter 46 to use oxytocin regulation as an example of *positive feedback*. Oxytocin stimulates uterine contractions and also stimulates production of prostaglandins by the placenta. Prostaglandins stimulate release of more oxytocin and more prostaglandins, stimulating more uterine contraction. Remember, in positive feedback, *more gets you more.*

Chapter 46: Animal Reproduction

YOU MUST KNOW

- The hormonal control of the menstrual cycle.
- How oogenesis and spermatogenesis differ.

Concept 46.1 Both asexual and sexual reproduction occur in the animal kingdom

▌ **Sexual reproduction** is the creation of offspring by the fusion of haploid gametes to form a zygote. The female gamete is the *ovum*, and the male gamete is the *sperm*.

▌ **Asexual reproduction** is reproduction in which all genes come from one parent; there is no fusion of egg and sperm. There are several modes of asexual reproduction.

  ■ **Fission** is the separation of a parent into two or more individuals of about the same size.

  ■ **Budding** occurs when new individuals arise from outgrowths of the parent. This is seen in certain cnidarians and tunicates.

  ■ **Fragmentation** occurs when an individual breaks into several pieces, all of which then may form complete adults. *Regeneration*, the regrowth of body parts, is a necessary part of fragmentation. This mode of reproduction can be seen in sea stars, sponges, and cnidarians.

  ■ **Parthenogenesis** is the process in which a female produces eggs that develop without being fertilized. Male bees are produced this way and are always haploid.

▌ Why sex? It may result in beneficial gene combinations arising through recombination that speed up adaptation, or the shuffling of genes during sexual reproduction might allow a population to rid itself of sets of harmful genes more readily.

▌ There are many triggers to reproduction and different patterns:

  ■ **Ovulation**, the release of mature eggs, may be cyclical, so that young are produced only at certain times of year when survival is most likely. These hormonal cycles may be influenced by changes in day length, seasonal temperature, rainfall, or lunar cycles.

  ■ In **hermaphroditism**, each individual has both male and female reproductive systems. This is common in sessile (stationary) animals such as barnacles and some parasites, including tapeworms.

  ■ In **sex reversal**, an individual changes its sex during its lifetime. An example is the bluehead wrasse, a reef fish. If a male dies, the largest female in the harem will undergo a sex change and begin producing sperm!

Concept 46.2 Mechanisms for fertilization bring together sperm and eggs of the same species

- **Fertilization** is the union of sperm and egg.
- **External fertilization** occurs when eggs are shed by the female and fertilized by the male outside the female's body, usually in water. The release of gametes must be synchronous and often involves environmental cues or courtship behaviors. Species with external fertilization in general produce very large numbers of gametes.
- **Internal fertilization** occurs when sperm are deposited in the female reproductive tract, and fertilization occurs within the tract. It is an adaptation that allows reproduction in a dry environment. Fewer gametes and fewer zygotes are often produced.
- When fertilized eggs are protected by shells or within the female's body, fewer zygotes are produced.
- **Gonads** are the organs that produce gametes in most animals.

Concept 46.3 Reproductive organs produce and transport gametes

Refer to Figure 9.10 as you review the parts of the male and female reproductive anatomy. Know the function of each structure.

Female Anatomy

- The **clitoris** and **labia** are the external female reproductive structures.
- The **ovaries** are the female gonads. They produce both the gametes (eggs) and the female sex hormones.
- **Follicles** are microscopic structures within the ovaries. They contain a partially developed egg (*oocyte*) surrounded by one or more layers of follicle cells, which help to develop, nourish, and protect the egg cell. One follicle matures and releases its egg cell during each menstrual cycle.
- The follicle cells also produce **estrogens**, the female hormones.
- **Ovulation** is the release of an egg from the follicle. The remaining follicle tissue heals and grows in the ovary to form a body called a **corpus luteum**, which secretes estrogen and progesterone. Progesterone helps to maintain the uterine wall during pregnancy. If the egg cell isn't fertilized, the corpus luteum disintegrates.
- The egg cell is released into the **oviduct**, and cilia lining the oviduct convey the egg cell down to the uterus.
- The **endometrium** is the inner lining of the uterus.
- The **cervix** is the neck of the uterus. It leads to the **vagina**, the canal through which a baby is born.

Male Anatomy

- The **scrotum** and **penis** are the external male reproductive structures. The scrotum is a sac that encloses the testes.
- The **testes** are the male gonads. They produce both the male gametes (sperm) and male sex hormones.
- The **seminiferous tubules** are highly coiled structures in the testes where sperm are made.

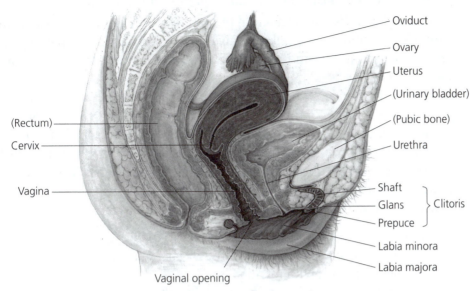

(a) Reproductive anatomy of female

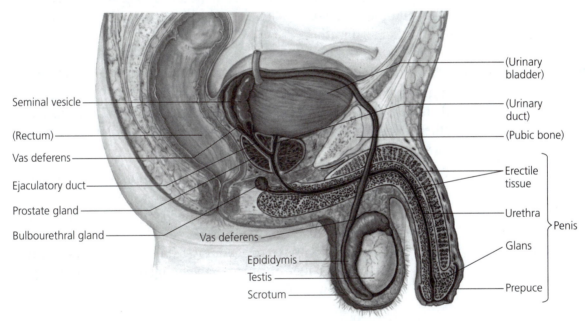

(b) Reproductive anatomy of male

Figure 9.10

- In between the tubules are **Leydig cells**, which produce testosterone and other androgens.
- The sperm pass from the seminiferous tubules into the **epididymis** where they will mature over a three week period.
- During ejaculation, sperm are propelled through the **vas deferens** to the urethra.
- The **seminal vesicles**, the **prostate gland**, and the **bulbourethral glands** all contribute secretions that make up semen. These secretions supply necessary nutrients and a medium for the sperm cells.

Concept 46.4 The timing and pattern of meiosis in mammals differ for males and females

▌ **Spermatogenesis** is the production of mature sperm cells, and it occurs in the seminiferous tubules. The cells that give rise to sperm are called *spermatogonia*. They undergo meiosis and differentiation eventually to form mature, motile sperm.

▌ **Oogenesis** is the development of mature ova. **Oogonia** are the cells that develop into ova; they multiply and begin meiosis, but they stop at prophase I of meiosis I. These egg cells are called **primary oocytes**, which are quiescent until puberty. From puberty onward, FSH periodically stimulates a follicle to grow and its egg cell to complete meiosis I and begin meiosis II. This forms the **secondary oocyte**.

▌ Spermatogenesis differs from oogenesis in several ways:

1. Production of four sperm cells versus a single egg.
2. Spermatogenesis continues throughout a mature male's entire life; oogenesis ends with menopause.
3. Spermatogenesis is continuous after puberty, whereas oogenesis has long interruptions. Recall that in humans, eggs are arrested in Prophase I prior to a female's birth! Meiosis is not completed until after fertilization.

Concept 46.5 The interplay of tropic and sex hormones regulates mammalian reproduction

▌ Humans and other primates have **menstrual cycles**. Menstruation occurs when the endometrium is shed from the uterus through the cervix and vagina. Other mammals have **estrous cycles** when the vagina is receptive to mating. There are three phases in the menstrual cycle:

1. In the **menstrual flow phase**, most of the endometrium is shed and menstrual bleeding occurs.
2. In the **proliferative phase**, the endometrium begins to regenerate and thicken.
3. In the **secretory phase**, the endometrium continues to thicken, and if an embryo has not implanted in the lining by the end of this phase, menstrual flow occurs.

> **A Tip from the Graders:**
> Because the hormonal control of menstruation is rich in examples of feedback and illustrates so well the circulation of hormones in the blood affecting target tissues, this is a good process to understand in detail.

▌ The **ovarian cycle** parallels the menstrual flow cycle. This cycling is under hormonal influence, and there is an interplay between pituitary hormones, whose target tissues are in the ovary, and ovarian hormones. Since you have probably just reviewed the chapter on hormones, let's look at what happens in this particular example. Note which hormone levels are increasing and which are decreasing and what these changes effect by studying Figure 9.11.

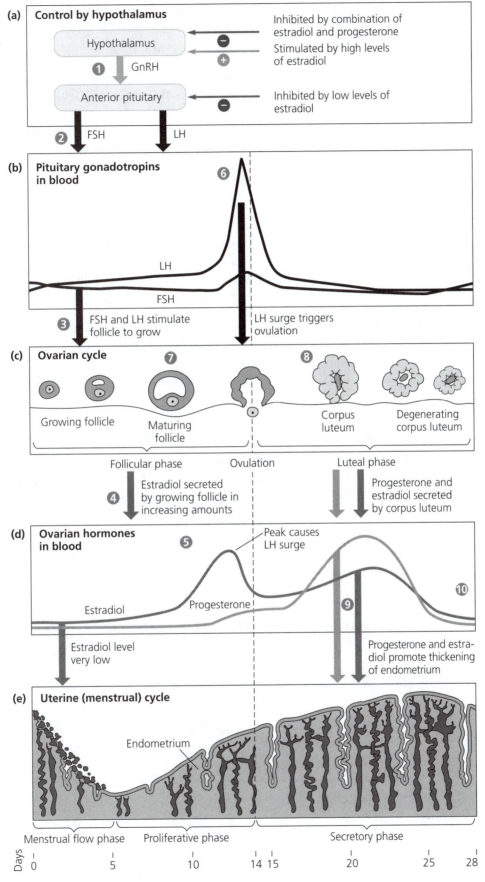

Figure 9.11

- Notice that the menstrual cycle is affected by two pituitary hormones and two ovarian hormones. We will first look at the pituitary hormones; since their target tissues are in the ovary, they are *gonadotropic hormones*:

  - **FSH** is secreted from the anterior pituitary. Its target is the follicle. As its level increases, the follicle enlarges and matures. FSH levels are affected by estradiol levels.
  - **LH** is also secreted from the anterior pituitary. It induces final development of the follicle, and a surge of LH triggers ovulation.

- Now, take a look at the ovarian hormones:

  - **Estradiol** is an estrogen secreted by the ovarian follicle. As the follicle develops, the level of estradiol increases and the endometrium thickens. Estradiol levels affect production of gonadotropins. (Study Figure 9.11.)
  - **Progesterone** levels increase markedly after the ovarian follicle ruptures in ovulation and becomes the corpus luteum. High progesterone and estradiol levels promote further endometrial development.

- If fertilization of the egg does not occur, the corpus luteum will disintegrate, and the sharp decline in levels of estradiol and progesterone will lead to shedding of the endometrium (menstruation).
- Spermatogenesis and androgen levels in males also involve feedback loops with varying FSH, LH, and testosterone levels.

Concept 46.6 *In placental mammals, an embryo develops fully within the mother's uterus*

- **Human chorionic gonadotropin (hCG)** is secreted by the developing embryo and acts like LH to maintain the secretion of progesterone and estrogens by the corpus luteum in early pregnancy. Its high level in the urine is used as the basis of a common early pregnancy test.
- **Pregnancy**, or **gestation**, is the condition of carrying one or more embryos in the uterus. It culminates in birth, or *parturition*, which is brought about by a series of strong rhythmic uterine contractions.

Chapter 47: Animal Development

YOU MUST KNOW

- What occurs in cleavage, gastrulation, and organogenesis.
- Two structures derived from each germ layer.
- The events that occur when a sperm contacts an egg.

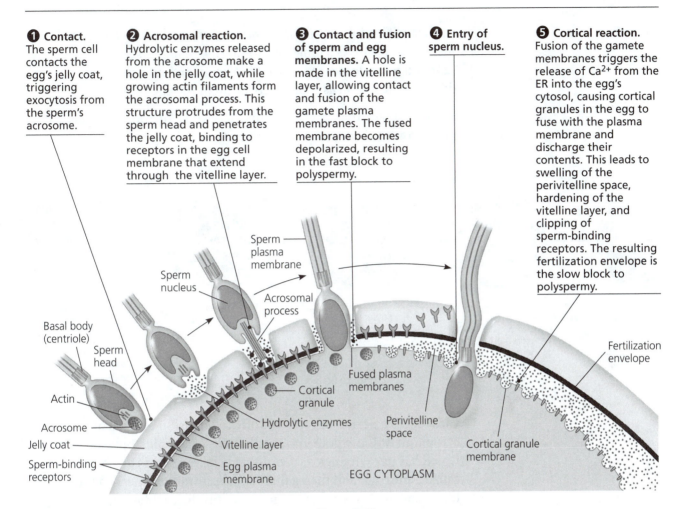

① Contact. The sperm cell contacts the egg's jelly coat, triggering exocytosis from the sperm's acrosome.

② Acrosomal reaction. Hydrolytic enzymes released from the acrosome make a hole in the jelly coat, while growing actin filaments form the acrosomal process. This structure protrudes from the sperm head and penetrates the jelly coat, binding to receptors in the egg cell membrane that extend through the vitelline layer.

③ Contact and fusion of sperm and egg membranes. A hole is made in the vitelline layer, allowing contact and fusion of the gamete plasma membranes. The fused membrane becomes depolarized, resulting in the fast block to polyspermy.

④ Entry of sperm nucleus.

⑤ Cortical reaction. Fusion of the gamete membranes triggers the release of Ca^{2+} from the ER into the egg's cytosol, causing cortical granules in the egg to fuse with the plasma membrane and discharge their contents. This leads to swelling of the perivitelline space, hardening of the vitelline layer, and clipping of sperm-binding receptors. The resulting fertilization envelope is the slow block to polyspermy.

Sperm plasma membrane

Sperm nucleus

Acrosomal process

Basal body (centriole)

Sperm head

Actin

Acrosome

Jelly coat

Sperm-binding receptors

Cortical granule

Hydrolytic enzymes

Vitelline layer

Egg plasma membrane

Fused plasma membranes

Perivitelline space

EGG CYTOPLASM

Cortical granule membrane

Fertilization envelope

Figure 9.12

Concept 47.1 After fertilization, embryonic development proceeds through cleavage, gastrulation, and organogenesis

▌ Fertilization combines haploid sets of chromosomes from two individuals into a single diploid cell, the *zygote*. Figure 9.12 provides an outline of the main events as seen in sea urchins.

There are a few minor differences in mammalian fertilization.

1. **The acrosomal reaction** occurs when the head of the sperm contacts the jelly coat of an egg. The *acrosome*, a specialized vesicle at the tip of the sperm, discharges hydrolytic enzymes, which digest the jelly coat. Molecules of the sperm membrane adhere to receptor proteins on the egg plasma membrane, and egg and sperm plasma membranes fuse.

2. **Ion channels open** in the egg's plasma membrane, and sodium ions flow in, causing depolarization of the membrane. This depolarization prevents other sperm from binding and serves as a *fast block to polyspermy*.

3. **The cortical reaction** occurs when molecules are secreted into the space between the plasma membrane and the vitelline layer and

harden it. This forms a *fertilization envelope* and resists additional sperm entry. This is a longer-term *slow block to polyspermy.*

4. **Calcium ion release** from the endoplasmic reticulum causes both cortical granule fusion as well as activation of the egg.

■ There are three stages that begin to build the body of most animals. Study Figure 9.13 as you read the accompanying text.

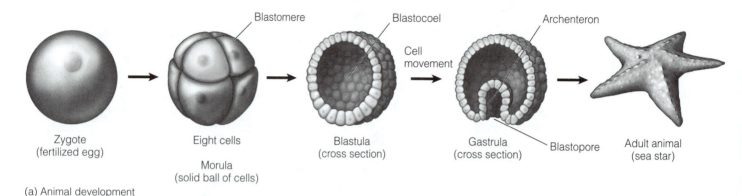

Figure 9.13

1. **Cleavage** is a period of rapid mitotic cell division that partitions the cytoplasm of the zygote into smaller cells called **blastomeres,** each of which has its own nucleus. Continued cleavage leads to a ball of cells called a **morula,** and then a fluid-filled central cavity called the **blastocoel** forms within the morula to produce a **blastula.**

2. **Gastrulation** is a drastic rearrangement of the cells in the blastula to form a three-layered embryo with a primitive gut (the *archenteron*). Invagination begins at the blastopore. In gastrulation, three germ cell layers are produced: the ectoderm, endoderm, and mesoderm. You should know at least two derivatives of each germ layer. Here is a chart to help.

ECTODERM	MESODERM	ENDODERM
• Skin, nails, teeth	• Skeletal, muscular systems	• Epithelial *linings* of digestive, respiratory, excretory tracts.
• Lens of eye	• Excretory, circulatory systems	
• Nervous System	• Reproductive system	• Liver, pancreas.
	• Blood, bone, and muscle	

3. **Organogenesis** is the development of the three germ layers into the rudiments of organs. Some additional notes about chordate development follow:
   • The **notochord** is a stiff dorsal skeletal rod characteristic of all chordate embryos. It forms from mesoderm.

- The **neural plate** forms from ectoderm above the notochord. It curves inward, rolling into a *neural tube* that will become the brain and spinal cord.
- **Neurulation** is the process in which the hollow dorsal nerve chord forms.
- **Somites** are blocks of mesoderm that are serially arranged along the notochord. They are a sign of segmentation.

▌ Birds and reptiles lay shelled eggs where the embryo is surrounded by fluid within the amnion; they are called **amniotes**. Notice the four *extraembryonic membranes* that form within the shelled egg and the function of each (see Figure 9.14).

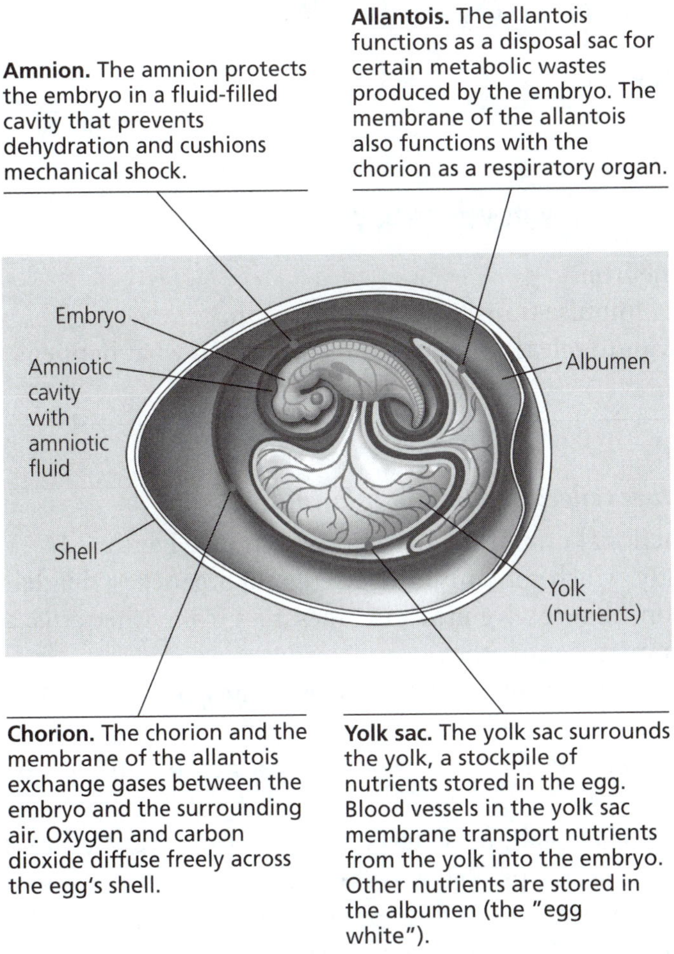

Amnion. The amnion protects the embryo in a fluid-filled cavity that prevents dehydration and cushions mechanical shock.

Allantois. The allantois functions as a disposal sac for certain metabolic wastes produced by the embryo. The membrane of the allantois also functions with the chorion as a respiratory organ.

Embryo

Amniotic cavity with amniotic fluid

Shell

Albumen

Yolk (nutrients)

Chorion. The chorion and the membrane of the allantois exchange gases between the embryo and the surrounding air. Oxygen and carbon dioxide diffuse freely across the egg's shell.

Yolk sac. The yolk sac surrounds the yolk, a stockpile of nutrients stored in the egg. Blood vessels in the yolk sac membrane transport nutrients from the yolk into the embryo. Other nutrients are stored in the albumen (the "egg white").

Figure 9.14

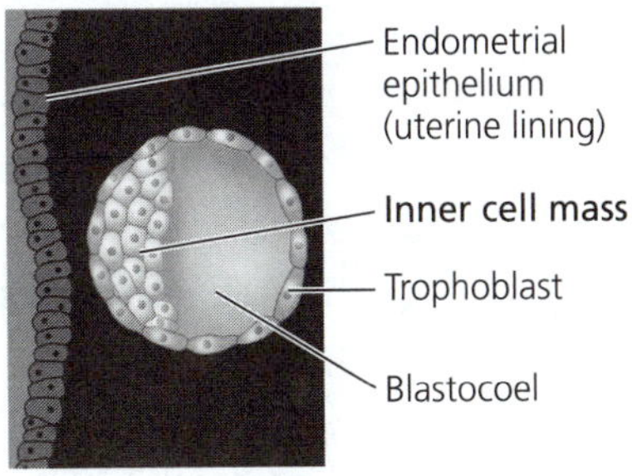

Endometrial epithelium (uterine lining)

Inner cell mass

Trophoblast

Blastocoel

Figure 9.15

▌ The **blastocyst** is the mammalian version of a blastula (see Figure 9.15).

- The **inner cell mass** is a group of cells that will develop into the embryo. Its cells are the source of embryonic stem cell lines.
- The **trophoblast** is the outer epithelium of the blastocyst. It initiates formation of the fetal portion of the placenta.

- The patterns of development are due to a combination of different **cytoplasmic determinants** and **inductive cell signals.** *Cytoplasmic determinants* are chemical signals such as mRNAs and transcription factors that may be parceled out unevenly in early cleavages. *Induction* is an interaction among cells that influences their fate, usually by causing changes in gene expression.
- The dorsal lip of the blastopore is an "organizer" which induces a series of events that result in formation of notochord and neural tube
- **Totipotent cells** are capable of developing into all the different cell types of that species. The cells of mammalian embryos remain totipotent until the 16-cell stage.

Chapter 48: Neurons, Synapses, and Signaling

YOU MUST KNOW
• The anatomy of a neuron.   • The mechanisms of impulse transmission in a neuron.   • The process that leads to release of neurotransmitter, and what happens at the synapse.

Concept 48.1 Neuron organization and structure reflect function in information transfer

- The **neuron** is the functional unit of the nervous system (See Figure 9.16). It is composed of a **cell body**, which contains the nucleus and organelles; **dendrites**, which are cell extensions that receive incoming messages from other cells; and **axons**, which transmit messages to other cells.
- Many axons are covered by an insulating fatty **myelin sheath**. This speeds the rate of impulse transmission.
- The **synapse** is a junction between two neurons (or a neuron and a muscle fiber or gland).
- **Neurotransmitters** are chemical messengers released from vesicles in the **synaptic terminals** into the synapse. They will diffuse across the synapse and bind to receptors on the neuron, muscle fiber, or gland across the synapse, effecting a change in the second cell. *Examples*: acetylcholine, dopamine, serotonin.

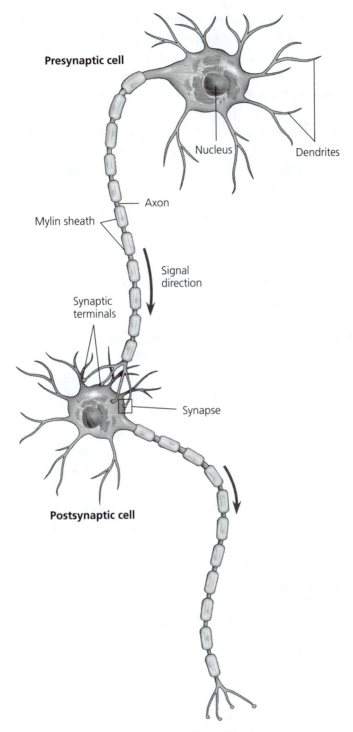

Presynaptic cell

Nucleus

Dendrites

Axon

Mylin sheath

Signal direction

Synaptic terminals

Synapse

Postsynaptic cell

Figure 9.16   Structure of a vertebrate neuron

- The **central nervous system (CNS)** consists of the brain and spinal cord, and the **peripheral nervous system (PNS)** consists of the nerves that communicate motor and sensory signals throughout the rest of the body.
- **Sensory receptors** collect information about the world outside the body as well as processes inside the body. *Examples*: the rods and cones of the eye; pressure receptors in the skin.
- **Sensory neurons** transmit information from eyes and other sensors that detect stimuli to the brain or spinal cord for processing.
- **Interneurons** connect sensory and motor neurons or make local connections in the brain and spinal cord.
- **Motor neurons** transmit signals to *effectors*, such as muscle cells and glands.
- **Nerves** are bundles of neurons. A nerve can contain all motor neurons, all sensory neurons, or be mixed.

Concept 48.2 *Ion pumps and ion channels maintain the resting potential of a neuron*

- **Membrane potential** describes the difference in electrical charge across a cell membrane.
- The membrane potential of a nerve cell at rest is called its **resting potential**. It exists because of differences in the ionic composition of the extracellular and intracellular fluids across the plasma membrane.
- The concentration of Na^+ is higher outside the cell, whereas the concentration of K^+ is higher inside the cell.
- Changes in the membrane potential of a neuron are what give rise to **nerve impulses**. A stimulus first affects the membrane's permeability to ions, and this is a graded potential with a magnitude proportional to the size of the stimulus.

Concept 48.3 *Action potentials are the signals conducted by axons*

- An **action potential** (nerve impulse) is an all-or-none response to depolarization of the membrane of the nerve cell (See Figure 9.17). A stimulus opens voltage-gated sodium channels, and Na^+ ions enter the cell, bringing the membrane potential to a positive value.
- In order to generate an action potential, a certain level of depolarization must be achieved, known as the **threshold**.
- The membrane potential is restored to its normal resting value by the inactivation of Na^+ channels and by opening voltage-gated K^+ channels, which increases K^+ leaving the cell.
- A *refractory period* follows the action potential, corresponding to the interval when the Na^+ channels are inactivated.
- Action potentials are propagated along the axon; **saltatory conduction**, which is the jumping of the nerve impulse between *nodes of Ranvier* (areas on the axon not covered by the myelin sheath), speeds up the conduction of the nerve impulse.

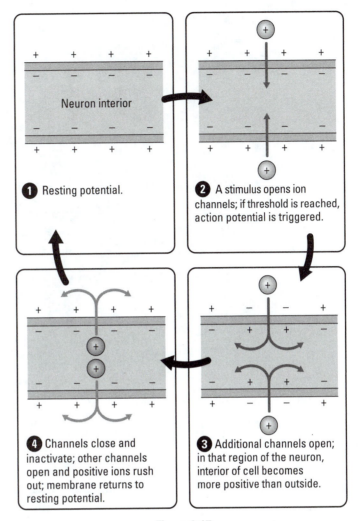

1 Resting potential.

2 A stimulus opens ion channels; if threshold is reached, action potential is triggered.

4 Channels close and inactivate; other channels open and positive ions rush out; membrane returns to resting potential.

3 Additional channels open; in that region of the neuron, interior of cell becomes more positive than outside.

Neuron interior

Figure 9.17

Concept 48.4 Neurons communicate with other cells at synapses

▌ The signal is conducted from the axon of a presynaptic cell to the dendrite of a postsynaptic cell via an **electrical** or **chemical synapse**.

▌ Study the steps involved in neurotransmitter release in Figure 9.18.

▌ Neurotransmitters are released by the presynaptic membrane into the synaptic cleft. They bind to receptors on the postsynaptic membrane and are then broken down by enzymes, or taken back up into surrounding cells.

▌ There are two categories of neurotransmitters: excitatory and inhibitory. Excitatory causes depolarization of the postsynaptic membrane, whereas inhibitory causes hyperpolarization of the postsynaptic membrane.

▌ **Acetylcholine** is a very common neurotransmitter; it can be inhibitory or excitatory. It is released by neurons at the neuromuscular junction. Other common **neurotransmitters** are epinephrine, norepinephrine, dopamine, and serotonin.

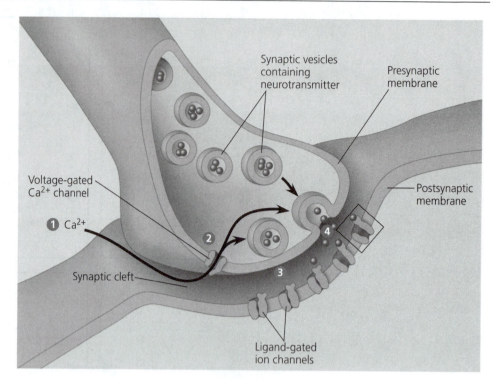

Figure 9.18

Chapter 49: Nervous Systems

YOU MUST KNOW

- The components of a reflex arc and how they work.
- The organization and function of the major parts of the nervous system.
- One function for each major brain region.

Concept 49.1 Nervous systems consist of circuits of neurons and supporting cells

▌ You may be asked to recall the evolution of the nervous system through representative organisms or phyla. Use this list as a quick check:

- ▪ Cnidarians have a **nerve net**.
- ▪ **Cephalization** is a trend toward clustering sensory neurons and interneurons at the anterior end.
- ▪ Flatworms show cephalization, with a small brain and longitudinal nerve cord. They have the simplest clearly defined *central nervous system*.
- ▪ Annelids such as the earthworm and arthropods have a *ventral nerve cord*.
- ▪ Vertebrates have a *hollow dorsal nerve cord*.

- A **reflex** is a simple automatic nerve circuit in response to a stimulus. Let's use a protective reflex, such as jerking your finger off a flame, as an example:

  - The *stimulus* is detected by a *receptor* in the skin, conveyed via a *sensory neuron* to an *interneuron* in the spinal cord, which synapses with a *motor neuron*, which will cause the *effector*, a muscle cell, to contract.

- Note that at its simplest level, conscious thought is not required in a reflex.
- **Cerebrospinal fluid** circulates through a central canal in the spinal cord and the ventricles of the brain, bathing cells with nutrients and carrying away wastes. It also cushions the brain and spinal cord.
- **Gray matter** consists of mainly neuron cell bodies and unmyelinated axons.
- **White matter** is white because of the myelin sheaths about the axons.
- **Glia** are cells that support neurons. Three important kinds of glia are **astrocytes**, which provide support for neurons; **oligodendrocytes**, which form myelin sheaths in the CNS; and **Schwann cells**, which form myelin sheaths in the PNS.
- Study Figure 9.19 as you review the organization of the nervous system.

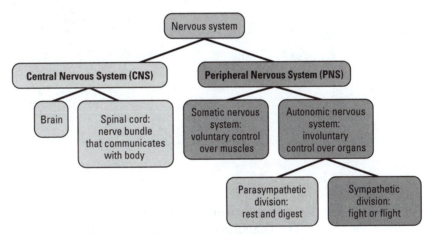

Figure 9.19

- The **central nervous system (CNS)** is the brain and spinal cord.
- The **peripheral nervous system (PNS)** consists of paired cranial and spinal nerves and associated ganglia. It is divided into

  - The *motor (somatic) nervous system*, which carries signals to skeletal muscles. It is a voluntary system.
  - The *autonomic nervous system*, which regulates the primarily automatic, visceral functions of smooth and cardiac muscles. This is the involuntary system.

- The **autonomic nervous system** transmits signals that regulate the internal environment by controlling smooth and cardiac muscle, including those in the gastrointestinal, cardiovascular, excretory, and endocrine systems. Its divisions are as follows:

    - The **sympathetic division**, which, when activated, causes the heart to beat faster and adrenaline to be secreted (with all its effects).
    - The **parasympathetic division**, which has the opposite effect when activated, slowing heartbeat and digestions.

Concept 49.2 *The vertebrate brain is regionally specialized*

- The **brainstem** is made up of the medulla oblongata, pons, and midbrain. The brainstem controls homeostatic functions such as breathing rate, conducts sensory and motor signals between the spinal cord and higher brain centers, and regulates arousal and sleep (See Figure 9.20.)

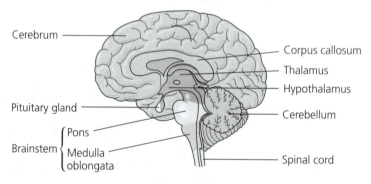

Figure 9.20

- The **cerebellum** helps coordinate motor, perceptual, and cognitive functions.
- The **thalamus** is the main center through which sensory and motor information passes to and from the cerebrum.
- The **hypothalamus** regulates homeostasis; basic survival behaviors such as feeding, fighting, fleeing, and reproducing; thermostat, appestat, thirst center, and circadian rhythms.
- The **cerebrum** has two hemispheres, each with a covering of gray matter over white matter. Information processing is centered here, and this region is particularly extensive in mammals.
- The **cerebral cortex** controls voluntary movement and cognitive functions.
- The **corpus callosum** is a thick band of axons that enables communication between the right and left cortices.

Chapter 50: Sensory and Motor Mechanisms

> **YOU MUST KNOW**
>
> - The location and function of several types of sensory receptors.
> - How skeletal muscle contracts.
> - Cellular events that lead to muscle contraction.

Concept 50.1 *Sensory receptors transduce stimulus energy and transmit signals to the central nervous system*

▎ **Mechanoreceptors** are receptors stimulated by physical stimuli, such as pressure, touch, stretch, motion, or sound.

▎ **Thermoreceptors** detect heat or cold and help maintain body temperature.

▎ **Chemoreceptors** transmit information about solute concentration in a solution. Taste and smell receptors are two types of chemoreceptors.

▎ **Electromagnetic receptors** detect various forms of electromagnetic energy such as visible light *(photoreceptors)*, electricity, and magnetism.

▎ **Pain receptors** respond to excess heat, pressure, or specific classes of chemicals released from damaged or inflamed tissues.

▎ **Reception** occurs when a receptor detects a stimulus. **Perception** occurs in the brain as this information is processed. As an example, when you view an optical illusion in which a figure seems to change, what is actually changing is your perception of the object.

Concept 50.2 *The mechanoreceptors involved with hearing and equilibrium detect moving fluid or settling particles*

▎ There are three regions in the mammalian ear (see Figure 9.21):

1. The **outer ear** is the external **pinna** and **auditory canal**. These collect sounds and direct them to the **tympanic membrane** (eardrum), which separates the outer ear from the middle ear.
2. In the **middle ear**, vibrations are conducted through three small bones (the malleus, incus, and stapes, or hammer, anvil, and stirrup) and through the **oval window**. The **Eustachian tube** allows air pressure to be equalized between the outer and middle ear.
3. Then the vibrations are conducted to the **inner ear**, which consists of fluid-containing channels lined by membrane situated in bone.

▎ The **cochlea** is involved in hearing. It is a snail-shaped structure containing the **organ of Corti**. Here, hair cells are distorted from the vibrations caused by sound waves. These hair cells in the organ of Corti are the actual receptors of sound.

▎ The **semicircular canals** are three fluid-filled chambers located at right angles to each other. Our sense of equilibrium results when hair cells within are distorted.

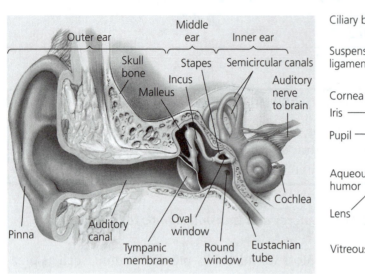

Figure 9.21

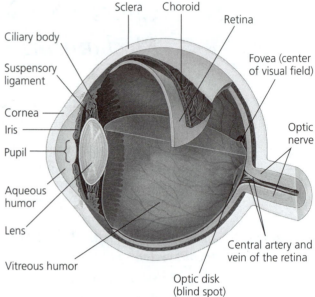

Figure 9.22

Concept 50.3 The senses of taste and smell rely on similar sets of sensory receptors

▌ Taste buds are modified epithelial cells situated on different parts of the tongue and mouth.

Concept 50.4 Similar mechanisms underlie vision throughout the animal kingdom

▌ **Compound eyes** (in insects and crustaceans) consist of up to several thousand light detectors called *ommatidia*, each of which has its own lens.

▌ **Single-lens eyes** are found in vertebrates and some invertebrates (See Figure 9.22).

▌ The **sclera** of the eye is tough and white; at the front it becomes the transparent **cornea**, which allows light into the eye and acts as a fixed lens.

▌ The **choroid** is a layer of pigmented cells inside the sclera.

▌ The **retina** is the innermost layer of the eyeball. It contains the photoreceptor cells (the rods and cones). Where it is pierced by the optic nerve, there are no photoreceptors. This is the *blind spot*.

▌ The **rods** are adapted for vision in dim light, and the **cones** for color vision.

▌ The **iris** is the pigmented area at the front of the eyeball. It is a muscular structure that regulates the amount of light admitted through the **pupil**, which is the hole at its center.

▌ **Aqueous humor** fills the anterior cavity of the eye, and the **vitreous humor** fills the posterior cavity of the eye.

▌ **Rhodopsin** is the light-absorbing pigment that triggers a signal transduction pathway that ultimately leads to sight.

Concept 50.5 The physical interaction of protein filaments is required for muscle function

▌ **Skeletal muscle** is attached to bones and responsible for the movement of bones. It consists of long fibers, each of which is a single muscle cell.

- **Skeletal muscle** is *striated* (striped in appearance).
- Each muscle fiber is a bundle of **myofibrils**, which in turn are composed of two kinds of myofilaments: **thin filaments** and **thick filaments**.

  - The thin filaments are *actin.*
  - The thick filaments are *myosin.*

- The **sarcomere** is the basic contractile unit of the muscle. Study the sarcomere in Figure 9.23 as you review these components.

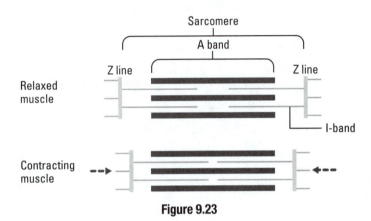

Figure 9.23

  - **Z lines** make up the border of sarcomeres. Actin is attached here.
  - The **I band** is the area near the end of the sarcomere where only the thin actin filaments are located.
  - The **A band** is the entire length of the thick myosin filaments.

- During **muscle contraction**, the length of the sarcomere is reduced. Actin filaments slide over the myosin.
- The **sliding-filament model** states that the thick and thin filaments slide past each other so that their degree of overlap increases.
- Study Figure 9.24 as you review what happens during depolarization of a muscle fiber.
- A **motor neuron** will cause a muscle fiber to contract when its depolarization causes the neurotransmitter *acetylcholine* to be released into the synapse of the *neuromuscular junction.*
- As acetylcholine binds to receptors on the muscle fiber, depolarization of the muscle cell leads to the following:

  - An action potential spreads along **T tubules (transverse tubules)** to the **sarcoplasmic reticulum.**
  - This depolarization causes the sarcoplasmic reticulum to release **calcium ions.**
  - The calcium ions bind to **troponin** and cause it to move, exposing the myosin sites on actin.

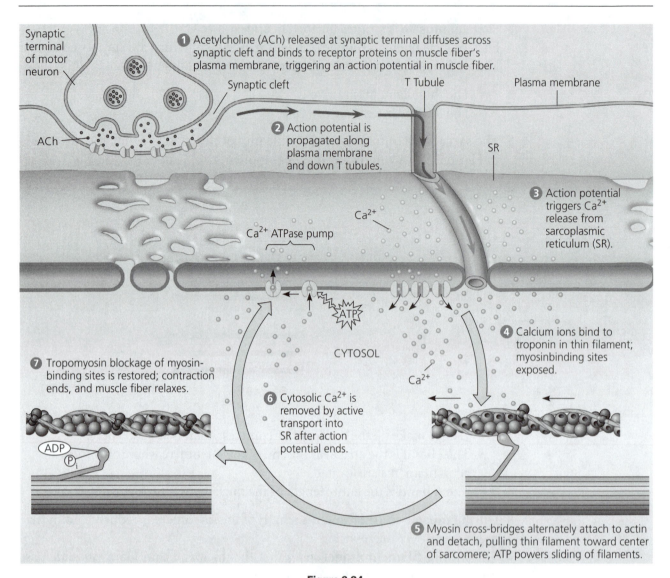

Figure 9.24

- Actin and myosin now interact as myosin heads attach to the actin filaments, after being phosphorylated by ATP.
- The muscle contracts, with actin filaments sliding over myosin.

A Hint from the Graders:
There are many possible topics to study in human systems! However, to focus your energies, know both the mechanism of impulse transmission in a neuron (48.2 and 48.3) and the mechanism of muscle contraction (50.5). Since these both involve cell signaling, specific activities of many organelles, and therefore integrate many areas of your course, this is a fertile ground for questions. Know them!

Concept 50.6 Skeletal systems transform muscle contraction into locomotion

- **Hydrostatic skeletons** consist of fluid held under pressure in a closed body compartment. This is seen in hydra, nematodes, and annelids.
- **Exoskeletons** are hard encasements on the surface of an animal, such as is found in the grasshopper.
- **Endoskeletons** consist of hard supporting elements buried within the soft tissues of an animal. An example is the human bony skeleton.

Chapter 51: Animal Behavior

YOU MUST KNOW

- The difference between a kinesis and a taxis.
- Various forms of animal communication.
- The role of altruism and inclusive fitness in kin selection.

Concept 51.1 Discrete sensory inputs can stimulate both simple and complex behaviors

- **Behavior** is what an animal does and how it does it. Behavior is a result of genetic and environmental factors, is essential for survival and reproduction, and is subject to natural selection over time.
- **Ethology** is the study of animal behavior.
- There are two fundamental levels of analysis, *proximate* and *ultimate*, in the study of behavior.

  - **Proximate** causes of behavior are the "how" questions and include the effects of heredity on behavior, genetic-environmental interactions, and sensory-motor mechanisms.
  - **Ultimate** causes are the "why" questions and include studies of the origin of a behavior, its change over time, and the utility of the behavior in terms of reproductive success.

- **Innate behaviors** are developmentally fixed. They are unlearned behaviors.
- A **fixed action pattern (FAP)** is a sequence of unlearned acts that is largely unchangeable and usually carried to completion once it is initiated. Fixed action patterns are triggered by *sign stimuli*.
- A classic example of an FAP was noted by *Nicholas Tinbergen* in male stickleback fish, which attack red objects. The red object is the sign stimulus; the attack is the FAP.
- A **kinesis** is a simple change in activity in response to a stimulus, whereas a taxis is an automatic movement toward or away from a stimulus. In AP Lab 11, the movement of pillbugs toward a moist habitat is a kinesis.
- **Migration** is a complex behavior seen in a wide variety of animals. Navigation may be by detection of the earth's magnetic field or visual cues.

- **Circadian rhythms** are those that occur on a daily cycle. Other rhythms occur over longer periods and can be triggered by differing day lengths or lunar cycles.
- A **signal** is a behavior that causes a change in the behavior of another individual and is the basis for animal communication. Examples of signals follow:

    - **Pheromones** are chemical signals that are emitted by members of one species that affect other members of the species.
    - **Visual signals** such as the warning flash of white of a mockingbird's wing.
    - **Auditory signals** such as the screech of a blue jay or song of a warbler.

- The **waggle dance** of the honeybee described by *Karl von Frisch* is a behavior in which the location and distance of a food source is communicated to the members of a hive by a foraging worker.

Concept 51.2 Learning establishes specific links between experience and behavior

- **Learning** is the modification of behavior based on specific experiences.
- **Imprinting** is a combination of learned and innate components that are limited to a *sensitive period* in an organism's life and is generally irreversible. Recall *Konrad Lorenz's* demonstration of imprinting in greylag geese. When hatchlings spent their first few hours with him, they followed him as though he were their mother!
- **Habituation** is a loss of responsiveness to stimuli that convey little or no information. It is a simple form of learning.
- A **cognitive map** is an internal representation of spatial relationships among objects in an animal's surroundings.
- **Associative learning** is the ability of many animals to associate one feature of their environment with another feature. **Classical conditioning** involves learning to associate certain stimuli with reward or punishment. **Operant conditioning** occurs as an animal learns to associate one of its behaviors with a reward or punishment.
- **Cognition** is the ability of an animal's nervous system to perceive, store, process, and use information from sensory receptors.

Concept 51.3 Both genetic makeup and environment contribute to the development of behaviors

- Twin studies in humans indicate that both environment and genetics contribute significantly to behaviors.
- Behavior can be directed by genes. For example, a single gene appears to control the courtship ritual in fruit flies.

Concept 51.4 Selection for individual survival and reproductive success can explain most behaviors

- **Foraging behavior** includes not only eating, but also mechanisms used in searching for, recognizing, and capturing food.
- **The optimal foraging model** proposes that it is a compromise between the benefits of nutrition and the cost of obtaining food.

■ **Mating systems** vary between species. The needs of the young are important constraints in the development of these systems. Various systems are

- ■ **Promiscuous** with no strong pair-bonds.
- ■ **Monogamous** with one male/one female.
- ■ **Polygamous** with one individual mating with several others.

■ **Agonistic behaviors** are often ritualized contests that determine which competitor gains access to a resource, such as food or mates.

Concept 51.5 Inclusive fitness can account for the evolution of altruistic social behavior

■ **Altruism** occurs when animals behave in ways that reduce their individual fitness but increase the fitness of other individuals in the population. *Example:* A blue jay giving an alarm call attracts attention to its location.

■ **Inclusive fitness** is the total effect an individual has on proliferating its genes by producing its own offspring and by providing aid that enables other close relatives to produce offspring. The natural selection that favors this kind of altruistic behavior by enhancing reproductive success of relatives is called **kin selection**.

For Additional Review

Consider how the immune, digestive, nervous, circulatory, and respiratory systems and the senses all contribute to homeostasis in animals. In doing so, connect the stimuli that engage these systems with the way the systems respond to those stimuli.

Multiple-Choice Questions

1. Which of the following is required in ALL living things in order for gas exchange to occur?
   (A) lungs
   (B) gills
   (C) moist membranes
   (D) blood
   (E) lymph

2. In animals, all of the following are associated with embryonic development EXCEPT
   (A) gastrulation.
   (B) cleavage.
   (C) depolarization.
   (D) organogenesis.
   (E) cell migration.

3. Which of the following is most likely to result in a release of epinephrine (adrenaline) from the adrenal glands?
   (A) falling asleep in front of the TV
   (B) watching a golf tournament
   (C) doing yoga
   (D) taking a test without having studied for it
   (E) being in the kitchen while dinner is being cooked

4. Oxygen is transported in human blood by which type of cell?
   (A) erythrocytes
   (B) leukocytes
   (C) phagocytes
   (D) B cells
   (E) platelets

5. The proximal tubules in the kidney reabsorb most of which of the following compounds?
   (A) H^+
   (B) proteins
   (C) H_2O
   (D) HCO_3^-
   (E) $C_6H_{12}O_6$

6. Salivary amylase, an enzyme secreted in saliva, begins the breakdown of which substance?
   (A) starches
   (B) proteins
   (C) lipids
   (D) nucleic acids
   (E) polypeptides

Directions: The group of questions below consists of five lettered choices followed by a list of numbered phrases or sentences. For each numbered phrase or sentence, select the one choice that is most closely related to it. Each choice may be used once, more than once, or not at all.

Questions 7–11
   (A) Ovary
   (B) Thyroid gland
   (C) Posterior pituitary gland
   (D) Adrenal medulla
   (E) Anterior pituitary gland

7. Releases hormones that raise blood glucose level, increase metabolic activities, and constrict blood vessels

8. Releases at least six different hormones, including several tropic hormones

9. Releases hormones that stimulate the mammary gland cells and contraction of the uterus

10. Releases hormones that stimulate growth of the uterine lining and promote the development of female secondary sex characteristics

11. Releases hormones that stimulate and maintain metabolic processes

12. Blood constitutes which of the following tissue types?
   (A) epithelial tissue
   (B) connective tissue
   (C) nervous tissue
   (D) vascular tissue
   (E) glandular tissue

13. The three types of muscle in the body are
   (A) skeletal, cardiac, and smooth.
   (B) skeletal, vascular, and smooth.
   (C) skeletal, cardiac, and rough.
   (D) skeletal, cardiac, and striated.
   (E) skeletal, vascular, and smooth.

14. Which of the following is an example of negative feedback?
   (A) the movement of sodium across a membrane through a transport protein and the movement of potassium in the opposite direction through the same transport
   (B) the pressure of the baby's head against the uterine wall during childbirth stimulates uterine contractions, which causes greater pressure against the uterine wall, which produces still more contractions
   (C) the growth of a population of bacteria in a petri dish until it has used all its nutrients and its subsequent decline
   (D) a heating system in which the heat is turned off when the temperature exceeds a certain point and is turned on when the temperature falls below a certain point
   (E) the progress of a chemical reaction until equilibrium is reached and then the cycling back and forth of reactant to product

15. Which of the following mechanisms does NOT help prevent the gastric juice from digesting the stomach lining?
    (A) Pepsin is stored and secreted as pepsinogen.
    (B) Mucus lines the inside surface of the stomach.
    (C) Mitosis generates enough new cells to replace the stomach lining every few days.
    (D) Gastric juice is not secreted continuously.
    (E) Pepsin activates pepsinogen by a chain reaction.

16. Which of the following has a diet that consists solely of autotrophs?
    (A) omnivore
    (B) carnivore
    (C) herbivore
    (D) detritovore
    (E) frugivore

17. The four stages of food processing are ingestion, digestion, absorption, and
    (A) incorporation.
    (B) circulation.
    (C) elimination.
    (D) excretion.
    (E) cellular uptake.

18. Hydras possess which of the following type of digestive system?
    (A) food vacuole
    (B) complete digestive tract
    (C) alimentary canal
    (D) lumen
    (E) gastrovascular cavity

19. Which of the following is the site of the production of bile?
    (A) gallbladder
    (B) small intestine
    (C) prostate
    (D) pancreas
    (E) liver

20. The primary sites of carbohydrate digestion are which of the following structures?
    (A) mouth and large intestine
    (B) mouth and stomach
    (C) stomach and small intestine
    (D) mouth and small intestine
    (E) small intestine and colon

21. Pepsin in the stomach is primarily responsible for the breakdown of which type of molecule?
    (A) starches
    (B) proteins
    (C) lipids
    (D) nucleic acids
    (E) glycogens

22. Which of the following structures is primarily responsible for reabsorbing water from the alimentary canal?
    (A) small intestine
    (B) nephron
    (C) glomerulus
    (D) colon
    (E) cecum

23. Which one of the following statements is NOT true about mammalian circulatory systems?
    (A) The pulmonary circuit carries blood between the heart and the lungs.
    (B) The systemic circuit carries blood between the heart and the rest of the body.
    (C) Mammals have two atria and two ventricles in their hearts.
    (D) A mammal uses about 10 times as much oxygen as a lizard of the same size.
    (E) The left side of a mammal's heart sends blood to the lungs.

24. A transport system in which blood bathes the organs directly is termed
(A) an open circulatory system.
(B) a closed circulatory system.
(C) a cardiovascular system.
(D) a gastrovascular system.
(E) a gastrovascular cavity system.

25. Which of the following carry blood away from the heart?
(A) venules
(B) veins
(C) arteries
(D) capillaries
(E) atria

26. In the mammalian heart, the sinoatrial (SA) node is responsible for which of the following functions?
(A) Delaying the nerve impulse to the walls of the ventricle
(B) Controlling the atrioventricular valve
(C) Controlling the semilunar valves
(D) Setting the rate and timing of cardiac muscle contraction
(E) Monitoring stroke volume

27. Fluid and proteins lost from the capillaries are returned to the blood via
(A) the venous system.
(B) the arteriole system.
(C) the lymphatic system.
(D) capillary beds.
(E) the hemolymph system.

28. All of the following are components of blood EXCEPT
(A) red blood cells.
(B) white blood cells.
(C) platelets.
(D) leukocytes.
(E) lymph.

29. Red blood cells are produced in which of the following structures?
(A) the heart
(B) the blood vessels
(C) bone
(D) muscles
(E) masses of other blood cells

30. Which of these is the organ of respiration for a spider?
(A) lungs
(B) the skin
(C) tracheal system
(D) Malpighian tubules
(E) book lungs

31. In the blood, carbon dioxide is primarily transported in what way?
(A) by hemoglobin
(B) by hemocyanin
(C) as carbon monoxide
(D) as bicarbonate
(E) in erythrocytes

32. All of the following are first-line barriers against infectious agents EXCEPT
(A) skin.
(B) nasal membranes.
(C) saliva.
(D) mucous secretions.
(E) phagocytes.

33. An immune response to a specific antigen generates the production of which type of cell that launches an attack the next time that same antigen infects the body?
(A) effector cells
(B) memory cells
(C) T cells
(D) B cells
(E) antibodies

34. All of the following are ways by which organisms exchange heat EXCEPT
    (A) transference.
    (B) conduction.
    (C) convection.
    (D) radiation.
    (E) evaporation.

35. Which of the following animals is most likely an ectoderm?
    (A) human
    (B) snake
    (C) bird
    (D) monkey
    (E) dolphin

36. The Malpighian tubules are the organs that constitute the excretory system of which of the following animals?
    (A) planaria
    (B) humans
    (C) fishes
    (D) insects
    (E) annelids

37. The ball of capillaries that is associated with the nephron and associated with filtration in the kidney is
    (A) the Bowman's capsule.
    (B) the loop of Henle.
    (C) the proximal tubule.
    (D) the glomerulus.
    (E) the distal tubule.

Directions: The group of questions below consists of five lettered choices followed by a list of numbered phrases or sentences. For each numbered phrase or sentence, select the one choice that is most closely related to it. Each choice may be used once, more than once, or not at all.

Questions 38–42
    (A) Vitreous humor
    (B) Cone cell
    (C) Eustachian tube
    (D) Cochlea
    (E) Taste bud

38. A photoreceptor sensitive to bright light and color

39. A fluid-filled coiled organ with sensory hair cells

40. Constitutes most of the volume of the eye

41. Equalizes the pressure between the middle ear and the atmosphere

42. A receptor that can be stimulated by a broad range of chemicals

43. Muscle cell contraction occurs via
    (A) contraction of the A band.
    (B) contraction of the I band.
    (C) contraction of the Z lines.
    (D) the sliding of the thin filaments by the thick filaments.
    (E) the contraction of the sarcoplasmic reticulum.

44. The succession of rapid cell division that follows fertilization is called
    (A) gastrulation.
    (B) cleavage.
    (C) morulation.
    (D) involution.
    (E) polarization.

45. The circuit of a sensory neuron, the spinal cord, a motor neuron, and an effector cell constitutes a
(A) presynaptic sequence.
(B) reflex arc.
(C) nerve circuit.
(D) nerve impulse.
(E) saltatory conduction system.

46. Which of the following is released into the synaptic cleft and acts as an intercellular messenger?
(A) sodium
(B) chloride
(C) neurotransmitter
(D) action potential
(E) voltage gradient

47. An egg cell surrounded by one or two layers of cells is called a
(A) follicle.
(B) corpus luteum.
(C) oviduct.
(D) endometrium.
(E) uterus.

48. Sperm are formed in the
(A) Leydig cells.
(B) prostate gland.
(C) seminal vesicles.
(D) seminiferous tubules.
(E) baculum.

49. The regulation of the internal environment in animals is referred to as
(A) equilibrium.
(B) stasis.
(C) homeostasis.
(D) regulation.
(E) feedback.

50. Fertilization—the fusion of egg and sperm cell—results in which of the following?
(A) embryo
(B) zygote
(C) gamete
(D) ovum
(E) follicle

51. Fixed action patterns (FAPs) are instigated by which of the following?
(A) mating behavior
(B) ritual behavior
(C) innate stimulus
(D) sign stimulus
(E) action potential

52. One morning, a woman who usually feeds her two cats in the morning passes by the food bowl without putting food in it. The cats usually run over to the bowl as she approaches it, but after four mornings of her passing the bowl without putting food in it, the cats no longer run over to the bowl. This is an example of
(A) maturation.
(B) imprinting.
(C) habituation.
(D) foraging.
(E) sensitivity.

53. Pavlov's dogs learned to salivate when they heard the ring of a particular bell; this is an example of
(A) classical conditioning.
(B) operant conditioning.
(C) sensitivity.
(D) imprinting.
(E) maturation.

54. The phenomenon in which young ducks follow their mother in a line is a result of which of the following?
(A) habituation
(B) imprinting
(C) maturation
(D) foraging
(E) conditioning

55. Altruism exists in populations because
(A) it deprives members of the species of territory and results in agonistic behavior.
(B) it can result in the passing on of the altruistic member's genes.
(C) it can result in the overall success of the ecosystem.
(D) it can result in a bond between the altruistic member and the recipient of the altruism, and the recipient might later reciprocate the altruism.
(E) it can result in the maximizing of the altruistic member's genetic representation in a population, if the altruistic member's behavior is directed toward a close relative.

Free-Response Question

1. *Muscle cells are responsible for moving parts of the skeleton by contracting. However, during muscle contraction, none of the muscle cells themselves actually contract.*

   (a) **Describe** how a muscle can contract without any of its cells contracting.
   (b) **Explain** the phenomenon of tetanus.
   (c) **Explain** why muscles become "sore" after exercise.

ANSWERS AND EXPLANATIONS

Multiple-Choice Questions

▌ **1. (C) is correct.** The only condition listed that is necessary in all organisms that breathe is the presence of moist membranes. The movement of O_2 and CO_2 across the membranes between the environment and the respiratory surface occurs by diffusion. Respiratory surfaces are generally thin and, since living animal cells must be wet in order to maintain their plasma membranes, these respiratory surfaces must be moist.

▌ **2. (C) is correct.** The three main stages of development in animals are cleavage, in which a multicellular embryo forms from the zygote through a series of mitotic cell divisions; gastrulation, in which cells migrate and rearrange to form three germ layers; and organogenesis, in which rudimentary organs are formed from the germ layers. It is important for you to know these key terms.

▌ **3. (D) is correct.** Epinephrine is a hormone that is secreted by the adrenal glands, specifically the adrenal medulla. It functions in raising the blood glucose level, increasing metabolic activities, and constricting blood vessels; all of this prepares the animal for the fight-or-flight response that is elicited in the body during stressful times.

4. (A) is correct. Erythrocytes are red blood cells, and they transport oxygen around the body. They are the most numerous blood cells, and are small and disk-shaped. In mammals, erythrocytes have no nuclei. Instead, they contain millions of molecules of hemoglobin, which is the iron-containing protein that transports oxygen. One molecule of hemoglobin can bind four oxygen molecules.

5. (D) is correct. The proximal tubule is the site of secretion and reabsorption that substantially changes the content and volume of the filtrate. It secretes hydrogen ions and ammonia to regulate the pH of the filtrate and also reabsorbs about 90% of the bicarbonate, which is an important buffer.

6. (A) is correct. Salivary amylase is contained in saliva; it is an enzyme that hydrolyzes starch, a glucose polymer found in plants, and glycogen, a glucose polymer found in animals. After hydrolysis, smaller polysaccharides and maltose remain.

7. (D) is correct. The adrenal medulla secretes epinephrine and norepinephrine. The regulation of these hormones is controlled by the nervous system, and these hormones act to raise the blood glucose level, increase metabolic activity in the cell, and change blood flow patterns.

8. (E) is correct. The anterior pituitary produces FSH, LH, TSH, ACTH, along with several other hormones. It is regulated by releasing hormones from the hypothalamus.

9. (C) is correct. The posterior pituitary gland releases two main hormones: oxytocin, which stimulates the contraction of the uterus and mammary gland cells, and antidiuretic hormone (ADH), which promotes the retention of water by the kidney. The actions of the posterior pituitary are regulated by the nervous system and the water/salt balance in the body.

10. (A) is correct. The ovaries secrete hormones called estrogens, which stimulate the growth of the uterine lining and promote the development of secondary sex characteristics in females. They are regulated by two other hormones: FSH and LH.

11. (B) is correct. The thyroid gland releases the hormone triiodothyronine, which stimulates and maintains metabolic processes, and calcitonin, which lowers the blood calcium levels. The thyroid gland secretions are regulated by TSH and by the level of calcium in the blood.

12. (B) is correct. Blood is a connective tissue. It functions very differently from the other connective tissues, but it has an extensive extracellular matrix, which is the criterion for being considered connective tissue. The matrix is plasma, which consists of water, salts, and dissolved proteins.

13. (A) is correct. The three types of muscle in the body are skeletal muscle (responsible for voluntary movements); cardiac muscle (which forms the contractile wall of the heart); and smooth muscle (found in the walls of the digestive tract, bladder, arteries, and other internal organs).

14. (D) is correct. The traditional example of a negative feedback system is the thermostat example. In the body, one very prominent example of negative feedback is the regulation of our body temperature at about 37°C. A section of the brain is responsible for keeping track of the temperature of the blood, and if the blood is too warm, for example, it tells the sweat glands to increase production. Remember, more gets you less!

15. (E) is correct. HCl converts pepsinogen to active pepsin. Other proteolytic enzymes such as trypsin and chymotrypsin are also secreted as zymogens and require cleavage of a portion of the molecule in order to be activated.

16. (C) is correct. Herbivores are animals that eat only autotrophs (plants and algae). Some examples of herbivores are gorillas and cows. Carnivores eat other animals, and omnivores eat animals as well as plants or algae. Detritovores feed on decaying matter. Frugivores feed on fruit.

17. (C) is correct. The four stages of food processing are ingestion (the act of eating), digestion (the process by which food is broken down into small particles), absorption (the uptake of nutrients by the body), and elimination (the release of undigested material). Excretion is generally used to refer to the removal of nitrogenous wastes.

18. (E) is correct. Hydras are simple animals that contain a gastrovascular cavity, which is a pouch that functions in both digestion and the distribution of nutrients throughout the body. The gastrovascular cavity's single opening acts as both mouth and anus. Review the evolution of digestive systems now.

19. (E) is correct. The liver is responsible for the production of bile, which contains bile salts (which act as detergents or emulsifying agents that facilitate the digestion of fats). Bile is stored in the gall bladder, not produced there. Bile contains pigments that are the byproducts of red blood cells destroyed in the liver. These are eliminated from the body along with feces.

20. (D) is correct. The digestion of carbohydrates, such as starch and glycogen, begins in the mouth through the action of salivary amylase. In the small intestine, pancreatic amylases hydrolyze starch, glycogen, and smaller polysaccharides into monosaccharides.

21. (B) is correct. Pepsin is an enzyme in the gastric juice of the stomach. It begins the hydrolysis of proteins by breaking peptide bonds between adjacent amino acids and by cleaving proteins into smaller polypeptides. The digestion of proteins continues in the small intestine by the enzymes trypsin and chymotrypsin.

22. (D) is correct. The large intestine, also known as the colon, is responsible for recovering water from the alimentary canal. It is also responsible for compacting the wastes into feces, which are stored in the rectum and then excreted.

23. (E) is correct. The left side of a mammal's heart pumps blood into the aorta for systemic circulation.

24. (A) is correct. In an open circulatory system, which exists in insects and other arthropods, the blood bathes the organs directly. In closed circulatory systems, blood is contained in vessels and is separate from the interstitial fluid. In closed circulatory systems, one or more hearts pump blood into vessels that branch and feed blood through the vessels.

25. (C) is correct. An artery is a kind of blood vessel that carries blood away from the heart, branching into arterioles and eventually into capillary beds. The capillary beds then converge into venules, which converge further into veins, which return blood to the heart.

26. (D) is correct. The role of the sinoatrial (SA) node, or pacemaker, is to control the rate and timing of the contraction of heart muscles. It generates nerve impulses just like the ones that occur in nerve cells, and the impulses spread rapidly through the walls of the atria, making them contract in unison.

27. (C) is correct. The lymphatic system collects fluid and proteins lost during regular circulation and returns them to the blood. This system is composed of a network of lymph vessels throughout the body, with lymph nodes, which are the sites at which lymph is filtered and viruses and bacteria are collected and killed.

28. (E) is correct. All of the answers listed, except lymph, are constituents of blood. White blood cells are leukocytes, and red blood cells are also called erythrocytes. Platelets are cell fragments that function in blood clotting. Lymph is found within its vessels only in the lymphatic system.

29. (C) is correct. Erythrocytes, leukocytes, and platelets all develop from stem cells in the red marrow of bones—primarily in the ribs, vertebrae, breastbone, and pelvis. The cells that develop into blood cells have the potential to develop into any type of blood cell; they are called pluripotent cells.

30. (E) is correct. Book lungs are the respiratory system of most arachnids. Insects have a tracheal system, which is made up of air tubes that branch throughout the body. The large tubes are called trachea, and they open to the outside, while the smallest branches reach the surface of every cell, where gas exchange takes place. Animals whose respiratory system involves diffusion across the skin must be kept moist. Examples include lungless salamanders and earthworms.

31. (D) is correct. Carbon dioxide is most commonly transported in the blood in the form of bicarbonate—it reacts with water to form carbonic acid, and a hydrogen dissociates from carbonic acid to produce bicarbonate. Less commonly, carbon dioxide is transported by hemoglobin, or transported in solution in the blood.

32. (E) is correct. All of the answers listed—except phagocytes—are examples of first-line barriers to infection by infecting agents that might attack the body. Phagocytosis constitutes the body's nonspecific internal mechanism for defending itself against infectious agents; it is the process by which invading organisms are ingested and destroyed by white blood cells.

33. (B) is correct. When a lymphocyte is activated by an antigen, it is stimulated to divide and differentiate, and it forms two clones. One clone is of effector cells that combat the antigen. One clone is of memory cells that stay in circulation, recognize the antigen if it infects the body in the future, and launch an attack against it.

34. (A) is correct. All of the answers except *A* constitute methods animals have for exchanging heat with the environment. Conduction is the transfer of heat between objects that are in direct contact. Convection is the transfer of heat by the movement of air past a surface. Radiation is the emission of electromagnetic waves by warm objects. Evaporation is heat loss through the loss of molecules as gas.

35. (B) is correct. Ectotherms are animals that have such low metabolic rates that the amount of heat they generate will not influence their body temperature. Their internal temperature is therefore determined by their environment. Ectotherms include most invertebrates; fishes; amphibians; and nonbird reptiles, including snakes. Endotherms have high metabolic rates, and this makes their bodies quite a bit warmer than the external environment. Endotherms include birds and mammals (humans, monkeys, and dolphins are mammals).

36. (D) is correct. Malpighian tubules are organs that remove the nitrogenous wastes of insects and other arthropods. They open into the digestive tract and dead-end at tips that are submerged in hemolymph. The tubules have an epithelial lining that secretes solutes into the lumen of the tubule, and water follows the solutes into the tubule by osmosis. You should know that the organs of excretion in a planaria (flatworm) are flame cells, kidneys for humans and fishes, and nephridia for annelids.

▌**37. (D) is correct.** The nephron, which is the functional unit of the kidney, is composed of a long tubule and the glomerulus, which is a ball of capillaries. The Bowman's capsule surrounds the glomerulus. The blood in the glomerulus is forced into the Bowman's capsule by blood pressure, and this process acts to filter the blood.

▌**38. (B) is correct.** Rods and cones are the two types of photoreceptors in the eye. They are contained in the retina and account for 70% of all the sensory receptors in the body. Cones can distinguish colors in daylight, whereas rods are sensitive to light but cannot distinguish colors.

▌**39. (D) is correct.** The cochlea is part of the inner ear that is involved in hearing. It is a coiled organ that has two large chambers—a vestibular canal and a lower tympanic canal—which are separated by a cochlear duct. The floor of the cochlear duct is home to the organ of Corti, which contains the receptors of the ear—hair cells.

▌**40. (A) is correct.** The vitreous humor is jellylike and fills the posterior cavity of the eye, constituting most of the eye's volume. The aqueous humor fills the anterior cavity of the eye and is clear and watery.

▌**41. (C) is correct.** On one end, the Eustachian tube connects to the middle ear, and on the other, the Eustachian tube connects with the pharynx. This enables it to equalize the pressure between the middle ear and the atmosphere.

▌**42. (E) is correct.** Taste buds are modified epithelial cells that act as receptors for taste. Most taste buds are on the surface of the tongue and mouth. Sweet, sour, salty, and bitter are taste perceptions detected by taste buds.

▌**43. (D) is correct.** The sliding-filament model of muscle contraction states that the thin and thick filaments do not shrink during muscle contraction. Instead, the filaments slide past each other so that the degree of their overlap increases; this sliding is based on the interactions of actin and myosin molecules that make up the filaments.

▌**44. (B) is correct.** Three successive stages of development follow fertilization. The first is cleavage, which is rapid cell division that produces a mass of new cells that share the cytoplasm of the original cell. The new cells all have their own nuclei and are called blastomeres. The second stage is gastrulation, and the third is organogenesis.

▌**45. (B) is correct.** The reflex arc is the simplest type of nerve circuit (automatic response), and it requires just two types of nerve cells. A sensory neuron receives information from a receptor and passes it to the spinal cord and then to a motor neuron, which signals an effector cell to respond to the stimulus.

▌**46. (C) is correct.** Neurotransmitters are excreted by the synaptic vesicles and act as intercellular messengers, transmitting the nerve impulse from one neuron to the next neuron or another cell. A single postsynaptic neuron can receive signals from many neurons that secrete different neurotransmitters.

47. (A) is correct. Each of the two ovaries in the female body contains many follicles. Follicles are composed of an egg cell surrounded by one or two layers of follicle cells; these serve to nourish and protect the cell.

48. (D) is correct. Sperm is produced in the seminiferous tubules, which are coiled tightly in the testes and surrounded by connective tissue. Production of sperm cannot take place at the high temperature of the body, so the testes are held in the scrotum of the male, outside the abdominal pelvic cavity, where it is about two degrees cooler. The seminal vesicles and prostate gland contribute to the fluid component of penis, the Leydig cells of the testes produce testosterone, and a baculum is a penis bone found in some mammalian species (but not humans).

49. (C) is correct. Homeostasis is the ability of many animals to regulate their internal environment. They do this through thermoregulation, which is the maintenance of internal temperature in a certain range, and osmoregulation, which is the regulation of solute balance within certain parameters.

50. (B) is correct. Fertilization is the fusion of egg cell (ovum) and the sperm cell, and it results in the formation of a zygote. The zygote is diploid, whereas the egg cell and the sperm cell, both the products of meiosis, are haploid.

51. (D) is correct. Fixed action patterns are a sequence of behavioral acts that are virtually unchangeable and usually carried to completion once they are initiated. Sign stimuli trigger fixed action patterns. These stimuli may be a feature of another animal, such as an aspect of its appearance, or some other event.

52. (C) is correct. Habituation is one type of learning. Learning is defined as the ability of an animal to modify its behavior as a result of specific experiences. Habituation is a very simple form of learning, in which there is a loss of responsiveness to stimuli that convey very little or no information.

53. (A) is correct. Classical conditioning is a form of associative learning (the ability of animals to learn to associate one stimulus with another). It specifically refers to an animal's ability to associate an arbitrary stimulus with a reward or a punishment.

54. (B) is correct. Imprinting is a form of learning that occurs during a sensitive period. Imprinting is generally irreversible, and the sensitive period is a limited phase in the animal's development when the learning of a particular behavior can take place.

55. (E) is correct. Altruism is thought to occur in populations because if parents sacrifice their own well-being for that of their offspring, this increases their fitness by better ensuring that the genes that they passed on will make it to the next generation. Likewise, helping other close relatives increases the chances that they will survive to pass on genes that are shared between them and the altruistic member.

Free-Response Question

(a) Skeletal muscle is fibrous, and each fiber is a single long cell. The fibers are composed of myofibrils, which are, in turn, composed of two kinds of myofilaments. These are thin filaments—which are made up of two actin strands and one regulatory protein strand, coiled—and thick filaments made up of myosin molecules. The sarcomere is the basic contracting unit of the muscle. During muscle contraction, the length of the sarcomere decreases. The sliding-filament model of muscle contraction states that the thick and thin filaments slide past each other horizontally, due to the interactions of actin and myosin.

The myosin molecules look like golf clubs arranged horizontally in a group, with the heads of the golf clubs pointing up. This head region is the center of the reactions that take place during muscle contraction. Myosin binds ATP and hydrolyzes it into ADP, and its structure is changed in the process, which causes it to bind to a specific site on actin and form a cross-bridge. Myosin then releases the stored energy and relaxes to its normal conformation. This changes the angle of attachment of the myosin head relative to its tail. As myosin bends inward upon itself, tension increases on the actin filament, and the filament is pulled toward the middle of the sarcomere.

(b) In the transmission of action potentials through muscle cells, in response to a nerve impulse, a single action potential will cause an increase in tension in the muscle cell, and if a second action potential arrives within a certain short period of time, the response will be greater; the two responses are summed. If a muscle cell receives action potentials from many nerve cells surrounding it, these, too, will be summed, and the level of tension will depend on how quickly the action potentials follow one another. If the rate of stimulation is sufficiently high, the muscle twitches will blur, and tetanus will result.

(c) When oxygen is scarce, as in situations in which a person is taking part in strenuous exercise, human muscle cells switch to lactic acid fermentation (which normally undergoes regular aerobic cellular respiration) to produce ATP. In lactic acid fermentation, pyruvate is reduced by NADH to form lactate with no release of CO_2. The lactate that accumulates as a result of this reaction can cause muscle fatigue and pain.

This response uses the following key terms in context, showing the writer's knowledge of their meanings and relatedness:

skeletal muscle	*cross-bridge*
myofibrils	*action potential*
myofilaments	*summation*
thin filaments	*tetanus*
actin strands	*aerobic cellular respiration*
thick filaments	*lactic acid fermentation*
myosin	*pyruvate*
sarcomere	*lactate*
sliding-filament model	

It also shows knowledge of how these important biological processes take place: how muscle cells contract and ultimately cause bones to move, how tetanus is reached, and how and why strenuous exercise produces muscle pain.

Ecology

Chapter 52: An Introduction to Ecology and the Biosphere

YOU MUST KNOW

- The role of abiotic factors in the formation of biomes.
- Features of freshwater and marine biomes.
- Major terrestrial biomes and their characteristics.

Concept 52.2 Interactions between organisms and the environment limit the distribution of species

▌ **Ecology** is the scientific study of the interactions between organisms and the environment.

▌ The ecological study of species involves **biotic** (living) and **abiotic** (nonliving) influences.

   ▪ **Biotic factors** may include behaviors as well as interactions with other species. Population and community ecology (Chapters 53 and 54) will explore many of the interactions between organisms.

   ▪ The **abiotic components** of an environment are the nonliving, chemical, and physical components. Some important abiotic factors include temperature, water, salinity, sunlight, and soil.

▌ The major components that make up the **climate** in a certain location are temperature, precipitation, sunlight, and wind. Climate patterns can be described on two scales: macroclimate and microclimate.

   ▪ **Macroclimate patterns** work at the global, regional, or local level.
   ▪ **Microclimates** are small-scale environmental variations—for example, under a log.

Concept 52.3 Aquatic biomes are diverse and dynamic systems that cover most of Earth

▌ **Biomes** are the major types of ecosystems that occupy very broad geographic regions. **Aquatic biomes** make up the largest part of the biosphere, because water covers roughly 75% of Earth's surface. These biomes are classified into **freshwater biomes** and **marine biomes.**

▌ All aquatic biomes display vertical stratification, which forms the following ecologically unique areas:

  ▪ The **photic zone,** in which there is enough light for photosynthesis to occur, and an **aphotic zone,** where very little light penetrates.
  ▪ The **benthic zone** is located at the bottom of the biome, where it is made up of sand, inorganic matter, and organic sediments. Organic sediments also include **detritus,** which is dead organic matter.
  ▪ **Thermoclines** are narrow layers of fast temperature change that separate a warm upper layer of water and cold deeper waters.

▌ The **two types of freshwater biomes** are standing bodies of water, such as lakes and wetlands, and moving bodies of water, such as streams and rivers.

  ▪ In lakes, communities are distributed according to the water's depth. The **littoral zone** (well-lit shallow waters near the shore) contains rooted and floating aquatic plants, whereas the **limnetic zone** (well-lit open surface waters farther from shore) is occupied by phytoplankton.
  ▪ **Oligotrophic lakes** are deep lakes that are nutrient-poor and oxygen-rich and contain sparse phytoplankton. **Eutrophic lakes** are shallower, and they have higher nutrient content and lower oxygen content with a high concentration of phytoplankton.
  ▪ The prominent physical attribute of **streams and rivers** is current. A great diversity of organisms inhabits unpolluted streams and rivers. These organisms are distributed in vertical zones and from the headwaters to the mouth.
  ▪ **Estuaries** are areas where freshwater streams or rivers merge with the ocean.

▌ **Marine biomes** include the following:

  ▪ The **intertidal zone,** where land meets the water, is periodically submerged and exposed by the twice-daily tides.
  ▪ The **neritic zone,** beyond the intertidal zone, is the shallow water over the continental shelves.
  ▪ The **pelagic biome** is a vast realm of open blue water found past the continental shelves.
  ▪ **A coral reef** is a biome created by a group of cnidarians that secrete hard calcium carbonate shells, which vary in shape and support the growth of other corals, sponges, and algae. Coral reefs are among the most productive ecosystems on Earth.

Concept 52.4 Climate largely determines the distribution and structure of terrestrial biomes

▋ The importance of climate, especially precipitation and temperature, are reflected in the climograph for the major biomes of North America featured in Figure 10.1.

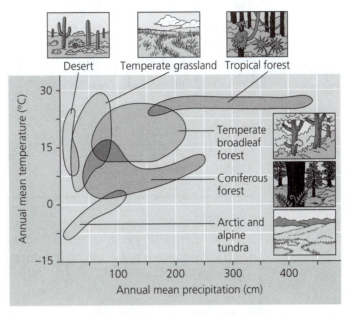

Figure 10.1   Climograph for some major biomes in North America

▋ **Savannas** are characterized by grasses and also some trees. The dominant herbivores are insects, such as ants and termites. Fire is a dominant abiotic factor, and many plants are adapted for fire. Plant growth is quite substantial during the rainy season, but large grazing mammals must migrate during regular seasons of drought.

▋ **Desert** is marked by sparse rainfall, and desert plants and animals are adapted to conserve and store water. Deserts contain many CAM plants and plants with adaptations that prevent animals from consuming them, such as the spines on cacti. Temperature (either hot or cold) is usually extreme.

▋ **Chaparral** is dominated by dense, spiny, evergreen shrubs. These are coastal areas with mild rainy winters and long, hot, dry summers. Plants are adapted to fires.

▋ **Temperate grassland** is marked by seasonal drought with occasional fires and by large grazing mammals. All these factors prevent the significant growth of trees. Grassland soil is rich in nutrients, making these areas good for agriculture.

▋ **Temperate broadleaf forest** is marked by dense stands of deciduous trees that require sufficient moisture. These forests are more open than (and not as tall as) rain forests. They are stratified—the top layer contains one or two strata of trees; beneath that are shrubs; and under that is an herbaceous stratum. **Canopy** refers to the upper layers of trees in a forest. These trees drop their leaves in fall, and many mammals enter hibernation. Many birds migrate to warmer climates.

- **Coniferous forest** is dominated by cone-bearing trees such as pine, spruce, and fir. The conical shape of conifers prevents much snowfall from accumulating on—and breaking—these trees' branches.
- **Tundra** is marked by permafrost (permanently frozen layer of soil), very cold temperatures, high winds, and little rainfall. Tundra supports no trees or tall plants. It accounts for about 20% of Earth's terrestrial surface.
- **Tropical forest** has pronounced vertical stratification. The canopy is so dense that little light breaks through. These forests are marked by epiphytes, which are plants that grow on other plants instead of the soil. Rainfall is varied. Biodiversity is greatest of all the terrestrial biomes.

Chapter 53: Population Ecology

YOU MUST KNOW

- How density, dispersion, and demographics can describe a population.
- The differences between exponential and logistic models of population growth.
- How density-dependent and density-independent factors can control population growth.

Concept 53.1 Dynamic biological processes influence population density, dispersion, and demographics

- A **population** is a group of individuals of a single species living in the same general area. **Population ecology** explores how biotic and abiotic factors influence the density, distribution, size, and age structure of populations.
- Three fundamental characteristics of the organisms in a population follow:

  - **Density** is the number of individuals per unit area or volume. The density of a population increases by births or immigration and decreases by deaths or emigration.
  - **Dispersion** is the pattern of spacing among individuals within the boundaries of the population.

    1. The most common pattern of dispersion is *clumped*, with individuals in patches, usually around a required resource. *Example:* Cottonwood trees along a stream in the arid southwest.
    2. A *uniform* dispersion pattern is often the result of antagonistic interactions. Animals that defend territories often show a uniform pattern. *Example:* Red-winged blackbirds during mating season.
    3. *Random* dispersion shows unpredictable spacing. This is not a common spacing in nature, as there is usually a reason for a pattern of spacing.

- **Demography** is the study of vital statistics of a population, especially birth and death rates. A graphic way to show birth and death rates in a population is survivorship curves, as shown in Figure 10.2.

  1. Note that Type I shows low death rates during early and midlife; then the death rate increases sharply in older age groups.
  2. Type II survivorship curves show a constant death rate over the organism's life spans.
  3. Type III curves show very high early death rates early, then a flat rate for the few surviving to older age groups.

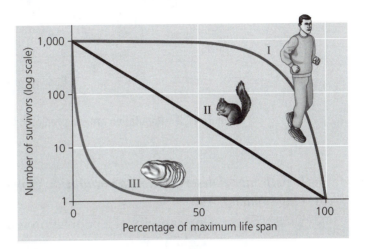

Figure 10.2

Concept 53.2 *Life history traits are products of natural selection*

- Traits that affect an organism's schedule of reproduction and survival make up its **life history.** Life histories entail three variables:

  1. When reproduction begins or the age of sexual maturation.
  2. How often the organism reproduces. Some organisms save their resources for one big reproductive event (*big-bang reproduction*), whereas others produce offspring in *repeated reproduction.*
  3. The number of offspring during each reproductive event.

- Life history traits are evolutionary outcomes, *not* conscious decisions by organisms.

Concept 53.3 *The exponential model describes population growth in an idealized, unlimited environment*

- **Exponential population** growth refers to population growth under ideal conditions. Figure 10.3 shows a graph of population growth as predicted by the exponential model. Any species, regardless of its life history, is capable of exponential growth if resources are abundant.

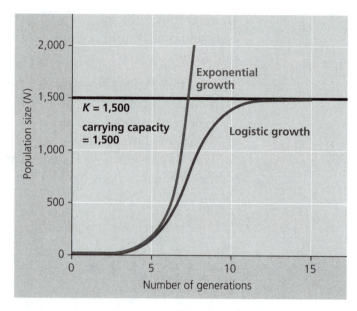

Figure 10.3 Population growth predicted by the exponential and logistic model.

Concept 53.4 **The logistic model describes how a population grows more slowly as it nears its carrying capacity**

▮ The **carrying capacity** of a population is defined as the maximum population size that a certain environment can support at a particular time with no degradation of the habitat.

▮ In the logistic growth model, the per capita rate of increase declines as carrying capacity is reached. Figure 10.3 shows a graph of population growth as predicted by the logistic growth model.

▮ Selection of life history traits that are sensitive to population density and carrying capacity are known as **K-selection.** K-selection operates in populations living close to the density imposed by the carrying capacity. By contrast, selection for life history traits that maximize reproductive success is called **r-selection.**

▮ The logistic growth model is sometimes associated with K-selection, whereas the exponential growth model is often associated with r-selection. Both K-selection and r-selection are two ends of a continuum of life history strategies.

Concept 53.5 *Many factors that regulate population growth are density dependent*

▌ A death rate that rises as population density rises and a birth rate that falls as population density rises are **density-dependent factors.** Examples of factors that reduce birth rates or increase death rates include the following:

1. *Competition for resources.* As population density increases, competition for resources intensifies. This might include competition for food, space, or essential nutrients.
2. *Territoriality.* Available space for territories or nesting may be limited, thus controlling the population.
3. *Disease.* Increasing densities allow for easier transmission of diseases.
4. *Predation.* As prey populations increase, predators may find the prey more easily.

▌ When a death rate does not change with increase in population density, it is said to be **density independent.** Natural disasters are examples of density-independent factors.

▌ All populations exhibit some size fluctuations. Many populations undergo regular boom-and-bust cycles that are influenced by complex interactions between biotic and abiotic factors.

Concept 53.6 *The human population is no longer growing exponentially but is still increasing rapidly*

▌ The exponential growth model in Figure 10.3 approximates the population explosion of humans since 1650. However, since about 1970 the rate of growth has fallen by nearly 50%.

▌ One reason for falling human population growths is demographic transition. **Demographic transition** occurs when a population goes from high birth rates and high death rates to low birth rates and low death rates. Demographic transition may regularly take 150 years to complete. First, death rate falls, usually due to increased medical care and sanitation; however, falling birth rates take much longer, thus delaying transition.

▌ Global carrying capacity for humans is not known. A concept termed the **ecological footprint** examines the total land and water area needed for all the resources a person consumes in a population. Currently, 1.7 hectares per person is considered sustainable. A typical person in the United States has a footprint of 10 hectares.

Chapter 54: Community Ecology

> **YOU MUST KNOW**
>
> • The difference between a fundamental niche and a realized niche.
> • The role of competitive exclusion in interspecific competition.
> • The symbiotic relationships of parasitism, mutualism, and commensalism.
> • The impact of keystone species on community structure.
> • The difference between primary and secondary succession.

Concept 54.1 *Community interactions are classified by whether they help, harm, or have no effect on the species involved*

▌ A **community** is a group of populations of different species living close enough to interact. **Interspecific interactions** may be positive for one species ($+$), negative ($-$), or neutral (0) and include competition, predation, and symbioses.

> *Study Tip:* The prefix *inter-* means between different groups, whereas *intra-* means within the same group. *Intraspecific competition* is competition within the same species, like two males fighting over a territory. *Interspecific competiton* is competition between two different species for resources, like food. Pay attention to the prefix! You could be asked to write about either type of competition in an essay.

▌ **Interspecific competitions** for resources occur when resources are in short supply. Competition is a $-/-$ interaction between the species involved. Central to the idea of competition and community structure are these two concepts:

1. The **competitive exclusion principle** states that when two species are vying for a resource, eventually the one with the slight reproductive advantage will eliminate the other.
2. An organism's **ecological niche** is the sum total of biotic and abiotic resources that the species uses in its environment. A species' **fundamental niche,** the niche potentially occupied by the species, is often different from the **realized niche,** the portion of the fundamental niche the species actually occupies.

▌ **Predation** is a $+/-$ interaction between two species in which one species (the **predator**) eats the other species (the **prey**). Defenses for predators include the following:

- ▪ **Cryptic coloration,** in which the animal is camouflaged by its coloring.
- ▪ **Aposematic** or **warning coloration,** in which a poisonous animal is brightly colored as a warning to other animals.

- **Batesian mimicry** refers to a situation in which a harmless species has evolved to mimic the coloration of an unpalatable or harmful species. In **Müllerian mimicry,** two bad-tasting species resemble each other, ostensibly so that predators will learn to avoid them equally.
- **Herbivory** is also a $+/-$ interaction in which an herbivore eats part of a plant or alga. It is advantageous for an animal to be able to distinguish toxic from nontoxic plants. A plant's main protective devices are chemical toxins, spines, and thorns.

■ **Symbiosis** occurs when individuals of two or more species live in direct contact with one another.

- **Parasitism** is a $+/-$ symbiotic interaction in which the parasite derives its nourishment from its host. Parasites may have a significant effect on the survival, reproduction, and density of their host population.
- **Mutualism** is an interspecific interaction that benefits both species $(+/+)$. Both pollinators and flowering plants benefit from their relationship.
- **Commensalism** benefits one of the species but neither harms nor helps the other species. A fern growing in the shade of another plant could be a commensal relationship.

Concept 54.2 Dominant and keystone species exert strong controls on community structure

■ **Species diversity** measures the number of different species in a community (species richness) *and* the relative abundance of each species. A community with an even species abundance is more diverse than one in which one or two species are abundant and the remainder are rare.

■ The **trophic structure** of a community refers to the feeding relationships among the organisms. **Trophic levels** are the links in the trophic structure of a community.

■ The transfer of food energy from plants through herbivores through carnivores through decomposers (from one trophic level to another) is referred to as a **food chain. Food webs** consist of two or more food chains linked together.

■ **Dominant species** in a community have the highest **biomass** (the sum weight of all the members of a population) or are the most abundant.

■ **Keystone species** exert control on community structure by their important ecological niches. Notice in Figure 10.4 the impact of the keystone predator *Pisaster* (a sea star) on the diversity of species present in a tidal pool.

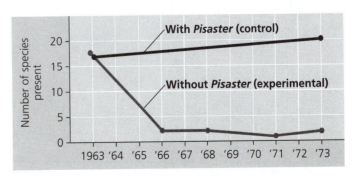

Figure 10.4 Impact of keystone predator on species diversity

Concept 54.3 Disturbance influences species diversity and composition

▌ A **disturbance**—storm, fire, flood, drought, or human activity—changes a community by removing organisms or changing resource availability. Disturbance is not necessarily bad for a community. The **intermediate disturbance hypothesis** states that moderate levels of disturbance create conditions that foster greater species diversity than low or high levels of disturbance.

▌ **Ecological succession** refers to transitions in species composition in a certain area over ecological time.

   ▪ In **primary succession,** plants and animals gradually invade a region that was virtually lifeless where soil has not yet formed. The gradual colonization of a newly formed volcanic island would be an example.

   ▪ **Secondary succession** occurs when an existing community has been cleared by a disturbance that leaves the soil intact. An abandoned farm will show secondary succession as it starts with the soil intact.

Concept 54.4 Biogeographic factors affect community biodiversity

▌ Two biogeographic contributions are especially important in community diversity:

   1. *The latitude of the community.* Plant and animal life is generally more abundant and diverse in the tropics, becoming less so moving toward the poles.
   2. *The area of the community.* If all other factors are held equal, the larger the geographic area of a community is, the more species it has.

▌ Because of their isolation and limited size, islands are natural laboratories for studying biogeographical factors. In addition to actual islands, this idea also pertains to islands of land, like national parks surrounded by development.

▌ **Island biogeography** is primarily influenced by two factors:

   1. Rates of immigration and extinction are influenced primarily by the *size* of the island and the *distance* of the island from the mainland. The greater the sizes of the island, the higher the *immigration* rates and lower the rates of *extinction*.
   2. As the distance from the mainland increases, the rate of *immigration* falls, whereas *extinction* rates increase.

Chapter 55: Ecosystems

> **YOU MUST KNOW**
>
> - How energy flows through the ecosystem by understanding the terms in bold that relate to food chains and food webs.
> - The difference between gross primary productivity and net primary productivity.
> - The carbon and nitrogen biogeochemical cycles.

Concept 55.1 Physical laws govern energy flow and chemical cycling in ecosystems

- An **ecosystem** is the sum of all the organisms living within its boundaries (biotic community) and all the abiotic factors with which they interact. Ecosystem ecology involves two unique processes: *energy flow* and *chemical cycling*.
- The flow of energy can be traced through the feeding or trophic levels in food chains and food webs. *Energy cannot be recycled*; therefore, energy must be constantly supplied to an ecosystem—in most cases by the sun.

  - **Primary producers** in an ecosystem are the **autotrophs** ("self-feeders"). They support all others organisms in the ecosystem.
  - Organisms that are in trophic levels above primary producers cannot make their own food and are therefore consumers or **heterotrophs** ("other feeders").
  - Herbivores eat primary producers and are called **primary consumers.**
  - Carnivores that eat herbivores are called **secondary consumers,** while carnivores that eat secondary consumers are termed **tertiary consumers.**
  - **Detritivores,** or **decomposers,** are consumers that get their energy from detritus, which is nonliving organic material such as the remains of dead organisms, feces, dead leaves, and wood. Detritivores convert organic materials from all trophic levels to inorganic compounds that can be used by producers. In this way nutrients cycle through ecosystems.

- It is not uncommon for a species to feed at more than one trophic level. An animal's diet might consist of berries and fish or algae and insects. The feeding level may also change as the stage in a species' life cycle changes.

Concept 55.2 Energy and other limiting factors control primary production in ecosystems

- The amount of light energy converted to chemical energy by autotrophs is an ecosystem's **primary production.** The amount of all photosynthetic production sets the spending limit for the energy budget of the entire ecosystem.

  - Total primary production in an ecosystem is known as that system's **gross primary production (GPP).**
  - GPP is not the amount of energy available to consumers, however. Some of the fuel molecules made by the producers must be used as fuel for

their own cellular respiration. **Net primary production (NPP)** is equal to gross primary production minus the energy used for respiration (R) by the producers:

$$NPP = GPP - R$$

Primary production in aquatic ecosystems is affected primarily by light availability and nutrient availability. In the photic zone, light—and therefore photosynthesis—decreases with depth. The nutrient most often limiting marine production is either nitrogen or phosphorus. A lake that is nutrient-rich and that supports a vast array of algae is said to be **eutrophic.**

> *Study Tip:* AP Lab 12 on Aquatic Primary Productivity should be reviewed with special attention paid to how net and primary productivity are determined as light intensity is reduced. You may want to use *The LabBench* to review the lab.

▌ Temperature and moisture are the key factors controlling primary production in terrestrial ecosystems. A measure of the amount of water transpired by plants and evaporated from the landscape, termed **evapotranspiration,** combines both key terrestrial factors.

Concept 55.3 *Energy transfer between trophic levels is typically only 10% efficient*

▌ If 10% of energy is transferred from primary producer to primary consumer to secondary consumer, only 1% of the net primary production (10% of 10%) is available to secondary consumers. The loss of energy from trophic level to trophic level is one of the factors that keep food chains so short.

▌ Pyramids of energy or biomass or numbers are sometimes used to give insight to food chains. Energy pyramids are never inverted. Number pyramids or aquatic biomass pyramids may be inverted, but never energy flow pyramids.

Concept 55.4 *Biological and geological processes cycle nutrients between organic and inorganic parts of an ecosystem*

▌ **Biogeochemical cycles** are nutrient cycles that contain both biotic and abiotic components. Understanding these cycles allows scientists to trace how nutrients flow through ecosystems and how humans may have altered the flow.

▌ The **carbon cycle** is a balance between the amount of CO_2 removed from ecosystems by photosynthesis and added by cellular respiration. The burning of fossil fuels has added significant amounts of additional CO_2 to the atmosphere. Examine Figure 10.5 to see the generalized flow of carbon while also considering the effects of CO_2 on global warming.

▌ The **nitrogen cycle** moves nitrogen from the atmosphere through the living world. Nitrogen is a common limiting factor for plant growth, making its movement through ecosystems especially important. Note the important role of bacteria in the nitrogen cycle while tracing nitrogen flow through Figure 10.6.

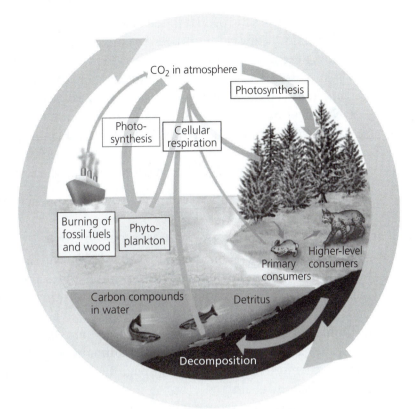

Figure 10.5   The carbon cycle

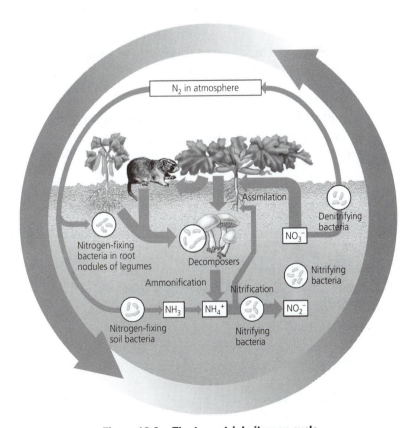

Figure 10.6   The terrestrial nitrogen cycle

- Most of Earth's nitrogen is in the form of N_2, which is unusable by plants. The major pathway for nitrogen to enter an ecosystem is **nitrogen fixation,** the conversion of N_2 by bacteria to forms that can be used by plants. Earlier we noted this relationship between plants that are legumes and the bacterium *Rhizobium* as an example of mutualism.
- **Nitrification** is the process by which ammonium (NH_4^+) is oxidized to nitrite and then nitrate (NO_3^-) by bacteria. Two inorganic nitrogen forms can be absorbed by plants: nitrates and ammonium.
- **Denitrification** by bacteria releases nitrogen to the atmosphere.

- Other important nutrient cycles involve water and phosphorus.

Concept 55.5 Human activities now dominate most chemical cycles on Earth

- **Acid precipitation** is defined as rain, snow, or fog with a pH less than 5.6. The burning of wood and fossil fuels releases sulfur oxides and nitrogen oxides into the atmosphere. These oxides react with water, forming sulfuric acid and nitric acid.
- In **biological magnification,** toxins become more concentrated in successive trophic levels of a food web. The toxins cannot be broken down biologically by normal chemical means, so they magnify in concentration as they move through the food chain.
- The **greenhouse effect** refers to the absorption of heat the Earth experiences due to certain atmospheric gases. Carbon dioxide and water vapor intercept and absorb much reflected infrared radiation, re-reflecting some back toward Earth.

  - Because of the burning of fossil fuels, CO_2 levels have been steadily increasing. One effect of this increase is that Earth is being warmed significantly (**global warming**).

- The **ozone layer** reduces the amount of UV radiation penetration from the sun through the atmosphere. Chlorine-containing compounds used by humans are eroding the ozone layer, allowing more DNA-damaging UV radiation to penetrate to the surface of the Earth.

Multiple-Choice Questions

1. All of the following statements about Earth's ozone layer are false EXCEPT
   (A) It is composed of O_2.
   (B) It increases the amount of ultraviolet radiation that reaches Earth.
   (C) It is thinning as a result of widespread use of certain chlorine-containing compounds.
   (D) It is a result of widespread burning of fossil fuels
   (E) It allows green light in but screens out red light.

2. Which of the following is the major primary producer in a savanna ecosystem?
   (A) lion
   (B) gazelle
   (C) grass
   (D) snake
   (E) diatom

3. The carrying capacity of a population is defined as
(A) the amount of time the parents in the population spend rearing and nurturing their offspring.
(B) the maximum population size that a certain environment can support at a particular time.
(C) the amount of vegetation that a certain geographic area can support.
(D) the number of different types of species a biome can support.
(E) the number of different genes a population can carry at a particular time.

4. Which of the following terms is used to describe major types of ecosystems that occupy broad geographic regions?
(A) biome
(B) community
(C) chaparral
(D) trophic level
(E) photic zone

5. A lake that is nutrient rich and that supports a vast array of algae is said to be
(A) oligotrophic.
(B) abyssal.
(C) littoral.
(D) eutrophic.
(E) limnetic.

6. Which of the following best describes an estuary?
(A) An area that is periodically flooded, causing its soil to be consistently damp
(B) An area where a river changes course after being diverted from its original course by an obstacle
(C) The area where a freshwater river merges with the ocean
(D) The area where a mass of cold water and a mass of warm water meet in the pelagic zone
(E) An outshoot of land that extends into the ocean

7. Which of the following is the term that refers to the layer of light penetration in aquatic ecosystems?
(A) littoral zone
(B) limnetic zone
(C) photic zone
(D) benthic zone
(E) aphotic zone

Directions: The group of questions below consists of five lettered choices followed by a list of numbered phrases or sentences. For each numbered phrase or sentence, select the one choice that is most closely related to it. Each choice may be used once, more than once, or not at all.

Questions 8–12
(A) Temperate grassland
(B) Tropical forest
(C) Temperate broadleaf forest
(D) Tundra
(E) Desert

8. Characterized by permafrost and few large plants

9. Characterized by epiphytes and significant canopy

10. Characterized by an understory of shrubs and trees that lose their leaves in the fall

11. Characterized by occasional fires, nutrient-rich soil, and large grazing animals

12. Characterized by sparse rainfall and extreme daily temperature fluctuations

13. A bacterial colony that exists in an environment displaying ideal conditions will undergo
(A) logistic growth.
(B) intrinsic growth.
(C) hyperactive growth.
(D) exponential growth.
(E) unbounded growth.

14. A species' specific use of the biotic and abiotic factors in an environment is collectively called the species'
    (A) habitat.
    (B) trophic level.
    (C) ecological niche.
    (D) placement.
    (E) partitioning.

15. In which type of camouflaging does a non-toxic animal mimic the appearance of a toxic animal?
    (A) Müllerian mimicry
    (B) cryptic coloration
    (C) aposematic coloration
    (D) Batesian mimicry
    (E) parasitoidism

16. The dominant species in a community is the one that
    (A) has the greatest number of genes per individual.
    (B) is at the top of the food chain.
    (C) has the largest biomass.
    (D) eats all other members of the community.
    (E) bears the most offspring in each mating.

17. Which statement best describes energy transfer in a food web?
    (A) Energy is transferred to consumers, which convert it to nitrogen compounds and use it to synthesize amino acids.
    (B) Energy from producers is converted into oxygen and transferred to consumers.
    (C) Energy from the sun is stored in green plants and transferred to consumers.
    (D) Energy is transferred to consumers that use it to synthesize food.
    (E) Energy moves from autotrophs to heterotrophs to decomposers, which convert it to a form producers can use again.

18. A fire cleared a large area of forest in Yellowstone National Park in the 1980s. When the first plants pioneered this burned area, this was an example of
    (A) primary succession.
    (B) secondary succession.
    (C) biological evolution.
    (D) a keystone species.
    (E) the top-down model.

19. In the nitrogen cycle, the process by which nitrogen in the atmosphere is made available for use by plants is known as
    (A) ammonification.
    (B) denitrification.
    (C) nitrogen fixation.
    (D) nitrogen cycling.
    (E) nitrogenation.

20. The process in which CO_2 in the atmosphere intercepts and absorbs reflected infrared radiation and re-reflects it back to Earth is known as
    (A) global warming.
    (B) atmospheric insulation.
    (C) stratospheric insulation.
    (D) biological magnification.
    (E) the greenhouse effect.

21. A Type I survivorship curve is level at first, with a rapid increase in mortality in old age. This type of curve is
    (A) typical of many invertebrates that produce large numbers of offspring.
    (B) typical of human and other large mammals.
    (C) found most often in r-selected populations.
    (D) almost never found in nature.
    (E) typical of all species of birds.

22. Which of the following would not be a density-dependent factor limiting a population's growth?
    (A) increased predation by a predator
    (B) a limited number of available nesting sites
    (C) a stress syndrome that alters hormone levels
    (D) a very early fall frost
    (E) intraspecific competition

23. The human population is growing at such an alarmingly fast rate because
    (A) technology has increased our carrying capacity.
    (B) the death rate has greatly decreased since the Industrial Revolution.
    (C) the age structure of many countries is highly skewed toward younger ages.
    (D) fertility rates in many developing countries are above the 2.1 children per female replacement level.
    (E) all of the above are true.

24. When one species was removed from a tide-pool, the species richness became significantly reduced. The removed species was probably
    (A) a strong competitor.
    (B) a potent parasite.
    (C) a resource partitioner.
    (D) a keystone species.
    (E) the species with the highest relative abundance.

25. Which of the following interspecific interactions is not an example of a $+/-$ interaction?
    (A) ectoparasite and host
    (B) herbivore and plant
    (C) honeybee and flower
    (D) pathogen and host
    (E) carnivore and prey

Free-Response Question

1. All of the organisms in a community are interrelated by the abiotic and biotic resources they use in the course of their lives.

   (a) Describe the relationships that exist among a hawk, a mouse, a plant, and soil in a particular ecosystem.
   (b) As unlikely as it may seem, biotic components of an environment do influence the abiotic components of an environment. Give two examples of this influence.

ANSWERS AND EXPLANATIONS

Multiple-Choice Questions

▮ **1. (C) is correct.** The ozone layer is located in the stratosphere and surrounds Earth. It is composed of O_3, and it absorbs UV radiation, preventing it from reaching the organisms in the biosphere. Researchers have been observing the thinning of the ozone layer since about 1975. The destruction of the ozone layer has been attributed to the widespread use of chlorofluorocarbons.

▮ **2. (C) is correct.** In a savanna, grass constitutes the primary producer. A primary producer traps the energy of sunlight and turns it into chemical energy through photosynthesis. Primary consumers (herbivores) consume primary producers, secondary consumers (carnivores) eat herbivores, and tertiary consumers eat carnivores.

3. (B) is correct. The carrying capacity of a population is defined as the maximum population size a particular environment can support at a particular time with no degradation of the habitat. It is fixed at certain times, but it varies over the course of time with the amount of resources that exist in an environment.

4. (A) is correct. Biomes are major types of ecosystems that occupy broad geographic regions. Some examples of biomes are coniferous forests, deserts, grasslands, and tropical forests.

5. (D) is correct. Lakes are classified according to how much organic matter they produce. Oligotrophic lakes are deep and generally poor in nutrients, and therefore, they have relatively little phytoplankton. Eutrophic lakes are usually shallower and have greater nutrient content, which allows the growth of more phytoplankton.

6. (C) is correct. An estuary is an area where a running freshwater source, such as a stream or river, meets the ocean. Often estuaries are bordered by large areas of coastal wetlands, and salinity varies with location within them, as well as with the rise and fall of the ocean tides. Estuaries are among the most biologically productive biomes, and they also are home to many of the fish and other animals that humans consume.

7. (C) is correct. All aquatic biomes display vertical stratification, which forms two ecologically unique areas, the photic zone in which there is enough light for photosynthesis to occur and an aphotic zone, where very little light penetrates.

8. (D) is correct. Tundra is characterized by having permafrost (which is a permanently frozen layer of soil), very cold temperatures, and high winds. These factors prevent tall plants from growing in the tundra. Tundra generally does not receive much rainfall throughout the year, and what rain does fall cannot soak into the soil because of the permafrost.

9. (B) is correct. Tropical forests generally have thick canopies that prevent much sun from filtering through. This means that in breaks in the canopy, other plants grow quickly to compete for sunlight. Tropical forests are home to epiphytes, and rainfall is frequent.

10. (C) is correct. Temperate broadleaf forests are characterized by dense populations of deciduous trees, which drop their leaves in the fall when the weather turns cold.

11. (A) is correct. Temperate grasslands are characterized by having thick grass, seasonal drought, occasional fires, and large grazing animals. Their soil is generally rich with nutrients, making them good areas for agriculture. Most of the temperate grassland in the United States is used today for agriculture.

12. (E) is correct. Deserts experience very little rainfall, so they are home to many plants and animals that have adaptations for storing and saving water. Deserts are marked by drastic temperature fluctuation; they can be very hot in the day but freezing at night. Many desert plants rely on CAM photosynthesis.

13. (D) is correct. A bacterial colony growing where it has limitless nutrients, and other ideal conditions, will experience what is called exponential growth. In exponential growth, all members are free to reproduce at their physiological capacity.

- **14. (C) is correct.** A species' ecological niche is defined as the sum of its use of the abiotic and biotic factors in an environment. For instance, a particular bird's niche refers to many things, including the food it consumes, where it builds its nest, the time of day it is active, and what climate it lives in.
- **15. (D) is correct.** In Batesian mimicry, a harmless or palatable animal evolves the same markings and/or colorings as a harmful or unpalatable animal and, in this way, can escape predation.
- **16. (C) is correct.** The dominant species in a community has the greatest biomass, or sum weight of all of the members of a population. Dominant species are also hypothesized to be the most competitive in exploiting the resources in an ecosystem.
- **17. (C) is correct.** Almost all organisms use solar energy stored in food to power life processes. Autotrophs convert solar energy to a form useful to both autotrophs and heterotrophs. At each successive trophic level, less energy is available because so much is converted to a form not useful to the organisms. Energy cannot be recycled.
- **18. (B) is correct.** Secondary succession refers to a situation in which a community has been cleared by a disturbance of some kind, but the soil is left intact. The area will begin to return to its original state through the process of plants invading the area and recolonizing.
- **19. (C) is correct.** Most of Earth's nitrogen is in the form of N_2, which is unusable by plants. The major pathway for nitrogen to enter an ecosystem is nitrogen fixation, the conversion of N_2 by bacteria to forms that can be used by plants.
- **20. (E) is correct.** The greenhouse effect is the process by which carbon dioxide and water vapor in the atmosphere intercept reflected infrared radiation from the sun and re-reflect it to Earth. Global warming is the process by which the amount of carbon dioxide in the atmosphere is increasing because of humans' combustion of fossil fuels, leading to higher temperatures on Earth.
- **21. (B) is correct.** Type I survivorship curve is the pattern described in the question. Type II curves show an equal chance of death throughout the life span, while Type III shows heavy mortality in early stages of life cycle.
- **22. (D) is correct.** An early fall frost is a density-independent factor—it occurs without regard to the density of the population. Density-dependent factors increasingly slow population growth as density increases.
- **23. (E) is correct.** All of the answers are contributing to increased human population growth, as much of the world is experiencing early demographic transition. Death rates have fallen, but birth rates are much slower to change.
- **24. (D) is correct.** Figure 10.4 shows the effect of the removal of a keystone species from a tide pool. In this case the removal of *Pisaster* allowed for the unrestrained growth of mussels, which eventually took over the rock faces of the tide pool and eliminated most other invertebrates.
- **25. (C) is correct.** A $+/-$ notation indicates one species benefits ($+$), whereas the other species is harmed ($-$). Answer C does not fit this notation because both the honeybee and the flower benefit, an example of mutualism ($+/+$).

Free-Response Question

(a) The hawk, mouse, and plant in this particular ecosystem are related by the passage of energy through them. Together they comprise a food chain—the mouse is a primary consumer, and it consumes the plant, which is a primary producer (the plant is an autotroph—capable of trapping the energy of the sun and converting it into chemical energy in the form of carbohydrates). The hawk then is a predator of the mouse—and a secondary consumer. Secondary consumers eat herbivores. All of these animals are dependent on the soil in which the plant has grown because the soil provides the plant with nutrients. The plant needs a variety of organic elements to produce carbohydrates, but it also needs mineral nutrients, such as nitrogen to make proteins and nucleic acids.

(b) One example in which the biotic factors of the biosphere impact the abiotic factors is seen in the case of global warming. We rely on the greenhouse effect (in which atmospheric carbon dioxide acts as an insulator, trapping infrared radiation from the sun and re-reflecting it) to help maintain the hospitable temperature of Earth. Yet, due to the burning of fossil fuels—beginning during the Industrial Revolution—the concentration of carbon dioxide in the atmosphere has increased significantly, and this has led to an increase in global temperatures.

The thinning of the ozone layer is another way in which humans (a biotic factor of the biosphere) impact abiotic processes. Organisms are protected from ultraviolet radiation from the sun by a protective layer of ozone that surrounds Earth. However, the ozone layer has been degraded by humans' use of chlorofluorocarbons, which are chemicals used in refrigeration and other industrial processes. Many countries have stopped using these chemicals, but chlorine molecules already in the atmosphere continue to have an effect on ozone.

This response uses the following key terms in context, showing the writer's knowledge of their meanings and relatedness:

ecosystem	*nitrogen fixation*
primary producer	*herbivore*
primary consumer	*global warming*
secondary consumer	*greenhouse effect*
food chain	*ozone layer*
autotroph	*chlorofluorocarbons*
predator	

This response also shows knowledge of the following important biological processes: food chains and the interaction of organisms in a community, global warming, and the depletion of the ozone layer.

Part III

The Laboratory

AP Biology is designed to be equivalent to a two-semester college course, and laboratory experience must be included in all AP Biology courses. There are 12 recommended laboratory activities; it is expected that your lab experience would include activities that cover the objectives for each lab, although your teacher may have selected different labs for your course than those published by the College Board.

Our students find that a review of the laboratory exercises is an excellent way to review many of the core concepts of their AP Biology course. A great resource for your review is The LabBench, available at www.biology.com. There, you will find exercises that take you through each lab, along with sample questions to test your understanding. Now, let's look at each lab.

Laboratory 1: Diffusion and Osmosis

YOU MUST KNOW

- Factors that affect diffusion across the membrane.
- How water potential is measured and its relationship to solute concentration and pressure potential of a solution.
- Water moves from a region where water potential is high to a region where water potential is low.
- The relationship of molarity to osmotic concentration.
- How to determine osmotic concentration of a solution from experimental data.

Overview of the Lab:

▮ In the first part of this lab, you put a solution of glucose and starch in a dialysis bag and place the bag in a beaker of water with iodine. Water and iodine enter the bag. Evidence of this is that the bag gains mass, and the clear solution in the bag turns black when iodine enters and binds with the starch. A test for glucose reveals that it is found outside the bag. The dialysis bag is selectively permeable and serves as a model for the cell membrane. Notice that each molecule moved down its respective concentration gradient. Factors that affect the tendency of particles to move across a membrane include size and relative concentration, demonstrated here.

▮ In the second part of the lab, you fill dialysis bags with varying molarities of a sucrose solution and place them into distilled water. By massing the bags before and after the experiment, you will note that water moves into the bags in direct relationship to the molarity of the solutions. Sucrose, a disaccharide, cannot pass through the bag pores.

▮ In the third part of this lab, potato cores are placed in varying molarities of a sucrose solution. The goal is to determine the molar solution of the potato cores by graphing their mass gain or loss. The point at which there is no net change in mass is the molarity of the potato tissue. This information can now be used to determine the *solute potential* and *water potential* of the potato cells.

Hints and Review

Get a firm fix on the terminology! This is one place where you will not get any credit if you understand the concept but garble the vocabulary.

▮ **Osmosis** is the movement of water from a region of high concentration to a region of low concentration through a selectively permeable membrane.

▮ In **dynamic equilibrium** molecules are in motion, but there is no net change in concentration.

▮ Study Figure 1.1 to review some important terms.

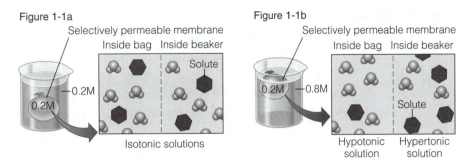

Figure 1-1a
Selectively permeable membrane
Inside bag   Inside beaker
Solute
—0.2M
0.2M
Isotonic solutions

Figure 1-1b
Selectively permeable membrane
Inside bag   Inside beaker
0.2M —0.8M
Solute
Hypotonic   Hypertonic
solution   solution

Figure 1.1

▌ In Figure 1.1a the two solutions are equal in their solute concentrations. We say that they are **isotonic** to each other.

▌ In Figure 1.1b, the solution in the bag contains less solute than the solution in the beaker. The solution in the bag is **hypotonic** (lower solute concentration) to the solution in the beaker. The solution in the beaker is **hypertonic** (higher solute concentration) to the one in the bag. Water will move from the hypotonic solution into the hypertonic solution.

▌ Obviously, you must have a solid knowledge of the principles of diffusion to understand this lab, but what many students find most difficult in this laboratory is the concept of water potential. Here's a quick review!

▌ **Remember this:** *Water moves from a region where water potential is high to a region where water potential is low.*

▌ **Water potential** (Ψ) involves *two* components: solute potential (Ψ_s) and pressure potential (Ψ_p).

▌ An increase in solute concentration will cause solute potential (Ψ_s) to decrease.

▌ In plant cells, pressure increases as water flows in due to the cell wall. This increases pressure potential (Ψ_p).

▌ A dehydrated potato slice does *not* have high water potential. Its water potential is low, and if placed in distilled water (which has high water potential), water will move into the potato cells.

Calculating Water Potential

Water potential is calculated using the following formula:

Water potential (Ψ) = pressure potential (Ψ_p) + solute potential (Ψ_s)	
Pressure potential (Ψ_p):	In a plant cell, pressure exerted by the rigid cell wall that limits further water uptake.
Solute potential (Ψ_s):	The effect of solute concentration. Pure water at atmospheric pressure has a solute potential of zero. As solute is added, the value for solute potential becomes more negative. This causes water potential to decrease also.
	In sum, as solute is added, the water potential of a solution drops, and water will tend to move into the solution.
In this laboratory we use bars as the unit of measure for water potential; 1 bar = approximately 1 atmosphere.	

Factors That Affect Water Potential

The water potential of pure water in an open container is zero because there is no solute and the pressure in the container is zero. Adding solute lowers the water potential. When a solution is enclosed by a rigid cell wall, the movement of water into the cell will exert pressure on the cell wall. This increase in pressure within the cell will raise the water potential.

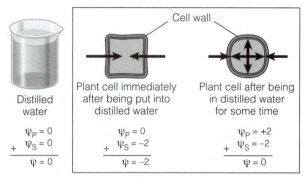

Figure 1.2

Questions

1. In beaker B, what is the water potential of the distilled water in the beaker, and of the beet core?
   (A) Water potential in the beaker = 0; water potential in the beet core = 0
   (B) Water potential in the beaker = 0; water potential in the beet core = −0.2
   (C) Water potential in the beaker = 0; water potential in the beet core = 0.2
   (D) Water potential in the beaker cannot be calculated; water potential in the beet core = 0.2
   (E) Water potential in the beaker cannot be calculated; water potential in the beet core = −0.2

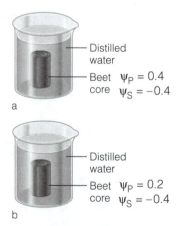

Figure 1.3

2. Which of the following statements is true for the diagrams?
   (A) The beet core in beaker A is at equilibrium with the surrounding water.
   (B) The beet core in beaker B will lose water to the surrounding environment.
   (C) The beet core in beaker B would be more turgid than the beet core in beaker A.
   (D) The beet core in beaker A is likely to gain so much water that its cells will rupture.
   (E) The cells in beet core B are likely to undergo plasmolysis.

Answers

1. The correct answer is b. Since water potential = solute potential (-0.4) + pressure potential (0.2), water potential = -0.2

2. The correct answer is a. The water potential for both the distilled water and the beet core in a is 0.

Laboratory 2: Enzyme Catalysis

YOU MUST KNOW

- The factors that affect the rate of an enzyme reaction such as temperature, pH, enzyme concentration.
- How the structure of an enzyme can be altered, and how pH and temperature affect enzyme function.
- How to name an enzyme, its substrate and products, and then design a controlled experiment to measure the activity of a specific enzyme under varying conditions.
- How to calculate the rate of a reaction.

Overview of the Lab

This experiment investigates the rate at which the enzyme catalase converts substrate to product. Catalase is allowed to react with hydrogen peroxide for varying amounts of time and then the reactions are stopped by adding H_2SO_4.

To determine the amount of hydrogen peroxide that remains after the reaction, you do a titration with $KMnO_4$, slowly adding it until a target color change is achieved.

Review the Procedure

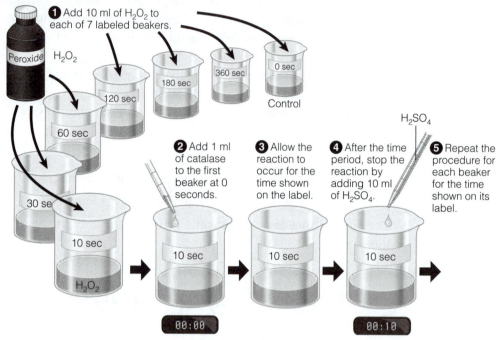

Figure 2.1

Hints and Review

- In this lab, the *enzyme* is catalase, the *substrate* is H_2O_2 (hydrogen peroxide), and the *products* are water and oxygen. When catalase is combined with peroxide, it fizzes as oxygen is released.

- Enzymes are large globular proteins; much of their three-dimensional shape is the result of interactions between the R (variable) groups of their amino acids. Anything that changes these interactions will change the shape of the enzyme and therefore alter the rate of reaction.

- Remember, *change the shape, change the function!*

- Enzyme activity is affected by pH and temperature because these affect the 3-D shape. Extremes of pH and temperature result in *denaturation* when the 3-D shape is so altered the enzyme can no longer function.

- Enzymes are not denatured by cold, but the rate of reaction is decreased as temperature decreases.

- Be able to calculate rate from graphed data using Figure 2.2.

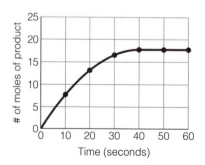

Figure 2.2

Enzyme Action Over Time

We can calculate the rate of a reaction by measuring, over time, either the disappearance of substrate (as in our catalase example) or the appearance of product (as in Figure 2.2). For example, on the graph above, what is the rate, in moles/second, over the interval from 0 to 10 seconds?

$$\text{Rate} = \frac{\Delta y}{\Delta x}$$

so for this example, the rate would be

$$\frac{7 \text{ moles} - 0 \text{ moles}}{(10 \text{ seconds} - 0 \text{ seconds})} = \frac{7}{10}$$

$$= 0.7 \text{ moles/second}$$

▮ Note that the slope of the graph is steepest during the *initial* time period; this is when the rate of a reaction is greatest and occurs because the substrate is most abundant. The rate of the reaction decreases as substrate is consumed and the slope of the graph flattens.

Questions

1. In order to keep the rate of reaction constant over the entire time course, which of the following should be done?
   (A) Add more enzyme.
   (B) Gradually increase the temperature after 60 seconds.
   (C) Add more substrate.
   (D) Add H_2SO_4 after 60 seconds.
   (E) Remove the accumulating product.

2. What is the role of sulfuric acid (H_2SO_4) in this experiment?
   (A) It is the substrate on which catalase acts.
   (B) It binds with the remaining hydrogen peroxide during titration.
   (C) It accelerates the reaction between enzyme and substrate.
   (D) It blocks the active site of the enzyme.
   (E) It denatures the enzyme by altering the active site.

3. A student was performing a titration for this laboratory and accidentally exceeded the endpoint. What would be the best step to obtain good data for this point?
   (A) Estimate the amount of $KMnO_4$ that was in excess and subtract this from the result.
   (B) Repeat the titration using the reserved remaining sample.
   (C) Obtain data for this point from another lab group.
   (D) Prepare a graph of the data without this point and then read the estimated value from the graph.

Answers

1. The correct answer is c. The rate of the reaction drops when the enzyme no longer has a maximum number of substrate molecules to interact with. Above the maximum substrate concentration, the rate will not be increased by adding more substrate; the enzyme is already working as fast as it can. An enzyme can catalyze a certain number of reactions per second, and if there is not sufficient substrate present for it to work at its maximum velocity, the rate will decrease. Therefore, to *keep* the enzyme working at its maximum, you must add more substrate.

2. The correct answer is e. H_2SO_4 lowers the pH so that the globular shape of the protein is altered. The active site is distorted to the point that the enzyme no longer functions.

3. The correct answer is b. You should not throw away any remaining samples until all titrations have been completed and the data analyzed. There is sufficient sample in this lab to repeat the titration. Any other response would provide a less accurate data point.

Laboratory 3: Mitosis and Meiosis

YOU MUST KNOW

- The events of mitosis and meiosis in plant and animal cells.
- How mitosis and meiosis differ.
- How to calculate the relative duration of each stage of mitosis.
- The roles of segregation, independent assortment, and crossing over in generating genetic variation.
- How to calculate map distance from experimental data.

Overview of the Lab:

This laboratory involves two distinct activities:

1. **Observing Mitosis**
   In this laboratory, you study and sketch the events of cell division in either plant or animal cells, using a microscope slide of cells arrested at various stages in the process of division. The data collected is used to calculate the relative duration of the phases of mitosis.
2. **Meiosis Simulation and Study of Crossing Over**
   Meiosis is studied using beads or other models to simulate the process. You will also observe the results of crossing over in a fungus, *Sordaria fimicola*.

Hints and Review

- Go back to Topic 2, Chapter 12 to review mitosis and the cell cycle. Go to Topic 3, Chapter 13 to review the important elements of meiosis.
- In this lab you calculate the time for each stage of mitosis. Here's how to do that:

> The average time for onion root tip cells to complete the cell cycle is 24 hours = 1440 minutes.
>
> To calculate the time for each stage:
>
> **% of cells in the stage × 1440 minutes = number of minutes in the stage**

Meiosis and Crossing Over in Sordaria

- When the growing filaments of two haploid strains of *Sordaria* that produce spores of different colors meet, fertilization occurs and zygotes form.
- Meiosis occurs within fruiting bodies to form four haploid *ascospores*, spores contained in *asci* (special sacs).
- Then one mitotic division doubles the number of ascospores to eight.

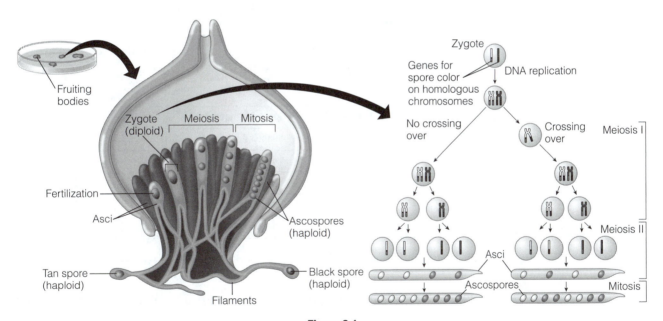

Figure 3.1

Hints and Review

- The number of **map units** between two genes or between a gene and the centromere is calculated by determining the percentage of recombinants that result from crossing over. The greater the frequency of crossing over, the greater the map distance.

- Calculate the percent of crossovers by dividing the number of crossover asci (These are the ones with spores arranged 2:2:2:2 or 2:4:2) by the total number of asci × 100.
- To calculate the map distance, divide the percent of crossover asci by 2. The percent of crossover asci is divided by 2 because only half of the spores in each ascus are the result of crossing over.

Questions

1. Which of the following statements is correct?
    - (A) Crossing over occurs in prophase I of meiosis and metaphase of mitosis.
    - (B) DNA replication occurs once prior to mitosis and twice prior to meiosis.
    - (C) Both mitosis and meiosis result in daughter cells identical to the parent cells.
    - (D) Karyokinesis occurs once in mitosis and twice in meiosis.
    - (E) Synapsis occurs in prophase of mitosis.

2. The cell cycle in a certain cell type has a duration of 16 hours. The nuclei of 660 cells showed 13 cells in anaphase. What is the approximate duration of anaphase in these cells?
    - (A) 2 minutes
    - (B) 13 minutes
    - (C) 19 minutes
    - (D) 32 minutes
    - (E) 647 minutes

3. A group of asci formed from crossing light-spored *Sordaria* with dark-spored produced the following results:

Number of Asci Counted	Spore Arrangement
7	4 light/4 dark spores
8	4 dark/4 light spores
3	2 light/2 dark/2 light/2 dark spores
4	2 dark/2 light/2 dark/2 light spores
1	2 dark/4 light/2 dark spores
2	2 light/4 dark/2 light spores

How many of these asci contain a spore arrangement that resulted from crossing over?
    - (A) 3
    - (B) 7
    - (C) 8
    - (D) 10
    - (E) 15

4. From this small sample, calculate the map distance between the gene and centromere.
    - (A) 10 map units
    - (B) 20 map units
    - (C) 30 map units
    - (D) 40 map units

Answers

1. The correct response is d. There are two nuclear divisions in meiosis, and only one in mitosis. Crossing over occurs only in meiosis; it does not occur at all in mitosis. Replication occurs only once in preparation for both mitosis and meiosis. The daughter cells of mitosis are identical to the parent cell, but in meiosis the daughter cells have only one of each homologous chromosome pair. Synapsis occurs only in prophase I of meiosis.

2. The correct response is c. Correct calculation:

   16 hours x 60 minutes/hour = 960 minutes.

   13 cells in anaphase ÷ 660 total cells = 2% of the cells in anaphase

   2% of 960 is approximately 19 minutes.

3. The correct response is d. Remember that if crossing over does not occur, the arrangement of spores will be 4 light and 4 dark. All other combinations are the result of crossing over.

4. The correct response is b. Map distance = number of crossovers divided by total number of asci $\times$ 100 divided by 2.

Laboratory 4: Plant Pigments and Photosynthesis

> **YOU MUST KNOW**
>
> - The equation for photosynthesis and understand the process of photosynthesis.
> - The principles of chromatography and how to calculate R_f values.
> - The relationship between light wavelength or intensity and photosynthetic rate.
> - How to determine the rate of photosynthesis and then be able to design a controlled experiment to test the effect of some variable factor on photosynthesis.

Overview of the Lab

There are two different exercises in this laboratory. In the first one, you use paper chromatography to separate the pigments in chlorophyll. In the second activity, you measure the rate of photosynthesis by evaluating the change in color that occurs when DPIP is reduced over time.

▌ In photosynthesis, plant cells convert light energy into chemical energy that is stored in sugars and other organic compounds.

> ▌ The equation for photosynthesis is
>
> $$6\,H_2O + 6\,CO_2 \rightarrow C_6H_{12}O_6 + 6\,O2$$

▌ **Chlorophyll** is a term often used to refer to the green material in leaves, but it is actually composed of several different pigments. These pigments can be separated by chromatography.

▌ When chromatography of photosynthetic pigments from spinach is done, you will get four or five distinct bands. These pigments include *carotene, xanthophyll, chlorophyll a, and chlorophyll b.*

▌ **Paper chromatography** is a technique used to separate a mixture into its component molecules. The molecules migrate, or move up the paper, at different rates because of differences in solubility, molecular mass, and hydrogen bonding with the paper.

▌ In paper chromatography the pigments are dissolved in a solvent that carries them up the paper. To separate the pigments of the chloroplasts, you must use an organic solvent.

▌ If you did a number of chromatographic separations, each for a different length of time, the pigments would migrate a different distance on each run. However, the migration of each pigment relative to the migration of the solvent would not change. This migration of pigment relative to migration of solvent is expressed as a constant, R_f (Reference front). It can be calculated by using the formula

> $$R_f = \frac{\text{distance pigment migrated}}{\text{distance solvent front migrated}}$$

▌ You should be able to calculate the R_f value from a pigment, given experimental results. There is a practice problem at the end of the lab review section.

▌ Pigments absorb in the visible range of the spectrum (380–760 nm). When chloroplasts are exposed to light, electrons from each photosystem will be boosted to a higher energy level, and this energy used to produce ATP and to reduce $NADP^+$ to NADPH.

▌ In this experiment, DPIP is substituted for $NADP^+$ and will be reduced when light strikes the chloroplasts.

▌ $DPIP_{ox}$ is blue. $DPIP_{re}$ is colorless.

▌ See Figure 4.1 for the experimental design.

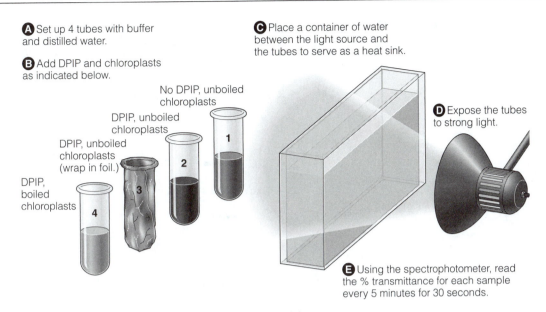

A Set up 4 tubes with buffer and distilled water.

B Add DPIP and chloroplasts as indicated below.

No DPIP, unboiled chloroplasts

DPIP, unboiled chloroplasts

DPIP, unboiled chloroplasts (wrap in foil.)

DPIP, boiled chloroplasts

C Place a container of water between the light source and the tubes to serve as a heat sink.

D Expose the tubes to strong light.

E Using the spectrophotometer, read the % transmittance for each sample every 5 minutes for 30 seconds.

Figure 4.1

Design of the Experiment II

▌ After illuminating the reaction tubes as shown above, we will use the spectrophotometer to measure the percentage of transmittance at wavelength 605 nm.

▌ If DPIP is in an oxidized state, it will be blue, and the percentage of light transmitted will be low. If, on the other hand, chlorophyll's electrons have been excited and move on to reduce the DPIP, the sample will become progressively paler, allowing more light energy to pass through the sample. We can measure this change over time until the sample has been completely reduced, is almost colorless, and the percentage of transmittance is very high.

▌ For this experiment, one tube (the blank) will contain all the solutions used in the reaction except the DPIP. Since the blank contains chloroplasts, it will be green; you will use this tube to calibrate the machine. The other tubes will be experimental and will contain either boiled or unboiled chloroplasts.

▌ The boiled chloroplasts and chloroplasts kept in the dark will show minimal change in transmittance of light.

▌ Another way you could measure the rate of photosynthesis would be to measure the rate of oxygen production. This can be done by submerging a leaf and then measuring the accumulated gas over time, for example.

▌ If you have a technique that will measure the rate of photosynthesis, such as DPIP reduction or oxygen accumulation, you could then design an experiment to test a variable such as exposure to different light intensities (vary the distance from a bulb, or vary the wattage of the bulb) or exposure to different wavelengths of light (use colored cellophanes or filters).

Questions

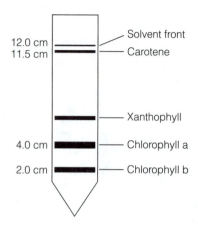

Figure 4.2

1. What is the R_f value for carotene calculated from the chromatogram above?
   (A) 1.09
   (B) 0.17
   (C) 0.96
   (D) 0.33
   (E) 0.50

2. For the chromatogram shown above,
   (A) The R_f for chlorophyll *b* can be determined by dividing the distance the green pigment migrated by the distance the solvent front migrated.
   (B) The R_f value of chlorophyll *b* will be higher than the R_f value for chlorophyll *a*.
   (C) The molecules of xanthophyll are not easily dissolved in this solvent, and thus are probably larger in mass than the chlorophyll *b* molecules.
   (D) If this same chromatogram were set up and run for twice as long, the R_f values would be twice as great for each pigment.

3. What is the role of DPIP in this experiment?
   (A) It mimics the action of chlorophyll by absorbing light energy.
   (B) It serves as an electron donor and blocks the formation of NADPH.
   (C) It is an electron acceptor and is reduced by electrons from chlorophyll.
   (D) It is bleached in the presence of light and can be used to measure light levels.

4. Some students were not able to get many data points in this experiment because the solution went from blue to colorless in only 5 minutes for the unboiled chloroplasts exposed to light. What modification to the experiment do you think would be most likely to provide better results?
   (A) Increase the number of drops of chloroplasts used from three to five.
   (B) Double the volume of DPIP so that the solution has a lower initial transmittance.
   (C) Modify the blank so that the initial transmittance is higher.
   (D) Use fresher spinach and prepare the chloroplast solution during the laboratory procedure.
   (E) Change the wavelength at which readings are taken.

5. If a student performed this experiment and got a flat curve, showing very little change in percent transmittance over time, which of these would be the most plausible explanation?
   (A) The DPIP was too pale at the beginning of the experiment.
   (B) The chloroplast solution was too concentrated.
   (C) The experimenter used chloroplasts that were damaged and could not respond to light.
   (D) The blank was not properly used to calibrate the spectrophotometer.

Answers

1. The correct answer is c. To calculate R_f remember that you take the distance the pigment migrated and divide this by the distance the solvent front migrated.

$$R_f = \frac{\text{distance pigment migrates}}{\text{distance solvent front migrates}} = \frac{11.5}{12} = 0.96.$$

2. The correct answer is a. Answer d is wrong because R_f values are relative and depend not on the total distance migrated, but on the relative distance migrated.

3. The correct answer is c. DPIP is reduced in this experiment as it receives high-energy electrons from chlorophyll.

4. The correct answer is b. When DPIP becomes colorless very rapidly either there are too many cholorplasts or the concentration of DPIP is so dilute that the quantity can be reduced rapidly.

5. The correct answer is c. A flat curve indicates that no DPIP is being reduced. This must mean the chloroplasts are not functioning.

Laboratory 5: Cell Respiration

YOU MUST KNOW

- The equation for cellular respiration.
- How a respirometer works.
- The relationship between movement of water in a respirometer and cellular respiration.
- The effect of temperature and increased metabolic activity on respiration.
- How to calculate the rate of respiration.

Overview of the Lab

In this experiment respirometers are used to measure the rate of cellular respiration in pea seeds. You investigate the rate of respiration in both germinating and nongerminating peas at two different temperatures.

Hints and Review

▌ The equation for cellular respiration is

$$C_6H_{12}O_6 + 6\,O2 \rightarrow 6\,H_2O + 6\,CO_2 + ATP$$

▌ How can the rate of cellular respiration be measured? When you study the equation for cellular respiration, you will see that there are at least three ways:

1. Measure the amount of glucose consumed.
2. Measure the amount of oxygen consumed.
3. Measure the amount of carbon dioxide produced.

In this experiment, a *respirometer* is used to measure the amount of oxygen consumed.

▌ A **respirometer** is an air-tight chamber except for one opening for gases to enter or leave. Potassium hydroxide (KOH) soaks a cotton ball and will combine with the CO_2 produced by the organism (peas in our AP Lab). A solid precipitate forms. Since CO_2 and O_2 are produced and consumed in equal amounts, any changes in the sealed container (assuming temperature and pressure remain constant) will be caused by a change in gas volume due to cellular respiration. With the CO_2 being removed as a solid precipitate, then it is O_2 that is consumed, lowering pressure within the respirometer, and allowing water to enter the pipette. Study Figure 5.1 to see the components of a respirometer.

▌ A **seed** contains an embryo plant and a food supply surrounded by a seed coat.

▌ The *germinating* (sprouting) seeds will show a higher rate of respiration than the *nongerminating* seeds. However, be sure that you know these dry, nongerminating seeds are not dead, but **dormant**. They can be stored for years and, when soaked in water, will germinate.

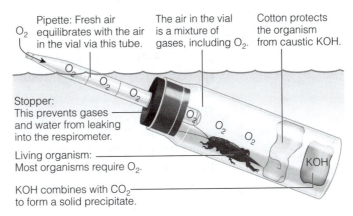

Pipette: Fresh air equilibrates with the air in the vial via this tube.

O_2

The air in the vial is a mixture of gases, including O_2.

Cotton protects the organism from caustic KOH.

Stopper: This prevents gases and water from leaking into the respirometer.

Living organism: Most organisms require O_2.

KOH combines with CO_2 to form a solid precipitate.

KOH

Figure 5.1

▌ There are actually two variables in this experiment. Besides germination versus nongermination, the second variable is the effect of temperature on metabolic rate. You run the two experiments simultaneously, with vials exposed to room temperature water (25°C), and vials that are exposed to cooler temperatures (10°C) by adding ice to the surrounding water.

▌ Notice that each experimental setup includes a vial of germinating peas; a vial of dry, nongerminating peas; and a vial of glass beads alone to act as a *control* for this experiment. The control will compensate for any change in pressure or temperature.

▌ Students are often confused by the difference between a *control* (the glass beads in this experiment) and *factors that are held constant*. In this experiment, you hold constant the volumes within the containers, the number of peas, and the temperature of the water for each set of three vials.

▌ Figure 5.1 shows a graph of typical results from this experiment.

▌ Be able to calculate the rate for each condition. Go back to Lab 2 on Enzymes to review this lesson.

Questions

1. What is the rate of oxygen consumption in germinating corn at 12°C as shown in the graph below?
   (A) 0.08 ml/min
   (B) 0.04 ml/min
   (C) 0.8 ml/min
   (D) 0.6 ml/min
   (E) 1.00 ml/min

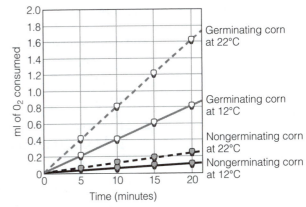

Figure 5.2

2. Using the figure above, which of the following is a true statement based on the data?
   (A) The amount of oxygen consumed by germinating corn at 22°C is approximately twice the amount of oxygen consumed by germinating corn at 12°C.
   (B) The rate of oxygen consumption is the same in both germinating and nongerminating corn during the initial time period from 0 to 5 minutes.
   (C) The rate of oxygen consumption in the germinating corn at 12°C at 10 minutes is 0.4 ml O_2/minute.
   (D) The rate of oxygen consumption is higher for nongerminating corn at 12°C than at 22°C.
   (E) If the experiment were run for 30 minutes, the rate of oxygen consumption would decrease.

3. Which of the following conclusions is supported by the data shown on the graph?
   (A) The rate of respiration is higher in nongerminating seeds than in germinating seeds.
   (B) Nongerminating seeds are not alive and show no difference in rate of respiration at different temperatures.
   (C) The rate of respiration in the germinating seeds would have been higher if the experiment were conducted in sunlight.
   (D) The rate of respiration increases as the temperature increases in both germinating and nongerminating seeds.
   (E) The amount of oxygen consumed could be increased if pea seeds were substituted for corn seeds.

4. What is the role of KOH in this experiment?
   (A) It serves as an electron donor to promote cellular respiration.
   (B) As KOH breaks down, the oxygen needed for cellular respiration is released.
   (C) It serves as a temporary energy source for the respiring organism.
   (D) It binds with carbon dioxide to form a solid, preventing CO_2 production from affecting gas volume.
   (E) Its attraction for water will cause water to enter the respirometer

Answers

1. The correct answer is b. To calculate this, it is easiest to find the change in y at 10 minutes (0.4 ml – 0 ml = 0.4) and divided by the change in x (10 minutes – 0 minutes = 10 minutes). 0.4 ml/10 minutes = 0.04 ml/min.

2. The correct answer is a. Study the graph carefully to see that at 10 minutes the 22°C germinating corn consumed 0.8 ml of oxygen, while the 12°C germinating corn consumed 0.04 ml of oxygen.

3. The correct answer is d. This is the only statement that is supported by information provided on the graph.

4. The correct answer is d. As carbon dioxide is released, it is removed from the air in the vial by this precipitation. Since oxygen is being consumed during cellular respiration, the total gas volume in the vial decreases. This causes pressure to decrease inside the vial, and water begins to enter the pipette.

Laboratory 6: Molecular Biology

YOU MUST KNOW

- The principles of bacterial transformation, including how plasmids are engineered and taken up by cells.
- Factors that affect transformation efficiency.
- The function of restriction enzymes and their role in genetic engineering.
- How gel electrophoresis separates DNA fragments.
- How to use a standard curve to determine the size of unknown DNA fragments.

Overview of the Lab

This laboratory is actually two different exercises. In the first part of the lab, **Bacterial Transformation**, you use antibiotic-resistance plasmids to transform *Escherichia coli*. In this part of the lab, a gene for resistance to the antibiotic ampicillin is introduced into a strain of *Escherichia coli* that is killed by ampicillin. If the susceptible bacteria incorporate the foreign DNA, they will become ampicillin resistant.

In the second part, **Gel Electrophoresis**, you separate fragments of DNA for further analysis.

Key Concepts I: Bacterial Transformation

▐ Genetic **transformation** occurs when a host organism takes in foreign DNA and expresses the foreign gene.

▐ Bacterial cells have a single main chromosome and circular DNA molecules called **plasmids** which carry genetic information. All of the genes required for basic survival and reproduction are found in the single chromosome.

▐ **Plasmids** are circular pieces of DNA that exist outside the main bacterial chromosome and carry their own genes for specialized functions including resistance to specific drugs. In genetic engineering, plasmids are one means used to introduce foreign genes into a bacterial cell. To understand how this might work, consider the plasmid in Figure 6.1.

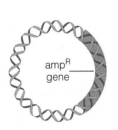

Figure 6.1

▐ Some plasmids have the amp^R gene, which confers resistance to the antibiotic ampicillin. *E. coli* cells containing this plasmid, termed "**+amp^R**" cells, can survive and form colonies on LB agar that has been supplemented with ampicillin.

- In contrast, cells lacking the amp^R plasmid, termed **"–amp^R"** cells, are sensitive to the antibiotic, which kills them. An ampicillin-sensitive cell (–amp^R) can be transformed to an ampicillin-resistant (+amp^R) cell by its uptake of a foreign plasmid containing the amp^R gene.

- **Competent cells** are cells that are most likely to take up extracellular DNA. Competent cells are in logarithmic growth, and chemical conditions are modified to induce the uptake of DNA. Study Figure 6.2 below to review the lab procedure used to prepare competent cells and get them to take up the amp^R plasmids.

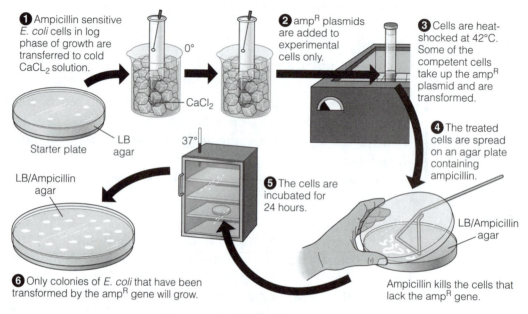

Figure 6.2

- Study Figure 6.3. It shows the expected results in this experiment.

- If there is no ampicillin in the agar, *E. coli* will cover the plate with so many cells it is called a "lawn" of cells (A + C).

- Only transformed cells can grow on agar with ampicillin. Since only some of the cells exposed to the amp^R plasmids will actually take them in, only some cells will be transformed. Thus, you will see only individual colonies on the plate (D).

- If none of the sensitive *E. coli* cells have been transformed, nothing will grow on the agar with ampicillin (B).

- **Restriction enzymes** or endonucleases are bacterial enzymes that will cut DNA at specific DNA sequences known as **recognition sites**. Often the enzymes cut the DNA so that the ends are single-stranded "sticky ends." A **gene of interest** (such as antibiotic resistance) can be introduced into a plasmid.

- Here are the general steps used to introduce a gene of interest into bacteria:
  1. Both the gene of interest and the plasmid are cut with the *same* restriction enzyme, so they have the same sticky ends.
  2. DNA ligase is used to anneal and seal the sticky ends.

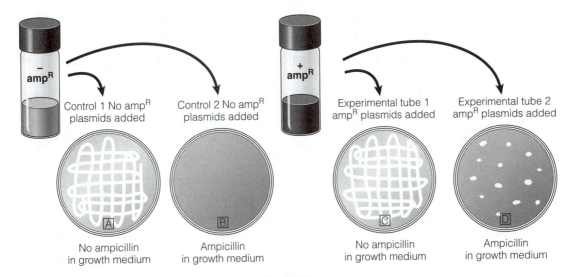

Control 1 No amp^R
plasmids added

Control 2 No amp^R
plasmids added

Experimental tube 1
amp^R plasmids added

Experimental tube 2
amp^R plasmids added

No ampicillin
in growth medium

Ampicillin
in growth medium

No ampicillin
in growth medium

Ampicillin
in growth medium

Figure 6.3

3. The recipient cells are transformed with the engineered plasmid.

4. Colonies carrying the plasmid are isolated.

▌ How do we know that transformation has been successful?

1. Use a **selection gene**, such as for antibiotic resistance. Only those cells that have incorporated the plasmid will have antibiotic resistance.

2. Use a **reporter gene** such as GFP (green fluorescent protein). Transformed cells will glow!

▌ **Transformation efficiency** is the number of transformed cells per microgram of the plasmid. High transformation efficiencies require cells that are in log phase of growth, suspended in ice-cold calcium chloride, have a rapid heat shock (this makes the membrane permeable to the plasmid), and plasmids that are not too large.

Key Concepts II: Restriction Enzyme Cleavage of DNA and Gel Electrophoresis

▌ **Gel electrophoresis** is a procedure that separates molecules on the basis of their rate of movement through a gel under the influence of an electrical field.

▌ The direction of movement is affected by the charge of the molecules, and the rate of movement is affected by their size and shape, the density of the gel, and the strength of the electrical field.

▌ DNA is a negatively charged molecule, so it will move toward the positive pole of the gel when a current is applied. When DNA has been cut by restriction enzymes, the different-sized fragments will migrate at different rates. Because the smallest fragments move the most quickly, they will migrate the farthest during the time the current is on. Keep in mind that the length of each fragment is measured in number of DNA base pairs.

▌ In your laboratory you will be given three samples of DNA obtained from a virus, the bacteriophage lambda. One sample will be uncut DNA, one will be incubated with the restriction enzyme HindIII, and one will be incubated with EcoRI. You will separate the fragments of DNA by electrophoresis, stain the DNA for visualization, and determine the fragment sizes formed in the EcoRI digest.

■ The samples are loaded into wells in the gel, and then electricity is applied. The DNA fragments will migrate. *Remember this!*

1. *DNA is negatively charged and will migrate toward the positive pole.*

2. *Smaller fragments of DNA will migrate faster than larger fragments.*

■ After electrophoresis, you must stain the DNA for visualization. You submerge the entire gel in methylene blue, which will bind to the DNA. You then rinse the gel repeatedly with water so that the dye washes off the gel. The DNA will appear as blue bands that are easily seen when a light is passed through the gel.

■ Each fragment of DNA is a particular number of nucleotides, or base pairs, long. When researchers want to determine the size of DNA fragments produced with particular restriction enzymes, they run the unknown DNA alongside DNA with known fragment sizes. The known DNA acts as a **marker**. In your laboratory, the DNA that has been cut with HindIII is the marker; you will use it to help you determine the fragment sizes in the EcoRI digest.

■ Figure 6.4 shows the results of electrophoresis. Semilog paper is used to plot the results of the HindIII digest. Since its fragments sizes are known, this is the *standard curve*. It can now be used to determine the other fragment sizes from DNA I and DNA II by interpolation.

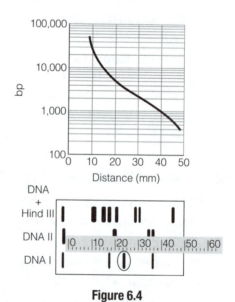

Figure 6.4

Questions

1. How many base pairs is the fragment circled in Figure 6.4?
   (A) 350
   (B) 22
   (C) 2200
   (D) 3500

2. Which of the following statements is correct?
   (A) Longer DNA fragments migrate farther than shorter fragments.
   (B) Migration distance is inversely proportional to the fragment size.
   (C) Positively charged DNA migrates more rapidly than negatively charged DNA.
   (D) Uncut DNA migrates farther than DNA cut with restriction enzymes.

   Here is a plasmid with restriction sites for BamHI and EcoRI. Several restriction digests were done using these two enzymes either alone or in combination. Use the figure to answer questions 3-4. **Hint:** Begin by determining the number and size of the fragments produced with each enzyme. "kb" stands for kilobases, or thousands of base pairs.

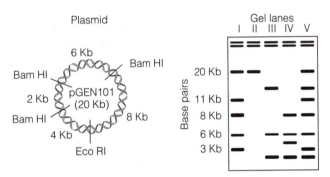

Figure 6.5

3. Which lane shows a digest with BamHI only
   (A) I
   (B) II
   (C) III
   (D) IV
   (E) V

4. Which lane shows a digest with both BamHI and EcoRI?
   (A) I
   (B) II
   (C) III
   (D) IV
   (E) V

In a molecular biology laboratory, a student obtained competent *E. coli* cells and used a common transformation procedure to induce the uptake of plasmid DNA with a gene for resistance to the antibiotic kanamycin. The results shown here were obtained.

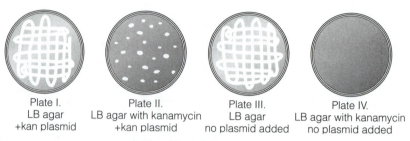

Plate I.	Plate II.	Plate III.	Plate IV.
LB agar	LB agar with kanamycin	LB agar	LB agar with kanamycin
+kan plasmid	+kan plasmid	no plasmid added	no plasmid added

Figure 6.6

5. On which petri dish do only transformed cells grow?
   (A) Plate I
   (B) Plate II
   (C) Plate III
   (D) Plate IV

6. Which of the plates is used as a control to show that nontransformed *E. coli* will not grow in the presence of kanamycin?
   (A) Plate I
   (B) Plate II
   (C) Plate III
   (D) Plate IV

7. If a student wants to verify that transformation has occurred, which of the following procedures should she use?
   (A) Spread cells from Plate I onto a plate with LB agar; incubate.
   (B) Spread cells from Plate II onto a plate with LB agar; incubate.
   (C) Repeat the initial spread of $-kan^R$ cells onto Plate IV to eliminate possible experimental error.
   (D) Spread cells from Plate II onto a plate with LB agar with kanamycin; incubate.
   (E) Spread cells from Plate III onto a plate with LB agar and also onto a plate with LB agar with kanamycin; incubate.

8. During the course of an *E. coli* transformation laboratory, a student forgot to mark the culture tube that received the kanamycin-resistant plasmids. The student proceeds with the laboratory because he thinks that he will be able to determine from his results which culture tube contained cells that may have undergone transformation. Which plate would be most likely to indicate transformed cells?
   (A) A plate with a lawn of cells growing on LB agar with kanamycin
   (B) A plate with a lawn of cells growing on LB agar without kanamycin
   (C) A plate with 100 colonies growing on LB agar with kanamycin
   (D) A plate with 100 colonies growing on LB agar without kanamycin

Answers

1. The correct answer is d. There are approximately 3500 base pairs in the circled fragment.

2. The correct answer is b. Small fragments migrate farther than large fragments. All DNA is negatively charged and so moves toward the positive electrode.

3. The correct answer is C.

4. The correct answer is D, Lane IV.

5. The correct answer is b, Plate II.

6. The correct answer is d, Plate IV.

7. The correct answer is d.

8. The correct answer is c.

Laboratory 7: Genetics of Organisms

<table>
<tr><td>

YOU MUST KNOW

- What is meant by degrees of freedom, critical value, the null hypothesis, and how to do chi-square analysis of data.
- How to use data to determine the mode of transmission and genetic make-up of the parents.
- How to use a Punnett square to verify your conclusions.

</td></tr>
</table>

Overview of the Lab

Over a period of 4 weeks three generations of flies are studied to determine the mode of inheritance of one or two mutations. The goal is determine the mutation, and whether it is transmitted on an autosome or on a sex chromosome. Chi-square analysis is used to evaluate the data gathered. While your class may have used yeast or *Brassica* or other organisms to collect this data, this review looks at fruit flies.

Hints and Review

▌ The **wild type** is the nonmutant form of the trait, such as normal wing shape and red eyes.

▌ To make this a little easier, this lab will use one of three modes of inheritance:

1. **Monohybrid:** A single contrasting pair of characteristics is involved, such as red eyes versus sepia eyes, or normal wings versus vestigial wings. In the F_2 generation, results of 3:1 are expected.

2. **Dihybrid:** Two pairs of contrasting characteristics are considered simultaneously, and in the F_2 generation, results of 9:3:3:1 are expected.

3. **Sex-linked:** If the mutant characteristic is associated with the X chromosome, the ratios obtained can be explained only if linkage to sex in considered.

▌ Here is a look at the procedure used in this lab. Note that you are not told what the mutations are, but have to determine them through observation.

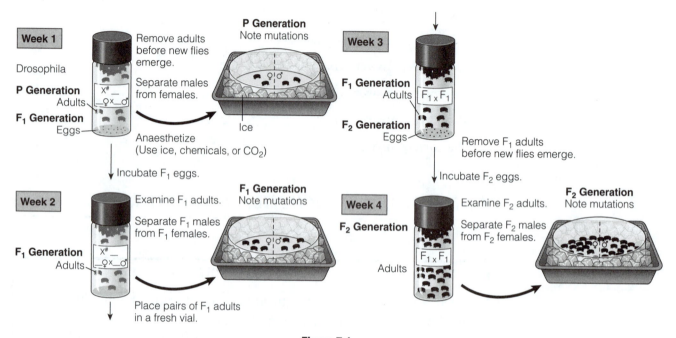

Figure 7.1

Chi-Square Analysis of Data

Let's review how to do this. Assume you obtained the following results in the F1:

F_1 RESULTS	OBSERVED PHENOTYPE AND NUMBERS
	Red eyes
♂ MALES	12
♀ FEMALES	8

F_2 RESULTS	OBSERVED PHENOTYPE AND NUMBERS	
	Red eyes	Sepia Eyes
♂ MALES	19	4
♀ FEMALES	12	9

Figure 7.2

▌ From the data presented, you can deduce that the F_1 cross was between individuals heterozygous for eye color: $+se$ x $+se$ (+ = red; se = sepia).

- The student kept data showing both males and females with the trait, because the unknown trait might be sex-linked. The data do not support a sex-linked trait, but do support an autosomal trait; thus the data are merged so that only the trait is considered in the cross.

- From this conclusion, you could write the following hypothesis: *If the parents are heterozygous for eye color, there will be a 3:1 ratio of red eyes to sepia eyes in the offspring.* Do your results support this hypothesis?

Calculating Chi-Square

The formula for chi-square is

$$\chi^2 = \text{the sum of } \frac{(o - e)^2}{e}$$

where:
o = observed number of individuals
e = expected number of individuals

- The actual results of an experiment are unlikely to match the expected results precisely. But how great a variance is significant? One way to decide is to use the chi-square (χ^2) test. This analytical tool tests the validity of a **null hypothesis**, which states that there is no statistically significant difference between the observed results of your experiment and the expected results. When there is little difference between the observed results and the expected results, you obtain a very low chi-square value; your hypothesis is supported.

Using the Chi-Square Critical Values Table

The chi-square critical values table provides two values that you need to calculate chi-square:

- **Degrees of freedom.** This number is one less than the total number of classes of offspring in a cross. In a monohybrid cross, such as our Case 1, there are two classes of offspring (red eyes and sepia eyes). Therefore, there is just one degree of freedom. In a dihybrid cross, there are four possible classes of offspring, so there are three degrees of freedom.

- **Probability.** The probability value (p) is the probability that a deviation as great as or greater than each chi-square value would occur simply by chance. Many biologists agree that deviations having a chance probability greater than 0.05 (5%) do not support the null hypothesis. Therefore, when you calculate chi-square, you should consult the table for the p value in the 0.05 row.

Critical Values Table

Probability (p)	Degrees of Freedom (df)				
	1	2	3	4	5
0.05	3.84	5.99	7.82	9.49	11.1
0.01	6.64	9.21	11.3	13.2	15.1
0.001	10.8	13.8	16.3	18.5	20.5

Steps to determining Chi-Square:

1. Complete the following chart.

Observed Phenotypes (o)	Expected(e)	(o-e)	(o-e)^2	$\dfrac{(o - e)^2}{e}$
Red Eyes 31	33	2	4	4/33 = 0.12
Sepia Eyes 13	11	2	4	4/11 = 0.36
				Total = χ^2 = 0.48

2. **Determine the degrees of freedom.** This is the number of categories (red eyes or sepia eyes) minus one. For this data, the number of degrees of freedom is 1.

3. **Find the probability (p) value for 1 degree of freedom in the 0.05 row.** This is the **critical value.** For this data, the critical value = 3.84.

4. **Accept or reject the null hypothesis.** Since the χ^2 value for this data is less than the critical value, you will accept the null hypothesis. This then supports your working hypothesis, *If the parents are heterozygous for eye color, there will be a 3:1 ratio of red eyes to sepia eyes in the offspring.*

▌ Don't forget this little nugget: **If the chi-square value is greater than the critical value, the null hypothesis is rejected,** and you must consider reasons for this variation, such as errors in sample size or data collection.

> **Hint from the Graders:** There was a chi-square problem on the AP Biology Exam a few years ago, and we found that a common error was that some students took the square root of the value they got for χ^2. Don't make this mistake! χ^2 is just the shorthand for the name of this mathematical technique, *chi-square.*

Questions

1. You have been given a vial contain a red-eyed male with normal wings and a red-eyed female with normal wings. These are the F1 generation. After 2 weeks, you collect the offspring from this pair and obtain the results shown in the following table:

F$_2$ RESULTS	OBSERVED PHENOTYPE AND NUMBERS			
	Red eyes normal wings	Red eyes no wings	Sepia eyes normal wings	Sepia eyes no wings
♂MALES	48 	13 	16 	4 
♀FEMALES	50 	9 	10 	10 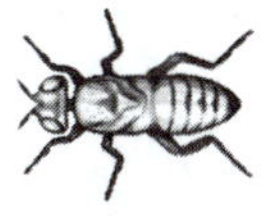

Figure 7.3

On the basis of the results shown in the table, which statement is most likely true?
   (A) The genes for red eyes and normal wings are linked.
   (B) The gene for no wings is sex-linked.
   (C) The gene for red eyes and the gene for no wings are both dominant.
   (D) The gene for eye color is inherited independently of the gene for wings.
   (E) The F$_1$ mates were both homozygous for both eye color and wings.

2. Based on the hypothesis that this is a dihybrid cross, with the two genes unlinked, calculate χ^2 using the data in the table of observed phenotypes.
   (A) 6.04
   (B) 7.81
   (C) 4.977
   (D) 24.0

3. Compare the chi-square value obtained in Question 3 with the Critical Values Table on the previous page for $p = 0.05$. Which of the following statements would be true?
   (A) Since the calculated value for chi-square is less than 7.82, the results support the hypothesis that the parents are heterozygous for two unlinked traits.
   (B) Since the calculated value for chi-square is less than 7.82, the results support the hypothesis that eye color and wings are linked.
   (C) Since the calculated value for chi-square is less than 7.82, the results are inconclusive. The experiment should be repeated.
   (D) Since the calculated value for chi-square is so large, the variations from the experiment are excessive. The experiment should be repeated.

Answers

1. The correct answer is d. There is no evidence for any type of linkage, since both males and females show the traits in approximately equal proportions, and eye color and wings appear to sort independently. If the parents were homozygous for these traits, the offspring would not show different phenotypes from both parents.

2. The correct answer is c. This is obtained by using the formula $\chi^2 =$ the sum of $\frac{(o - e)^2}{e}$.

	o	e	$o - e$	$\frac{(o - e)^2}{e}$
Red eyes/normal wings	98	90	8	0.7111
Red eyes/no wings	22	30	8	2.1333
Sepia eyes/normal wings	26	30	4	0.5333
Sepia eyes/no wings	14	10	4	1.6
				$\chi^2 = 4.977$

3. The correct answer is a. The expected results are based on obtaining a 9:3:3:1 ratio from two heterozygous parents. There are 3 degrees of freedom. Since the chi-square value is below 7.82, the results support the null hypothesis.

Laboratory 8: Population Genetics and Evolution

> ### YOU MUST KNOW
>
> - The Hardy-Weinberg equation and be able to use it to determine the frequency of alleles in a population.
> - Conditions for maintaining Hardy-Weinberg equilibrium.
> - How genetic drift, selection and the heterozygote advantage affect Hardy-Weinberg equilibrium.

Overview of the Lab

This lab used a class as a sample population in a simulation of mating events to generate offspring, and looked at conditions that would affect the genetic makeup of subsequent generations. Given the initial frequency of two forms of an allele (A and a), changes in gene frequency were evaluated after the simulated passage of several generations.

Hints and Review

▌ **The Hardy-Weinberg Law of Genetic Equilibrium** provides a mathematical model for studying evolutionary changes in allelic frequency within a population. It predicts that the frequency of alleles and genotypes in a population will remain constant from generation to generation if the population is stable and in genetic equilibrium.

> **Five conditions are required in order for a population to remain at Hardy-Weinberg equilibrium:**
>
> 1. A large breeding population
> 2. Random mating
> 3. No change in allelic frequency due to mutation
> 4. No immigration or emigration
> 5. No natural selection

▌ **A Large Breeding Population** helps to ensure that chance alone does not disrupt genetic equilibrium. In a small population, only a few copies of a certain allele may exist. If for some chance reason the organisms with that allele do not reproduce successfully, the allelic frequency will change. This random, nonselective change is what happens in **genetic drift** or a bottleneck event.

- **Random Mating**—In a population at equilibrium, mating must be random. In assortative mating, individuals tend to choose mates similar to themselves; for example, large blister beetles tend to choose mates of large size and small blister beetles tend to choose small mates. Though this does not alter allelic frequencies, it results in fewer heterozygote individuals than you would expect in a population where mating is random.

- **No Change in Allelic Frequency Due to Mutation**—Any mutation in a particular gene could change the balance of alleles in the gene pool. Mutations may remain hidden in large populations for a number of generations, but may show more quickly in a small population.

- **No Immigration or Emigration**—No new alleles can come into the population, and no alleles can be lost. Both immigration and emigration can alter allelic frequency.

- **No Natural Selection**—No alleles are selected over other alleles. If selection occurs, those alleles that are selected for will become more common. For example, if resistance to a particular herbicide allows weeds to live in an environment that has been sprayed with that herbicide, the allele for resistance may become more frequent in the population.

- To estimate the frequency of alleles in a population, we can use the Hardy-Weinberg equation. According to this equation:

---

p = the frequency of the dominant allele (represented here by A)

q = the frequency of the recessive allele (represented here by a)

For a population in genetic equilibrium:

$p + q = 1.0$ (The sum of the frequencies of both alleles is 100%.)

$$(p + q)^2 = 1$$

so

$$p^2 + 2pq + q^2 = 1$$

The three terms of this binomial expansion indicate the frequencies of the three genotypes:

p^2 = frequency of AA (homozygous dominant)
$2pq$ = frequency of Aa (heterozygous)
q^2 = frequency of aa (homozygous recessive)

- This information in the box contains all the information you need to calculate allelic frequencies when there are two different alleles. Make sure that you can use it!

Sample Problem #1:

Consider a population of pigs where B = tan pigs and b = black pigs. We can use the Hardy-Weinberg to determine the percent of the pig population that is heterozygous for tan coat and so arrive at the frequency of the two-coat color alleles.

1. Calculate q^2.

Count the individuals that are homozygous recessive in the illustration above. Calculate the percent of the total population they represent. This is q^2.

Answer: Four of the sixteen individuals show the recessive phenotype, so the correct answer is 25% or 0.25.

2. Find q.

Take the square root of q^2 to obtain q, the frequency of the recessive allele.

Answer: $q = 0.5$

3. Find p.

The sum of the frequencies of both alleles = 100%, $p + q = 1$. You know q, so what is p, the frequency of the dominant allele?

Answer: $p = 1 - q$, so $p = 0.5$

4. Find $2pq$.

The frequency of the heterozygotes is represented by $2pq$. This gives you the percent of the population that is heterozygous.

Answer: $2pq = 2(0.5)(0.5) = 0.5$, so 50% of the population is heterozygous.

> **Hints from the Graders:** There are a couple of ways you can get confused when doing these problems. Here's what to watch for:
>
> 1. If you are given the number of individuals showing the dominant trait, remember that this is *not* p^2 because it includes heterozygotes. However, you can use it to determine q^2 and from that get a value for q.
> 2. You also may be given a problem where you are told the frequency of an allele. You are being directly given p or q!

Sample Problem #2:

In a certain population of 1000 fruit flies, 360 have red eyes, while the remainder have sepia eyes. The sepia eye trait is recessive to red eyes. How many individuals would you expect to be homozygous for red eye color?

Hint: The first step is always to calculate q^2! Start by determining the number of fruit flies that are homozygous recessive.

Answer: You should expect 40 to be homozygous dominant.

Calculations:

q^2 for this population is $640/1000 = 0.64$

$q = \sqrt{0.64} = 0.8$

$p = 1 - q = 1 - 0.8 = 0.2$

The homozygous dominant frequency $= p^2 = (0.2)(0.2) = 0.04$

Therefore, you can expect 4% of 1000, or 40 individuals, to be homozygous dominant.

Questions

1. If the frequency of two alleles in a gene pool is 90% *A* and 10% *a*, what is the frequency of individuals in the population with the genotype *Aa*?
    (A) 0.81
    (B) 0.09
    (C) 0.18
    (D) 0.01
    (E) 0.198

2. If a population experiences no migration, is very large, has no mutations, has random mating, and there is no selection, which of the following would you predict?
   (A) The population will evolve, but much more slowly than normal.
   (B) The makeup of the population's gene pool will remain virtually the same as long as these conditions hold.
   (C) The composition of the population's gene pool will change slowly in a predictable manner.
   (D) Dominant alleles in the population's gene pool will slowly increase in frequency, whereas recessive alleles will decrease.
   (E) The population probably has an equal frequency of A and a alleles.

3. In a population that is in Hardy-Weinberg equilibrium, the frequency of the homozygous recessive genotype is 0.09. What is the frequency of individuals that are homozygous for the dominant allele?
   (A) 0.7
   (B) 0.21
   (C) 0.42
   (D) 0.49
   (E) 0.91

Answers

1. The correct answer is c. The question tells you that $p = 0.9$ and $q = 0.1$. From this, you can calculate the heterozygotes: $2pq = 2\,(0.9)\,(0.1) = 0.18$. If you selected e as your response, you may have confused the allelic frequency with genotypic frequency. This problem gives you the allelic frequency of a, which is 10%.

2. The correct answer is b. The conditions described all contribute to genetic equilibrium, where it would be expected for initial gene frequencies to remain constant generation after generation. If you chose e, remember that genetic equilibrium does not mean that the frequency of A = the frequency of a.

3. The correct answer is d. $q^2 = 0.09$, so $q = 0.3$.

$$p = 1 - q, \text{ so } p = 1 - 0.3 = 0.7$$
$$AA = q^2 = 0.49$$

Laboratory 9: Transpiration

> ### YOU MUST KNOW
>
> - The role of water potential and transpiration in the movement of water from roots to leaves.
> - The effects of various environmental conditions on the rate of transpiration.
> - How to identify xylem and phloem and relate their structure to their function.

Overview of the Lab

Transpiration is the major mechanism that drives the movement of water through a plant. In the first section of this laboratory, you investigate, factors that influence the rate of transpiration. In the second section, you study plant anatomy as it relates to transport.

Hints and Review

▌ Review **hydrogen bonding!** (Topic 1, Chapter 3) In water, a hydrogen bond is a weak bond between the hydrogen of one water molecule and the oxygen of another, and it accounts for the unique properties of water, including adhesion and cohesion.

▌ Water enters a plant through the root hairs, passes through the tissues of the root into the xylem, and travels up through the xylem vessels into the leaves.

▌ **Transpiration** is the evaporation of water from the leaves through the stomates. It is the major factor that pulls the water up through the plant.

▌ When water enters the roots, hydrogen bonds link each water molecule to the next (*cohesion*) so the molecules of water are pulled up the thin xylem vessels like beads on a string. The water molecules also cling to the thin walls of the xylem cells (*adhesion*). The water moves up the plant, enters the leaves, moves into air spaces in the leaf, and then evaporates (transpires) through the *stomata* (singular, stoma).

▌ **Stomata** are the pores in the epidermis of a leaf. There are hundreds of stomata in the epidermis of a leaf. Most are located in the lower epidermis. This reduces water loss because the lower surface receives less solar radiation than the upper surface. Each stoma allows the carbon dioxide necessary for photosynthesis to enter, while water evaporates through each one in transpiration.

- **Guard cells** are cells surrounding each stoma. They help to regulate the rate of transpiration by opening and closing the stomata. To understand how they function, study the following figures. As you look at the figures, keep in mind that an increase in solute concentration lowers the water potential of the solution, and that water moves from a region with higher water potential to a region of lower water potential.

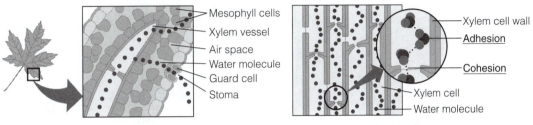

Figure 9.1a

Mesophyll cells
Xylem vessel
Air space
Water molecule
Guard cell
Stoma

Figure 9.1b

Xylem cell wall
Adhesion
Cohesion
Xylem cell
Water molecule

- Notice that in Figure 9.2a the guard cells are turgid, or swollen, and the stomatal opening is large. This turgidity is caused by the accumulation of K^+ (potassium ions) in the guard cells. As K^+ levels increase in the guard cells, the water potential of the guard cells drops, and water enters the guard cells.

- In Figure 9.2b, the guard cells have lost water, which causes the cells to become flaccid and the stomatal opening to close. This may occur when the plant has lost an excessive amount of water. In addition, it generally occurs daily as light levels drop and the use of CO_2 in photosynthesis decreases.

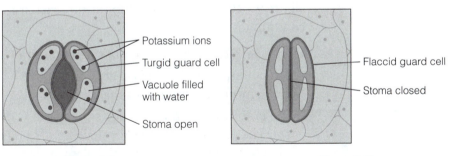

Figure 9.2a

Potassium ions
Turgid guard cell
Vacuole filled with water
Stoma open

Figure 9.2b

Flaccid guard cell
Stoma closed

- A leaf needs carbon dioxide and water for photosynthesis. For carbon dioxide to enter, the stomata on the surface of the leaf must be open. Transpiration draws water from the roots into the leaf mesophyll. However, the plant must not lose so much water during transpiration that it wilts. The plant must strike a balance between conserving water and bringing in sufficient amounts of CO_2 for photosynthesis.

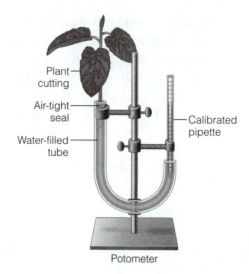

Plant cutting

Air-tight seal

Water-filled tube

Calibrated pipette

Potometer

Figure 9.3

▌ One way to measure water loss from a plant is to use a *potometer*, a device that measures the rate at which a plant draws up water. Since the plant draws up water as it loses it by transpiration, you are able to measure the rate of transpiration. The basic elements of a potometer are

   ▌ A plant cutting
   ▌ A calibrated pipette to measure water loss
   ▌ A length of clear plastic tubing
   ▌ An air-tight seal between the plant and the water-filled tubing

▌ Figure 9.4 shows a typical setup for this exercise. The rate of water loss in each condition is measured.

   ▌ The *control* is the potometer in room conditions.
   ▌ The potometer in *bright light* generally shows a higher rate of water loss, indicating more photosynthesis than the control.
   ▌ The potometer in the *fan* shows increased water loss compared to the control. The reason is that the air movement results in greater evaporation, lowering water potential outside the plant surface, and more water is pulled from the roots.
   ▌ The plant cutting inside the plastic bag is in a situation of *high humidity*. Because of high water potential in the area surrounding the leaves, the rate of transpiration will be low.

▌ Go to Topic 8, Chapter 35, to review the plant tissue types. Refer to Figure 9.5 as you read this section. You should be able to recognize and know the function of the following cell and tissue types:

   1. **Parenchyma** is the most abundant cell type, and cells are relatively unspecialized. The mesophyll of leaves (where most photosynthesis occurs), and root cortex (starch storage) are parenchyma.

❶ Assemble 4 potometers.

❷ Place each potometer in a different environment: room conditions, mist, wind, and bright light.

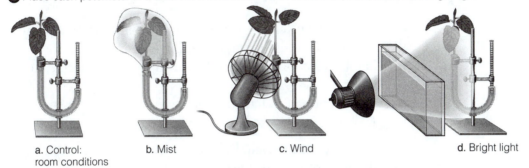

a. Control: room conditions

b. Mist

c. Wind

d. Bright light

❸ Measure water loss in each potometer every 3 minutes for 30 minutes.

00:00

30:00

Figure 9.4

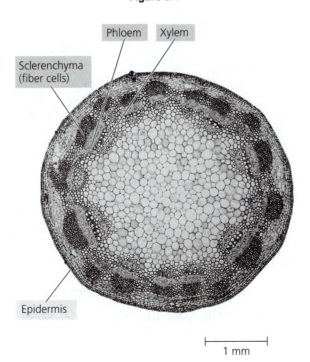

Phloem Xylem

Sclerenchyma (fiber cells)

Epidermis

1 mm

Figure 9.5   Dicot stem

2. **Sclerenchyma** cells make up fibers, have thick secondary cell walls, and often serve a support function.
3. **Collenchyma cells** have thickened cell walls and function in support.
4. **Xylem** cells (tracheids and vessel elements) are dead at maturity. Their function is water transport.
5. **Phloem** (sieve tubes and companion cells) transports solutes.
6. **Epidermal cells** are outermost and serve a protective function.

Questions

1. All of the following enhance water transport in terrestrial plants EXCEPT
   (A) Hydrogen bonds linking water molecules
   (B) Capillary action due to adhesion of water molecules to the walls of xylem
   (C) Evaporation of water from the leaves
   (D) K^+ being transported out of the guard cells

2. Under conditions of bright light, in which part of a transpiring plant would water potential be lowest?
   (A) Xylem vessels in the leaves
   (B) Xylem vessels in the roots
   (C) Root hairs
   (D) Spongy mesophyll of the leaves

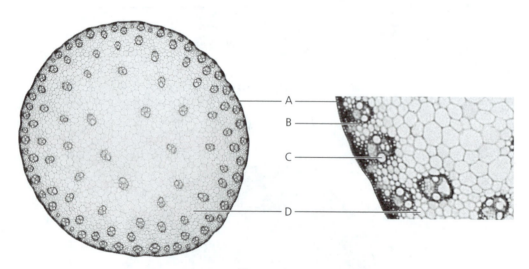

Figure 9.6

Identify each of the structures in the micrograph of a monocot stem above by answering the following questions using the appropriate choices from the list.

Choices for Tissue Type: xylem, phloem, parenchyma, epidermis

Choices for Function: solute transport, water transport, food storage, protection

3. Name tissue type A and describe its function.

4. Name tissue type B and describe its function.

5. Name tissue type C and describe its function.

6. Name tissue type D and describe its function.

Answers

1. The correct answer is d. When K^+ leave the guard cells, water follows, the guard cells become flaccid, and the stomata close. As a result, transpiration doesn't take place.

2. The correct answer is d. Water moves from a region of high water potential to a region of low water potential. The region of lowest water potential would be at the farthest end of the transpiration pathway.

3. The correct tissue type: epidermis. The correct function: protection.

4. The correct tissue type: phloem. The correct function: transport food.

5. The correct tissue type: xylem. The correct function: transport water.

6. The correct tissue type: parenchyma. The correct function: store food.

Laboratory 10: Physiology of the Circulatory System

> ## YOU MUST KNOW
>
> - How exercise changes heart rate and be able to relate this to fitness.
> - How changes in temperature affect the rate of respiration or circulation.

Overview of the Lab

This lab has two parts. In the first, you learn to take pulse and blood pressure in a human subject and gather data for a fitness index. In the second part, you measure the effect of temperature on the heart rate of a small invertebrate, *Daphnia magna*.

Hints and Review

Key Concepts I: Pulse, Blood Pressure and Fitness

▌ When the ventricles of the heart contract (*systole*), blood is pumped through the body, and pressure is increased in the arteries. During the relaxation phase of the heart (*diastole*), the pressure in the arteries decreases.

▌ A **sphygmomanometer** is a device used to measure these pressures.

- In a blood pressure reading of **110/80**, 110 = systolic pressure and 80 = diastolic pressure. The numbers refer to the number of millimeters the pressure will raise a column of mercury. The blood pressure of teenagers is frequently in the range of 120/70.

- **Pulse** is measured as the distension of an artery that can be felt each time the heart contracts. Pulse is measured in number of beats per minute. You can measure pulse anywhere an artery passes close to the skin. Clinically, it is most common to measure heart rate in the radial artery on the inside of the wrist.

- The rate of circulation can be affected by external conditions such as exercise or other activities that affect metabolism, and the efficiency of the circulatory system is often related to the health of the individual.

- In your laboratory you do five different tests of cardiovascular fitness and assign fitness points based on the results of each test. In general, in a fit individual pulse and blood pressure are lower and will return more quickly to the resting condition after exercise than in a less fit individual.

Key Concepts II: Temperature and Metabolic Activity

- In the Cellular Respiration (Lab 5), you experimented with peas and saw how the rate of oxygen consumption during cellular respiration varied with temperature. In animals, an increase in cellular respiration triggers homeostasis mechanisms that increase both breathing and heart rate, resulting in more oxygen being available to cells.

- In this part of the lab, you study the relationship between temperature and metabolic activity in an ectothermic animal, *Daphnia*. It is possible to measure metabolic rate indirectly by measuring heart rate.

- An **ectotherm** is an animal whose body temperature is much the same as its surroundings, such as a frog, a cricket, or a snake.

- The rate of metabolism in ectothermic animals increases as the environmental temperature increases. This rise occurs because the reactants in the cell have greater thermal energy, and many cellular enzymes are more active as temperature increases. This effect is noticeable in a range from approximately 5°C to 35°C; at temperatures much higher than this, enzymes become denatured.

- The relationship between temperature and metabolic rate is often measured as Q_{10}. For most ectotherms, each 10°C increase in temperature results in a doubling of the metabolic rate. If the metabolic rate doubles with a 10°C increase in temperature, then $Q_{10} = 2$.

Questions

1. Which of the following has the least effect on blood pressure in a young adult?
   (A) Temperature of the room
   (B) Position of the body
   (C) Level of conditioning
   (D) Supplemental vitamins

2. An individual's blood pressure is reported as 110/50. Which of the following is correct?
   (A) The pressure during the contraction phase of the heart is 50, and the pressure during the relaxation phase is 110.
   (B) Systolic pressure is 110 and diastolic pressure is 50.
   (C) The pulse is 110 during exercise and 50 when at rest.
   (D) The individual shows possible borderline high blood pressure.

3. Which of the following organisms would show the greatest fluctuation in body temperature hour by hour?
   (A) dolphin
   (B) mouse
   (C) lake trout
   (D) rattlesnake

4. What is the relationship between metabolic rate and body temperature in *Daphnia*?
   (A) As the body temperature increases, the metabolic rate decreases.
   (B) An increase of 10°C results in a doubling of metabolic rate.
   (C) Heart rate increases as body temperature decreases.
   (D) Cellular enzymes are less active at 35°C than at 20°C, resulting in decreased metabolic rate.

5. If Q_{10} = 2, then an enzymatic reaction that takes place at a given rate at 7°C would take place approximately how many times faster at 27°C?
   (A) Twenty times
   (B) Eight times
   (C) Four times
   (D) Three times
   (E) Two times

Answers

1. The correct answer is a. All other factors affect blood pressure.

2. The correct answer is b.

3. The correct answer is d. Although a lake trout is also ectothermic, its aquatic habitat results in fewer temperature fluctuations than the snake's.

4. The correct answer is b. It is the only statement to correctly link an increase in body temperature with an increase in metabolic rate.

5. The correct answer is c. Each 10°C increase in temperature results in a doubling of the metabolic rate.

Laboratory 11: Animal Behavior

Overview of the Lab

Animals exhibit a variety of behaviors, both learned and innate, that promote their survival and reproductive success in a variety of ways.

In this laboratory, detailed observations of an organism's behavior are made, and a controlled experiment is designed to test a hypothesis about a specific case of animal behavior.

Hints and Review

▌ This lab is an opportunity to make detailed observations and learn about some interesting animal behaviors. Because the topic is so broad and there are so many local organisms and possibilities for your teacher to choose, we will only make a few comments about animal behavior here. This is a wonderful laboratory opportunity to teach experimental design, so this is where we will focus our discussion. Refer to the Introduction where we have given some hints about writing the lab essay. For any of these 12 labs, be prepared to design your own experiment to test a variable. Laboratory 11 is an excellent guide to the specific steps of experimental design.

▌ There are hundreds of species of fruit flies—so how do members of the same species find each other and signal willingness to mate? Each species has evolved a complex series of behaviors that appear to be genetically programmed. Students who do the lab written in the AP Biology Lab Manual investigate this behavior in *Drosophila melanogaster*.

▌ In another behavior activity found in the Lab Manual, an experiment is done to observe how pillbugs respond to their environment. Pillbugs are placed in a choice chamber, half in the side lined with dry filter paper and the other half in the side lined with wet filter paper. Because pillbugs are crustaceans, they respire through gills and are generally found in a moist habitat. Because of this, most students hypothesize that more pillbugs will be found in the moist chamber. This is often what occurs, but not always!

■ This brings us to an important consideration: The pillbug exercise is not a controlled experiment. Could there be more light at one end of the choice chamber? More activity and vibration? A chemical residue on one side? Any of these conditions and more could possibly influence the organism's behavior. Without a control, it is very risky to state a conclusion.

■ What's needed here is a **controlled experiment**. A controlled experiment begins with a *hypothesis*, a proposed solution for the problem being investigated. A hypothesis is often written as an IF, THEN statement that predicts the outcome we should expect if the hypothesis is correct. A hypothesis should not only predict results, but must be testable.

■ In a controlled experiment, *all variables are held constant* except the one being tested or manipulated. For instance, if the goal is to test response to wet versus dry conditions, the light, temperature, chemicals in the filter paper or on the dish surface, and movement of the table must all remain constant. In addition, all the experimental organisms must be of the same approximate age, size, and state of health. It is not enough to say you will hold all variables constant; you must be explicit in your explanation of how you will do this.

■ To be meaningful, the experiment must include a *large sample size* to be representative of a general condition.

■ The *results must be measurable!* Are you going to count, measure, find the mass? Some way to quantify the results must be devised.

■ Several *repetitions (or replicates)* of the experiment must be done. Like a large sample size, this lets you verify your result.

■ Before you design an experiment, it may be useful to *search the literature* (including the World Wide Web) to see what has already been done and to help develop ideas for a reasonable study.

■ Finally, *statistical analysis of your data* (such as the Chi-square test used in Lab 7, Genetics) should be done to validate experimental results.

Questions

1. A student wanted to study the effect of nitrogen fertilizer on plant growth, so she took two similar plants and set them on a window sill for a 2-week observation period. She watered each plant the same amount, but she gave one a small dose of fertilizer with each watering. She collected data by counting the total number of new leaves on each plant and also measured the height of each plant in centimeters.
   Which of the following is a significant flaw in this experimental setup?
   (A) There is no variable factor.
   (B) There is no control.
   (C) There is no repetition.
   (D) Measurable results cannot be expected.
   (E) It will require too many days of data collection.

2. Students placed five pillbugs on the dry side of a choice chamber and five pillbugs on the wet side. They collected data as to the number on each side every 30 seconds for 10 minutes. After 6 minutes, eight or nine pillbugs were continually on the wet side of the chamber, and several were under the filter paper. Which of the following is *not* a reasonable conclusion from these results?
   (A) It takes the pillbugs several minutes to explore their surroundings and select a preferred habitat.
   (B) Pillbugs prefer a moist environment.
   (C) Pillbugs prefer a dark environment.
   (D) Pillbugs may find chemicals in dry filter paper irritating.
   (E) Pillbugs demonstrate no significant habitat preference.

3. If a student wanted to determine whether pillbugs prefer a moist or a dry environment, what would be the best way to analyze data from the experiment?
   (A) Total the number of pillbugs on the dry side throughout the entire experiment and compare this with the number on the wet side throughout the experiment.
   (B) After waiting 5 minutes for the pillbugs to acclimate, count the number of pillbugs on the dry side every 30 seconds for 5 minutes. Total and average the results, and compare this with the number of pillbugs on the wet side during this same time interval.
   (C) Compare the number of pillbugs on the dry side at the end of 10 minutes with the number of pillbugs on the wet side at the end of 10 minutes.
   (D) Divide the number of pillbugs on the dry side throughout the experiment by the number on the wet side throughout the experiment.

4. Which of the following hypotheses is stated best?
   (A) If pillbugs are allowed free movement, then more will be found in a moist environment than in a dry environment.
   (B) If pillbugs like a moist environment, then they will move to the wet side of a choice chamber.
   (C) If an experiment with pillbugs is run for 10 minutes, then more pillbugs will be found in the most favorable environment.
   (D) Pillbugs are found in moist habitats, so I predict that more will be found where it is wet.

Answers:

1. The correct answer is c. The amount of fertilizer is the variable factor, and the use of two similar plants in the same location receiving equal amounts of water implies the concept of a control. However, with such a small sample size (two plants) and no repetition, this is a flawed experiment.

2. The correct answer is e. However, this experiment has no control, so any of the other choices are reasonable conclusions from the data. In order to reach a reliable single conclusion, you must design a new controlled experiment.

3. The correct answer is b. Much of the early movement is simply exploration, so data analysis should not begin until the pillbugs have acclimated. Choice

c may seem reasonable, but it is not as good as b because it looks only at a single point in time.

4. The correct answer is a. It is the only statement that has the IF, THEN format, and gives both the conditions and a measurable predicted result

Laboratory 12: Dissolved Oxygen and Aquatic Primary Productivity

---

YOU MUST KNOW

- Factors that affect the amount of oxygen available in an aquatic ecosystem.
- The relationship between dissolved oxygen, photosynthesis, respiration, and how these processes relate to primary productivity.
- How to measure primary productivity based on changes in dissolved oxygen.
- The effect of changing light intensity on primary productivity.

Overview of the Lab

In the first part of the laboratory, you measure dissolved oxygen in water and note the effect of temperature on DO. Next, using screens to vary the amount of light exposure on an algal culture, primary productivity is measured, as the effect of varying amounts of light on primary productivity is investigated.

Hints and Review

▌ To begin, you absolutely, positively must have a firm grip on the summary equations for both cellular respiration and photosynthesis to understand this lab! When this lab was tested on the AP Biology Exam several years ago, we found many students lacked this fundamental understanding. So, here are the summary equations for both:

Photosynthesis $6 \, CO_2 + 6 \, H_2O + \text{Light energy} \rightarrow C_6H_{12}O_6 + 6 \, O_2$

Cellular Respiration $C_6H_{12}O_6 + 6 \, O_6 \rightarrow 6 \, CO_2 + 6 \, H_2O + \text{Energy}$

▌ In an aquatic environment, oxygen availability is influenced by a variety of chemical and physical factors. Some of the factors that affect the amount of oxygen dissolved in water are

1. **Temperature:** As water becomes warmer, its ability to hold oxygen decreases.

2. **Photosynthetic activity:** In bright light, aquatic plants are able to produce more oxygen.

3. **Decomposition activity:** As organic material decays, microbial processes consume oxygen.

4. **Mixing and turbulence:** Wave action, waterfalls, and rapids all aerate water and increase the oxygen concentration.

5. **Salinity:** As water becomes more salty, its ability to hold oxygen decreases.

▌ To measure how much oxygen water can hold (saturation), you need to be able to read a **nomograph**. In the lab, the amount of oxygen at different temperatures is determined by laying a straight edge to connect water temperature with the oxygen level of the sample. Where the straight edge crosses the % saturation line gives the correct reading. On the figure, you will see that at 7 mg/l O_2 at 10°C, % Saturation = 46%. Practice by making the measurements necessary to answer the questions at the end of this lab review.

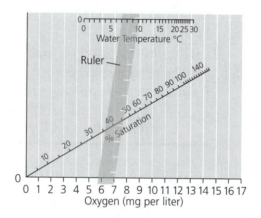

▌ In the second part of the laboratory you monitor the effect of varying light levels on dissolved oxygen in an algae-rich water culture by determining changes in productivity. So now, let's review what you need to know about productivity.

▌ **Primary productivity** is a term used to describe the rate at which plants and other photosynthetic organisms produce organic compounds in an ecosystem. There are two aspects of primary productivity:

1. **Gross productivity** = the entire photosynthetic production of organic compounds in an ecosystem.

2. **Net productivity** = the organic materials that remain after photosynthetic organisms in the ecosystem have used some of these compounds for their cellular energy needs (cellular respiration).

▌ Since oxygen is one of the most easily measured products of both photosynthesis and respiration, a good way to gauge primary productivity in an aquatic ecosystem is to measure dissolved oxygen. We cannot measure gross productivity directly because respiration, which uses up oxygen and organic compounds, is always occurring simultaneously with photosynthesis—but we can measure it indirectly. Let's see how to do this.

▌ Primary productivity can be measured in three ways:

1. The amount of carbon dioxide used

2. The rate of sugar formation

3. The rate of oxygen production

- In this laboratory the third method is used to measure primary productivity. Do you understand why the rate of oxygen production will reveal primary productivity? If you aren't sure, look again at the definition of primary productivity and the equation for photosynthesis.

- *Net productivity* is determined directly by measuring oxygen production in the light, when photosynthesis is occurring.

- *Respiration* without photosynthesis is obtained by measuring O_2 consumption in the dark, when photosynthesis does not occur.

- Since **Net Productivity = Gross Productivity − Respiration**, we can calculate *gross productivity.*

- The amount of light available for photosynthesis drops off sharply with increasing depth in an aquatic environment. You model this condition by wrapping water-sample bottles with increasing layers of screen, as shown in the overview diagram below.

1 Fill seven BOD bottles with aquatic culture, being careful not to agitate the sample.

2 Carefully seal the bottles with caps which are designed to prevent air entrapment.

3 On each cap write the % light the sample will receive.

4 Wrap bottles in screen or foil as indicated by the label.

5 Test the DO of the initial bottle.

(Wrapped in foil) (Wrapped 8 times) (Wrapped 5 times) (Wrapped 3 times) (Wrapped 1 time)

6 Place all the remaining bottles on their sides in a tray under a fluorescent light for 24 hours.

(Wrapped in foil) (Wrapped 8 times) (Wrapped 5 times) (Wrapped 3 times) (Wrapped 1 time)

Figure 12.1

- On the first day of the laboratory, the amount of oxygen present in the original sample is measured. After 24 hours of photosynthesis and respiration under the light, the DO is measured for each bottle.
- To analyze and review the results of the experiment, look at the illustration here.

Figure 12-2a

Initial DO — Dark DO = Respiration

Figure 12-2b

Light DO — Dark DO = Gross productivity

Figure 12-2c

Light DO — Initial DO = Net productivity

Figure 12.2

- The screens wrapped around the bottles simulate the attenuation of light with increasing water depth. Data from this experiment show that photosynthesis (and therefore productivity) drops off steeply as light is attenuated. Both gross and net productivity are graphed, and from this we see the point at which there is so little light available for photosynthesis that respiration exceeds photosynthesis. See Question 4 to review this.

Questions

Refer to the nomograph on page 358 to answer questions 1 and 2.

1. What is the percent oxygen saturation for a water sample at 25°C that has 7 mg O_2/l?

2. In which aquatic environment would you expect dissolved oxygen to be highest?
   (A) A mountain lake that is clear and cold
   (B) A bog where the water is shallow and warm and there is a mat of aquatic plants
   (C) A marine tidepool
   (D) A cold mountain stream dropping over a series of small rock falls
   (E) A coral reef in a still lagoon

3. Study this graph to answer the following question:

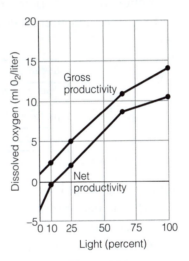

Figure 12.3

At what light intensity do you expect there to be no net productivity?
   (A) Any intensity below 100%
   (B) Only at intensities of 0% and 2%
   (C) Any intensity below 10%
   (D) Any intensity above 25%

4. What is meant by "net productivity" and how is it calculated in a sample aquatic environment?
   (A) It is a measure of the organic products of photosynthesis that accumulate after cellular respiration by those organisms is taken into account, and it is calculated by subtracting the amount of oxygen in the dark bottle from the amount in the light bottle.
   (B) It is a measure of the amount of respiration in a test area, and it is calculated by subtracting the amount of oxygen present in the light bottle from the amount in the dark bottle.
   (C) It is the total amount of carbon fixed, and it is calculated by measuring the amount of oxygen present in a bottle kept in the light.
   (D) It is the amount of oxygen produced during the day, and it is calculated by subtracting the amount of oxygen in the light bottle from the amount in the dark bottle.

5. A biology class used two aquatic cultures as described below for the experiment with screens that reduce light. They measured dissolved oxygen initially and then after 24 hours.

Culture A	Culture B
Little phytoplankton	Rich in phytoplankton
Rich in zooplankton	Rich in zooplankton
Low initial dissolved oxygen	High initial dissolved oxygen

What results would you predict for this experiment?
   (A) The net productivity in culture A will be much higher than in that in culture B.
   (B) Culture B will have both higher gross productivity and higher net productivity than culture A.
   (C) The net productivity for culture A will be negative at greater light intensity than that for culture B.
   (D) Cultures A and B will show similar results because of the comparable quantities of zooplankton.
   (E) Net productivity in culture B will exceed gross productivity in high light intensity.

Answers:

1. The answer should be approximately 65%. Even though the dissolved oxygen level is the same as at the cooler temperature, the percent oxygen saturation has increased. Warmer water has less capacity to hold oxygen. What might be one environmental hazard from thermal pollution of waterways?

2. The correct answer is d. Turbulence such as that seen in a rapidly moving stream aerates the water, so there would be more oxygen than in either a lake or a still lagoon. Salt water and warm temperature decrease the ability

of water to hold oxygen. A mat of aquatic plants blocks light to lower portions of the water, and the high rate of decomposition of organic matter consumes much oxygen.

3. The correct answer is c. This is the point were the line crosses the x axis and where respiration therefore exceeds photosynthesis.

4. The correct answer is a. Productivity refers to the amount of organic material that is fixed, and it can be measured several ways. In an aquatic environment, it is common to measure oxygen as a byproduct of photosynthesis.

5. The correct answer is b. Since culture A has little phytoplankton, gross productivity will be lower than that of culture B, and this will result in lower net productivity as well.

Part IV

Sample Tests with Answers and Explanations

On the following pages are two sample examinations that approximate the actual AP Biology Examination in format, types of questions, and content. Set aside three hours to take each test. To best prepare yourself for actual AP exam conditions, use only the allowed time for Section I and Section II.

Practice Test 1

Biology
Section 1

Time—1 hour and 20 minutes

Directions: Each of the questions or incomplete statements below is followed by five suggested answers or completions. Select the one that is best in each case, and then fill in the corresponding oval on the answer sheet.

Gene	Probability of Appearing in Gamete
P	1/4
Q	1/4
R	1/4

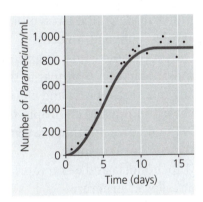

***Paramecium* Population in Lab**

1. In the diagram above, what number of paramecia best represents the carrying capacity of the environment for the population shown?
   (A) 200
   (B) 500
   (C) 600
   (D) 900
   (E) 1,000

2. Three genes—*P*, *Q*, and *R*—are not linked. The probability of each gene appearing in a gamete is shown in the table above. Which of the equations below represents the probability that all three genes will appear in the same gamete?
   (A) ¼ × ¼ × ¼
   (B) ¼ + ¼ + ¼
   (C) ¼ ÷ ¼ × ¼
   (D) (¼)⅓
   (E) 0

3. Which cellular organelle is the site of cellular respiration?
   (A) Golgi apparatus
   (B) Chloroplast
   (C) Mitochondria
   (D) Endoplasmic reticulum
   (E) Ribosomes

GO ON TO THE
NEXT PAGE

4. Which term best describes how goslings recognize their mothers after they are born?
   (A) Habituation
   (B) Imprinting
   (C) Reasoning
   (D) Instinct
   (E) Maturation

5. A student using a light microscope observes a relatively small, rod-shaped cell that has no observable nucleus or other membrane-bounded organelle. What type of cell is this most likely to be?
   (A) Viral
   (B) Eukaryote
   (C) Gamete
   (D) Prokaryote
   (E) Plant

6. In a testcross, which of the following must be true?
   (A) One of the individuals is homozygous dominant.
   (B) One of the individuals is homozygous recessive.
   (C) Both individuals are heterozygous.
   (D) Both individuals are homozygous.
   (E) Both individuals have an unknown phenotype.

7. Diatoms are the major primary producers in which of the following ecosystems?
   (A) Marine
   (B) Desert
   (C) Temperate broadleaf forest
   (D) Chaparral
   (E) Tropical rain forest

8. In deer, fur length is controlled by a single gene with two alleles. When a homozygous deer with long fur is crossed with a homozygous deer with short fur, the offspring all have fur of medium length. If these offspring with medium-length fur mate, what percentage of their offspring will have long fur?
   (A) 100%
   (B) 75%
   (C) 50%
   (D) 25%
   (E) 0%

9. All of the following are types of wastes excreted by animals EXCEPT
   (A) ammonia
   (B) urea
   (C) uric acid
   (D) carbon dioxide
   (E) nitrate

10. Which of the following statements best supports the idea that certain cell organelles are evolutionarily derived from symbiotic prokaryotes living in host cells?
    (A) The process of cellular respiration in certain prokaryotes is similar to that occurring in mitochondria and chloroplasts.
    (B) Mitochondria and chloroplasts have DNA and proteins that are very similar to those in eukaryotes.
    (C) Mitochondria and eukaryotes have similar cell wall structures.
    (D) Like prokaryotes, mitochondria have a double membrane.
    (E) Mitochondria and chloroplasts have DNA and ribosomes that are similar to those of prokaryotes.

11. During the course of which type of reaction is energy consumed?
    (A) Hydrolysis
    (B) Catabolic
    (C) Oxidation-reduction
    (D) Endergonic
    (E) Exergonic

12. Consumption of CO_2 can be used as a measure of photosynthetic rate because carbon dioxide is
    (A) consumed during the light reactions of photosynthesis.
    (B) consumed during the dark reactions of photosynthesis.
    (C) used to trap photons, the form of energy in sunlight.
    (D) necessary for the production of ATP in oxidative phosphorylation.
    (E) produced when fermentation takes place.

13. Which of the following cell organelles is not bound by a membrane?
    (A) Centrosome
    (B) Golgi apparatus
    (C) Cell nucleus
    (D) Mitochondrion
    (E) Peroxisome

14. Which of the following statements is NOT part of Darwin's theory of natural selection?
    (A) Individuals survive and reproduce with varying degrees of success.
    (B) Because there are more individuals than the environment can support, this leads to a struggle for existence in which only some of the offspring survive in each generation.
    (C) The unequal ability of individuals to survive and reproduce leads to a gradual change in the population.
    (D) Members of the population that are physically weaker than others will be eliminated first by forces in the environment.
    (E) Individuals in a population vary in their characteristics, and no two individuals are exactly alike.

15. In animal development, all of the following occur EXCEPT
    (A) cleavage, a succession of rapid cell divisions, occurs just after fertilization
    (B) the zygote develops polarity
    (C) the zygote eventually develops into a hollow ball of cells called a blastula
    (D) as cleavage continues a solid ball of cells called a morula is produced
    (E) all of the genes in the zygote are activated

16. Which of the following describes a protein capable of converting related proteins to an infectious form?
    (A) Virus
    (B) Retrovirus
    (C) Prion
    (D) Spirochete
    (E) Prokaryote

17. A difference between prokaryotic and eukaryotic cells is the presence of
    (A) a membrane-bounded nucleus
    (B) genetic material in the form of DNA
    (C) cytoplasm
    (D) ribosomes
    (E) a cell membrane

18. Which plant hormone is responsible for root growth and differentiation, cell division, germination, and delaying senescence?
    (A) Auxin
    (B) Cytokinins
    (C) Gibberellins
    (D) Abscisic acid
    (E) Ethylene

GO ON TO THE NEXT PAGE

19. A farmer selects one green pepper plant that has all of the most desirable traits of the species. The farmer then produces a group of offspring plants using only genetic material from this ideal parent plant. The resulting plants are genetically identical to the parent and are said to be
(A) a community
(B) a family
(C) clones
(D) a phylum
(E) a genus

20. The domain Archaea contains prokaryotic organisms that
(A) possess a nuclear envelope
(B) have plantlike features
(C) are capable of nitrogen fixation
(D) live in extreme heat or acid environments
(E) reproduce sexually

21. Which statement best describes the action of the hormone oxytocin in humans?
(A) it stimulates growth
(B) it stimulates the secretion of epinephrine
(C) it raises blood glucose levels
(D) it lowers blood glucose levels
(E) it stimulates contraction of the uterus

22. In dogs, the trait for long tail is dominant (L), and the trait for short tail is recessive (l). The trait for yellow coat is dominant (Y), and the trait for white coat is recessive (y). Mating two dogs gives a litter of 3 long-tailed, yellow dogs and 1 long-tailed, white dog. Which of the following is most likely to be the genotype of the parent dogs?
(A) $LLYY \times LLYY$
(B) $LLyy \times LLYy$
(C) $LlYy \times LlYy$
(D) $LlYy \times LLYy$
(E) $LlYY \times Llyy$

23. Which of the following can be viewed with a light microscope?
(A) Ribosomes
(B) Golgi apparatus
(C) Nucleus
(D) Lipids
(E) Proteins

24. Which of the following areas is a site of active cell division at the tips of plant roots and shoots?
(A) Lateral meristems
(B) Apical meristems
(C) Sclerenchyma cells
(D) Cortex
(E) Pericycle

25. Which substances are components of the plasma membrane of a cell?
(A) Glycoproteins
(B) Cytochromes
(C) Nucleic acids
(D) Phosphatidic acid
(E) Lipoproteins

26. Mitosis in vertebrate cells occurs just after which of the following phases of the cell cycle?
(A) G_1
(B) S
(C) DNA synthesis
(D) G_2
(E) M phase

27. Certain cells of all of the following organisms undergo meiosis EXCEPT
(A) ferns
(B) sponges
(C) fungi
(D) bacteria
(E) nematodes

28. Water and minerals flow up through a plant through the
    (A) sieve tubes of phloem
    (B) sieve tubes of xylem
    (C) tracheids and vessel elements of phloem
    (D) tracheids and vessel elements of xylem
    (E) only vessel elements of xylem

29. In plants, change in the level of which of the following substances causes the stomata to close and conserve water during drought?
    (A) Brassinosteroids
    (B) Abscisic acid
    (C) Auxin
    (D) Cytokinins
    (E) Ethylene

30. O_2 and CO_2 diffuse from regions where their partial pressures
    (A) are higher to regions where they are lower
    (B) are lower to regions where they are higher
    (C) are zero to regions of higher partial pressure
    (D) are zero to regions of lower partial pressure
    (E) are influenced by external atmosphere changes into the cell

31. A DNA molecule that can carry foreign DNA into a cell and then replicate is called a
    (A) probe
    (B) restriction fragment
    (C) restriction enzyme
    (D) vector
    (E) transcriptase

32. It is theorized that when organisms that were capable of self-replicating originated, Earth's atmosphere contained a low concentration of
    (A) gaseous oxygen.
    (B) water.
    (C) carbon dioxide.
    (D) nitrogen.
    (E) hydrogen.

33. The composition of lymph in lymph vessels is roughly the same as which of the following?
    (A) Blood
    (B) Interstitial fluid
    (C) Glomerular filtrate
    (D) Bile
    (E) Chyme

34. In photosynthesis, most ATP is produced as a result of which of the following processes?
    (A) The light reactions
    (B) Carbon fixation
    (C) Noncyclic photophosphorylation
    (D) The dark reactions
    (E) The Calvin cycle

35. Which of the following organisms is not usually considered alive because of its dependence on other organisms for reproduction?
    (A) Nematode
    (B) Tapeworm
    (C) Mold
    (D) Virus
    (E) Lichen

36. In guinea pigs, black fur (B) is dominant to brown fur (b). No tail (T) is dominant over tail (t). What fraction of the progeny of the cross $BbTt \times BbTt$ will have black fur and tails?
    (A) 1/16
    (B) 3/16
    (C) 3/8
    (D) 9/16
    (E) 1/4

GO ON TO THE
NEXT PAGE

37. A migrating flock of Canadian geese is nearly decimated by a severe storm. Only four members of the flock, which constitute the entire population of a specific region, survive and return north the next spring. These four start a new colony. This phenomenon is known as
(A) the founder effect
(B) natural selection
(C) migration
(D) polymorphism
(E) the bottleneck effect

38. Which of the following describes a drawing showing the evolutionary history among a particular species or a group of related species?
(A) Food web
(B) Punnett square
(C) Pedigree
(D) Phylogenetic tree
(E) Graph

39. Which of the following characteristics is common to all bryophytes?
(A) Large, independent gametophytes
(B) Monoecious plants
(C) Haploid spores
(D) Seed production
(E) Vascular tissue

40. The rate of flow of sugar and nutrients through the phloem is regulated by
(A) diffusion from source to sink
(B) hydrostatic pressure in the sieve tube
(C) the force of transpirational pull
(D) active transport by tracheid and vessel cells
(E) passive transport by the pith

41. Allolactose stimulates the cells of the human body to produce mRNAs that code for the enzyme β-galactosidase, which breaks down lactose into glucose and galactose. In this case, the role of allolactose can best be described as that of a
(A) DNA replication stimulator
(B) translation inhibitor
(C) stimulator of β-galactosidase secretion
(D) regulator of gene activity
(E) translation activator

42. Which of the following types of data can be used to map the locations of genes on chromosomes?
(A) Segregation frequency
(B) Rate of gene regulation
(C) Dominance patterns
(D) Rate of gene recombination
(E) Rate of gene expression

43. When the stomata of a plant leaf open, which of the following occurs?
(A) There is a decrease in CO_2 intake by the leaf.
(B) The plant shifts from C_3 photosynthesis to C_4 photosynthesis.
(C) The rate of transpiration decreases.
(D) There is an increase in the concentration of CO_2 in mesophyll cells.
(E) There is an increase in the rate of production of nucleic acids.

44. All of the following are evidence for evolution EXCEPT
(A) the presence of anatomical homologies
(B) vestigial organs
(C) the existence of molecular homologies
(D) the fossil record
(E) the existence of homologies in diet among species

45. In a certain group of iguanas, the presence of brown skin is the result of a homozygous recessive condition in the biochemical pathway producing skin pigment. If the frequency of the allele for this condition is 0.35, which of the following is closest to the frequency of the dominant allele in this population? (Assume that the population is in Hardy-Weinberg equilibrium.)
    (A) 0.15
    (B) 0.45
    (C) 0.55
    (D) 0.65
    (E) 0.85

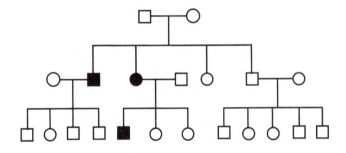

46. In the pedigree above, squares represent males, and circles represent females. Shaded figures represent individuals who possess a particular trait. Which of the following patterns of inheritance best explains how this trait is transmitted?
    (A) Partially dominant
    (B) Autosomal dominant
    (C) Autosomal recessive
    (D) Sex-linked recessive
    (E) Sex-linked dominant

47. The movement of H^+ across the inner mitochondrial membrane during chemiosmosis of cellular respiration, is an example of what type of movement across a membrane?
    (A) Active transport
    (B) Facilitated diffusion
    (C) The work of a symport
    (D) The work of an antiport
    (E) Cotransport

48. Which of the following is the most direct result of the presence of protein in the small intestine?
    (A) The secretion of bile by the gallbladder
    (B) The secretion of pepsin by the lining of the small intestine
    (C) The activation of the inactive form of trypsin and chymotrypsin
    (D) The activation of the inactive form of lipase
    (E) Peristalsis along the walls of the small intestine

49. During part of its life cycle, a tapeworm lives as an adult in a human's intestine. The tapeworm attaches to the intestinal lining, absorbs nutrients digested by the host, and releases eggs that are excreted in the human's feces. The feces happen to contaminate the food given to a pig, and larvae encyst in the muscles of the pig. The pig is later consumed by humans. The tapeworm is an example of
    (A) mutualistic symbiotic partner to humans
    (B) commensalistic symbiotic partner to humans
    (C) parasitic symbiotic partner to humans
    (D) mutualistic symbiotic partner to pigs
    (E) commensalistic symbiotic partner to pigs

50. Which of the following is a major food source for organisms that live in the benthic zone?
    (A) Floating aquatic plants
    (B) Phytoplankton
    (C) Zooplankton
    (D) Detritus
    (E) Cyanobacteria

GO ON TO THE NEXT PAGE

51. Which of the following cellular processes is most closely coupled with active transport?
    (A) The addition of H^+ to H_2O to produce a hydronium ion
    (B) The hydrolysis of ATP
    (C) The phosphorylation of ADP
    (D) The synthesis of G3P
    (E) The formation of peptide bonds between amino acids

52. Which of the following cells would most likely have the greatest concentration of mitochondria in its cytoplasm?
    (A) A cell lining the digestive tract
    (B) An active skeletal muscle cell
    (C) A cell in the liver
    (D) A cell in the lung
    (E) A cell in the epidermis

53. In which of the following pairs are the organisms most closely related taxonomically?
    (A) Mushroom; tulip
    (B) *E. coli*; euglenid
    (C) Lobster; spider
    (D) Shark; crayfish
    (E) Dolphin; sea star

Directions: Each group of questions below consists of five lettered choices (or five lettered items in a graph) followed by a list of numbered phrases or sentences. For each numbered phrase or sentence, select the one choice (or item) that is most closely related to it. Each choice may be used once, more than once, or not at all in each group.

Questions 54–56 refer to the following graph. Each of the curves represents one pathway for the same reaction, but one pathway is catalyzed by an enzyme.

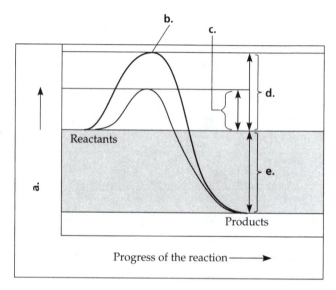

54. Represents the activation energy of the uncatalyzed reaction

55. Represents the activation energy of the catalyzed reaction

56. Represents the transition state of the uncatalyzed reaction

Questions 57–61 refer to the following diagram of the structure of a flower.

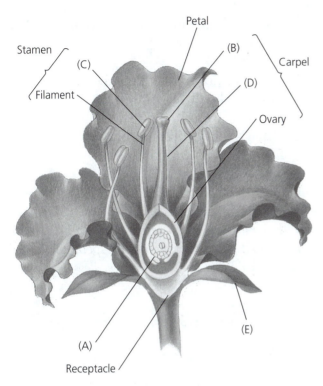

Structure of a Flower

57. Develop into seeds after fertilization

58. The site of pollen production

59. Receives pollen

60. Enclose the flower prior to its opening

61. Connects the stigma to the ovary

Questions 62–65
    (A) Follicle-stimulating hormone (FSH)
    (B) Growth hormone (GH)
    (C) Melatonin
    (D) Androgens
    (E) Endorphins

62. Secreted by the anterior pituitary gland, stimulates growth and metabolism

63. Secreted by the anterior pituitary gland, stimulates the production of ova and sperm

64. Secreted by the testes, promotes the development of secondary sex characteristics

65. Secreted by the pineal gland, involved in regulating biological rhythms

Questions 66–69
    (A) Savannah
    (B) Temperate broadleaf forest
    (C) Tundra
    (D) Chaparral
    (E) Coniferous forest

66. Possesses permafrost and is dominated by short shrubs and grasses; endures long, dark winters

67. Dominated by dense evergreen shrubs and other plants adapted to periodic fires

68. Has long, cold winters and short summers; a biome that is dominated by gymnosperms

69. Home to large herbivores and their predators, marked by grasses and scattered trees

Questions 70–73

(A) Inner mitochondrial membrane
(B) The cytosol
(C) Thylakoid membranes
(D) Ribosome
(E) Nucleus

70. Where mRNA is translated into proteins

71. Where DNA is replicated prior to cell division

72. Where chlorophyll is located

73. The location of glycolysis

Questions 74–78

(A) Rotifera
(B) Porifera
(C) Nematoda
(D) Platyhelminthes
(E) Chordata

74. Possess a notochord, a dorsal hollow nerve cord, and bilateral symmetry

75. Possess a tough exoskeleton called a cuticle, are not segmented, have a complete digestive tract but no circulatory system

76. Possess a complete digestive tract, are pseudocoelomates with a crown of cilia surrounding their mouths

77. Are hermaphrodites and suspension feeders, have no nerves or muscles, have a sac-like body

78. Possess a gastrovascular cavity with only one opening, are acoelomates, and include many parasitic species

Questions 79–82

(A) Telomere
(B) DNA polymerase
(C) Helicase
(D) Primer
(E) DNA ligase

79. DNA not made up of genes, but of multiple repetitions of short nucleotide sequences

80. Joins the sugar-phosphate backbones of the Okazaki fragments to make a complete DNA strand

81. Catalyzes elongation of new DNA at a replication fork

82. Catalyzes the unwinding of double-stranded DNA prior to transcription

Questions 83–84 A scientist is studying the cell cycles of various organisms to learn about their metabolic activities and division patterns. She kept track of the amount of time each type of cell spent in the cell cycles and collected them in the table below.

TOTAL MINUTES SPENT IN EACH CELL CYCLE PHASE				
Cell Type	G_1	S	G_2	M
Monkey liver	20	23	10	18
Plant stem	98	0	0	0

83. From the data in the table above, which of the following is the most likely conclusion about the cell from the plant stem?
(A) It is dead.
(B) This cell contains no DNA.
(C) This cell contains no mRNA.
(D) This cell has entered the G_0 phase.
(E) This cell is continually growing.

84. How long did the entire process of mitosis in the monkey liver cell last?
(A) 70 minutes
(B) 23 minutes
(C) 18 minutes
(D) 10 minutes
(E) 0 minutes

Questions 85–86

In a study of the development of chicken embryos, groups of cells in the early germ layers were stained with five different-colored dyes. After the organs of the chick developed, the location of the dyes were marked down as shown below.

Tissue	Color
Brain	Blue
Liver	Red
Mucous membranes	Green
Nerve cord	Yellow
Heart	Orange

85. Mesoderm would eventually give rise to tissues containing which of the following colors?
    (A) Yellow and purple
    (B) Red and blue
    (C) Orange and yellow
    (D) Orange and green
    (E) Red and green

86. Tissues that were stained blue were derived from
    (A) mesoderm.
    (B) ectoderm.
    (C) mesoderm and ectoderm.
    (D) ectoderm and endoderm.
    (E) mesoderm and endoderm.

Questions 87–88 refer to the following chromosome map. Letters represent gene loci and numbers represent map units.

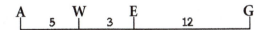

87. Considering the possibilities of recombination, which two genes on this chromosome are most likely to segregate together into a daughter cell?
    (A) A and W
    (B) A and E
    (C) A and G
    (D) W and E
    (E) W and G

88. If the rate of recombination between gene A and W is 5%, what is the rate of recombination between gene W and gene G?
    (A) 0%
    (B) 5%
    (C) 10%
    (D) 15%
    (E) 20%

Questions 89–91 refer to the following figure, which shows a food web in a particular ecosystem. Each letter represents a species in this ecosystem, and the arrows show the flow of energy.

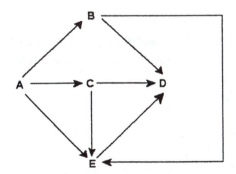

89. Which of the species in the food web is the primary producer?
    (A) A
    (B) B
    (C) C
    (D) D
    (E) E

90. Species B and C represent which of the following?
    (A) Primary producers
    (B) Primary consumers
    (C) Secondary consumers
    (D) Tertiary consumers
    (E) Omnivores

91. Which of the following most accurately describes species E?
    (A) An herbivore, and a secondary consumer
    (B) An omnivore, and a secondary consumer
    (C) An omnivore, and both a primary and secondary consumer
    (D) An herbivore, and both a primary and secondary consumer
    (E) A cannibal

Questions 92–94 refer to the information about five organisms shown in the table below.

Environment Inhabited by Animal	Body Length	Features of Gas Exchange System	Features of Gas Exchange Surface	Percentage of Oxygen Extracted from Air
1 Terrestrial	0.01 m	Branching air tubes, large tracheae that open to the outside	Moist epithelium lining the terminal ends of the tracheal system	53%
2 Terrestrial	0.02 m	Branching air tubes, large tracheae, ventilates with rhythmic body movements	Moist epithelium lining the terminal ends of the tracheal system	48%
3 Aquatic	0.5 m	Outfoldings in the body surface suspended in water	Uses countercurrent exchange and ventilation	73%
4 Terrestrial	1.0 m	Lungs that work in conjunction with circulatory system	Gas exchange occurs across epithelium of alveoli	67%
5 Terrestrial	2.0 m	Lungs that work in conjunction with circulatory system	Gas exchange occurs across epithelium of alveoli	78%

92. Which of the above organisms is most likely to have hemolymph as its main circulatory fluid?
    (A) 1 and 2
    (B) 2 or 3
    (C) 3 or 4
    (D) 4 and 5
    (E) None of these organisms

93. In which of these organisms is hemoglobin used to transport oxygen through the blood?
    (A) 5
    (B) 4 and 5
    (C) 3, 4, and 5
    (D) 2, 3, 4, and 5
    (E) All of them

94. In which of these animals can the process of gas exchange occur without physical movement of some part of the animal?
    (A) 1 and 2 only
    (B) 1 and 3 only
    (C) 1, 2, and 3 only
    (D) 3, 4, and 5 only
    (E) All of them

Questions 95–97 refer to the following gel, which was produced from four samples of a radioactively labeled single strand of DNA that were cut with one type of restriction enzyme. The samples were separated by gel electrophoresis. Answer the questions on the basis of the bands you can visualize below.

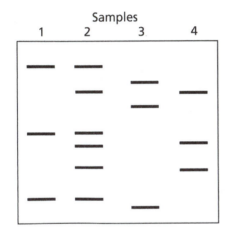

Samples
1   2   3   4

95. The DNA fragments in the gel were separated when an electric field was applied across the gel and they migrated at different speeds. The differential migration speed of the different DNA fragments was due to the
    (A) amount of radioactivity in the samples
    (B) degree to which the samples were negatively charged
    (C) degree to which the samples were positively charged
    (D) size of the fragments within the samples
    (E) polarity of the samples

96. Which of the following is true about the DNA samples that were loaded onto the gel?
    (A) The DNA strand of sample 2 was originally the longest.
    (B) The DNA strand of sample 4 was originally the shortest.
    (C) Samples 2 and 4 are the same DNA sample.
    (D) Sample 2 was cut at more restriction sites than was sample 4.
    (E) Sample 4 was cut at more restriction sites than was sample 2.

97. What was the purpose in radioactively labeling the DNA fragments in this experiment?
    (A) To visualize them
    (B) To make them travel through the gel
    (C) To hydrolyze them into fragments
    (D) To get rid of contaminants
    (E) To destroy their polarity

Questions 98–100 refer to an experiment in which there is an initial setup of a U-tube with its two sides separated by a membrane that permits the passage of water and NaCl but not molecules of glucose. The U-tube is filled on one side with a solution of 0.4 M glucose and 0.5 M NaCl, and on the other, 0.8 M glucose and 0.4 M NaCl.

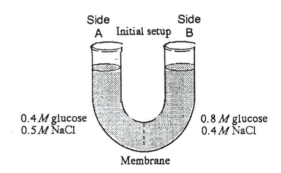

Side A   Initial setup   Side B

0.4 M glucose
0.5 M NaCl

0.8 M glucose
0.4 M NaCl

Membrane

GO ON TO THE NEXT PAGE

98. When this U-tube was set up, at time = 0 in the experiment, which of the following was true?
    (A) The solution on side A was more concentrated than the solution on side B.
    (B) The solution on side B was more concentrated than the solution on side A.
    (C) The two solutions had equal concentration.
    (D) There was more salt on side B than on side A.
    (E) There was more glucose on side A than on side B.

99. Which of the following is most likely to occur after two hours of the U-tube being undisturbed? (Assume that both sides are at atmospheric pressure.)
    (A) The water levels of sides A and B will remain the same.
    (B) The amount of NaCl on side B will have increased.
    (C) The amount of NaCl on side A will have increased.
    (D) The amount of glucose on side B will have increased.
    (E) The amount of glucose on side A will have increased.

100. After two hours, which of the following would probably be true of the level of water on each side of the U-tube?
    (A) There would be no change in the water levels on either side of the U-tube.
    (B) The water column in side A would be slightly higher.
    (C) The water column in side B would be slightly higher.
    (D) The water columns on both sides would be slightly lower.
    (E) The water columns on both sides would be slightly higher.

Biology
Section II

Time—10 minutes to plan responses; 1 hour and 30 minutes for writing

Answer all questions. Number your answer as the question is numbered below.

   Answers must be in essay form. Outline form is NOT acceptable. Labeled diagrams may be used to supplement discussion, but in no case will a diagram alone suffice. It is important that you read each question completely before you begin to write. After reading the questions thoroughly, allow yourself 10 minutes to organize your thoughts and plan your responses.

1. Birth control pills are chemical contraceptives that are made up of estrogen and progestin (which is a progesterone-like substance). They act through a negative feedback loop to stop the secretion of GnRH by the hypothalamus, and of FSH and LH by the pituitary.
   (a) **Explain** how a negative feedback loop works.
   (b) **Explain** how the effects of the birth control pill described above make pregnancy highly unlikely when taken as prescribed.

2. Gene expression in a cell is influenced by a variety of factors. Not all genes on the eukaryotic chromosome are expressed, and in fact, only a small fraction of the genes are transcribed into working proteins.
   (a) **Discuss** three ways in which gene control works in the cell.
   (b) **Describe** three laboratory procedures you could employ in order to determine how much transcription and translation is going on in a cell at a given time.

3. It has been determined that, evolutionarily, the closest relative of humans is the chimpanzee. Other somewhat close relatives are the gibbon and the orangutan.
   (a) **Describe** the relationships among these four species—taxonomically and through phylogeny.
   (b) **Describe** three kinds of evidence that were used to determine the relationship among these four species.
   (c) **Describe** the general structure of the ancestor of *Homo sapiens*, relative to that of other anthropoids.

4. A flowering plant in a ceramic pot is placed in a window that has light shining through most of the day, and it is given adequate water and soil nutrients.
   (a) **Describe** the daily and nightly events in the plant's metabolism.
   (b) **Describe** the changes in the plant that would be induced by rotating the plant 180°.

END OF EXAMINATION

ANSWERS AND EXPLANATIONS

Multiple-Choice Questions

1. **(D) is correct.** The carrying capacity of a population is defined as the maximum population size that a certain environment can support without itself being degraded. If you look at the graph, you can see that the population increases in number until it reaches about 900 members, and at that point it stabilizes. The number 900 therefore represents the carrying capacity.

2. **(A) is correct.** This is a simple probability question. In order to calculate the chance that two or more independent events will occur together in a specific combination, you can use the multiplication rule. Take the probability that gene P will segregate into a gamete ($\frac{1}{4}$), and multiply it by the probability that gene Q will segregate into a gamete ($\frac{1}{4}$). Then multiply that by the probability that gene R will segregate into a gamete to get $\frac{1}{4} \times \frac{1}{4} \times \frac{1}{4} = \frac{1}{64}$.

3. **(C) is correct.** The mitochondria of the cell is the site of cellular respiration and produces the most ATP by extracting energy from sugars, fats, and other fuels. Mitochondria are found in almost all eukaryotic cells, and are enclosed by two membranes—a smooth outer membrane and an inner membrane with many infoldings called cristae. The inside of the inner membrane is called the mitochondrial matrix. There are two processes that occur in the mitochondria to produce energy. One is the citric acid cycle, and the other is chemiosmosis.

4. **(B) is correct.** Imprinting is defined as a type of learning that is generally irreversible and that is limited to a certain period in an animal's life (usually when the animal is very young). The phenomenon of mother-offspring bonding in geese is an example of this type of learning. If this imprinting does not happen, the mother will not take care of the offspring, and the goslings will die.

5. **(D) is correct.** This is most likely a prokaryotic cell. Most prokaryotes are about one-tenth the size of eukaryotic cells, and the most common shapes of prokaryotic cells are spherical, rod-shaped, and helical. Finally, prokaryotes lack membrane-bound organelles, including nuclear membranes; instead they have nucleoid regions, which are a complex of DNA and fibers in a certain region of the cell.

6. **(B) is correct.** A testcross is the breeding of an organism that has an unknown genotype with one that is homozygous recessive, in order to determine the genotype of the unknown parent. The phenotypic ratio of the offspring will reveal the unknown genotype.

7. **(A) is correct.** Diatoms are single-celled algae that reproduce asexually by mitosis. They live in both freshwater and marine environments, and they are very numerous in those environments. They are major primary producers in marine environments.

8. **(D) is correct.** If you consider that the homozygous long-hair deer is *HH*, and the homozygous short-hair deer is *hh*, then crossing them would give all offspring with the genotype *Hh* and medium-length hair phenotype. Crossing the heterozygotes would give you offspring in the ratio of 1:2:1—*HH:Hh:hh*. This means that 25% of the offspring would have long hair (*HH*), 25% of

them would have short hair (*hh*), and 50% of them would have medium-length hair (*Hh*).

9. (E) is correct. Nitrate is the only answer listed that does not represent a form of waste secreted by some kind of animal. Ammonia is the waste product secreted by many aquatic species; urea is the common waste form of mammals, most amphibians, and many fishes; and birds and reptiles secrete uric acid. Carbon dioxide is a waste product of respiration.

10. (E) is correct. Among these answers, the one that best supports the idea that certain cell organelles, such as mitochondria and chloroplasts, were once symbiotic prokaryotes living inside larger cells is answer E, which states that mitochondria, chloroplasts, and bacteria have similar DNA and chromosomes.

11. (D) is correct. An endergonic reaction is a nonspontaneous chemical reaction; in order for the reaction to begin, free energy must be absorbed from the surroundings. During the course of an endergonic reaction, energy is released.

12. (B) is correct. The Calvin cycle—often referred to as the dark reactions of photosynthesis—occurs in the stroma of the chloroplast. In the course of the cycle, the enzyme rubisco combines carbon dioxide with a five-carbon sugar, consuming NADPH and ATP, and ultimately producing glyceraldehyde-3-phosphate. Since carbon dioxide is consumed in the course of the Calvin cycle, the rate of its consumption can be used to determine the rate of photosynthesis.

13. (A) is correct. All of the cell organelles listed, except centrosomes, are contained by a membrane composed of phospholipids. Centrosomes are areas that are present in the cell cytoplasm only during cell division; they are the sites at which the centrioles containing microtubules comprising the spindle apparatus are organized.

14. (D) is correct. The only one of the statements not included in Darwin's theory of natural selection is answer *D*. Darwin's theory stated that the individuals that are least well suited will not leave behind as many offspring, but physical weakness does not necessarily lead to an individual's being unsuited for its environment.

15. (E) is correct. All of the answers list events in the embryonic development of an animal except *E*. During development, certain genes are turned on at certain points in the process, but at no point are all of the genes in a cell activated.

16. (C) is correct. Prions are proteins that are the cause of diseases such as "mad cow" disease; they are misfolded proteins that are capable of converting normally folded proteins into misfolded forms (like themselves), triggering a chain reaction that vastly increases their numbers. These proteins exist primarily in the brain, which is why their misfolding has such serious negative effects.

17. (A) is correct. Prokaryotic cells are very simple cells that have no membrane-bound nucleus—instead they have a nucleoid region, at which the genetic material is concentrated. They do, however, share the remaining characteristics with eukaryotes—they are bound by a membrane, they have genetic material that is translated into proteins on ribosomes, and they have cytoplasm.

18. **(B) is correct.** Cytokinins enhance the growth and development of plant cells; they are produced primarily in actively growing tissues such as roots, embryos, and the fruits of a plant. They act in concert with auxins to cause cell division and differentiation. They are also responsible for retarding the aging process of some plant organs.

19. **(C) is correct.** Asexual reproduction is a form of reproduction in which just one parent contributes genetic material to the offspring, which are clones of the parent and of each other. On the other hand sexual reproduction, in which two parents contribute genetic material to the offspring, results in off-spring that differ from each other and their parents.

20. **(D) is correct.** The domain Archaea, possesses members that are known for their extreme hardiness. Many Archaea are extremophiles: they thrive in extreme environments such as hot-water geysers in Yellowstone Park. Extreme halophiles live in very saline places; extreme thermophiles live in very hot environments.

21. **(E) is correct.** Oxytocin is produced in the posterior pituitary gland and is regulated by the nervous system. It stimulates the powerful contractions of the smooth muscles in the uterine wall that occur during childbirth. Oxytocin also stimulates the placenta to secrete prostaglandins, which also contribute to contractions.

22. **(D) is correct.** The ratio of offspring produced is 3:1. Answer *A* can be eliminated because it would produce all dogs with long tails and yellow coats, since only the dominant alleles are present. First, figure out the gametes that the resulting parents in each answer would produce, and do a Punnett square to figure out the resulting offspring proportions. For answer *D*, the gametes produced by parent 1 are *LY*, *Ly*, *lY*, and *ly*. For parent two, the possible gametes are *LY* and *Ly*.

23. **(C) is correct.** A light microscope can be used to view structures that are no smaller than 0.2 μm. This means that they can be used to see most plant and animal cells, cell nuclei, and certain large organelles such as mitochondria. They can also be used to view some bacterial cells.

24. **(B) is correct.** Apical meristems are located at the roots and shoots of plants, and they supply cells that enable the plant to grow in length. The elongation of plants is called primary growth (as opposed to secondary growth, when plants grow in diameter). Meristems are perpetually embryonic tissues that exist in a plant's areas of growth.

25. **(A) is correct.** Glycoproteins are complexes of carbohydrate and protein that are associated with the cell membrane and that function in cell-cell recognition—a cell's ability to determine the function of a neighboring cell—for example. Glycoproteins vary from species to species and from individual to individual.

26. **(D) is correct.** Mitosis, or the M phase of the cell cycle, occurs just after the G_2 phase in the cell cycle. In the G_1 phase, the cell grows. In the S phase, the cell continues to grow and also copies its chromosomes. In the G_2 phase, again the cell grows, and it begins to prepare for cell division. It divides finally in the M phase.

27. (D) is correct. Bacteria, which are the only prokaryotes listed, reproduce by an asexual reproductive process called binary fission. In binary fission, the bacterial cell chromosome (which consists of a single circular DNA molecule) replicates itself, and the cell grows with the plasma membrane growing inward and eventually pinching off to form two cells. The organisms in all of the other answers produce gametes, and gametes are produced by meiosis.

28. (D) is correct. The water-conducting elements of xylem (which is the plant tissue that carries water upward from the roots toward the shoots of the plant) are the tracheids and vessel elements. These cells are elongated and dead when they reach functional maturity. When the cell dies, its interior disintegrates. This leaves only the hard cell wall, which forms a conduit for the movement of water and dissolved minerals.

29. (B) is correct. Abscisic acid is a plant hormone that prompts seeds to enter a dormant phase until conditions are favorable for germination. Abscisic acid also acts as the primary hormonal signal in times of drought. When a plant begins to wilt, abscisic acid accumulates in leaves and causes the stomata to close, which prevents water loss through transpiration.

30. (A) is correct. Gases in general, including carbon dioxide and oxygen, diffuse down a pressure gradient from regions where their partial pressure is higher to regions where it is lower. This means that if cells in one location are depleting their supply of oxygen, the partial pressure of oxygen in that area will drop, and more oxygen will be unloaded at the site of the depletion.

31. (D) is correct. A vector is an important component of genetic engineering; it is a plasmid (a circular piece of DNA) in which a foreign piece of DNA can be inserted. The vector can then be injected into a bacterial cell, causing transformation, and can then be replicated (along with the DNA fragment that it carries) by the cell's machinery.

32. (A) is correct. It is theorized that diatomic oxygen levels in the atmosphere were quite low around the time when the first self-replicating organisms were appearing on Earth. Oxygen began to accumulate in the atmosphere when photosynthesis evolved.

33. (B) is correct. Lymph is the fluid that is inside the lymphatic system, and the composition of lymph is similar to that of interstitial fluid. Fluids and proteins that are lost as they pass through the capillaries are picked up by the lymph vessels and returned to the blood. The lymphatic system drains into the circulatory system near the junction of the venae cavae and the right atrium.

34. (C) is correct. In noncyclic photophosphorylation, ATP is generated by chemiosmosis; the redox reactions of the electron transport chain create an H^+ gradient across the thylakoid membrane, and this gradient is used to power an ATP synthase. This ATP synthase makes ATP.

35. (D) is correct. Viruses are usually not considered to be alive, because they depend on host cells in order to reproduce. Invading viruses use the host cell's machinery to produce its proteins and replicate itself.

36. (B) is correct. Crossing guinea pigs with genotype *BbTt* gives a ratio of offspring of 9:3:3:1; this is a dihybrid cross between two independently assorting characters. Nine-sixteenths of the offspring will display both of the dominant traits; 3/16 of the offspring will display one of the dominant traits and one recessive one; 3/16 of the offspring will display the other dominant trait and the other recessive trait; 1/16 of the offspring will display both recessive traits. This question asks how many of the offspring will display one of the dominant traits (black fur) and one of the recessive traits (tail); the answer is 3/16.

37. (E) is correct. This is an example of the bottleneck effect—a type of genetic drift. Genetic drift is defined as a change in the allelic frequencies of a population due to chance, and the bottleneck effect occurs when a large part of a certain population is destroyed by a disaster such as an earthquake, drought, or fire. The surviving members may not be representative of the population's gene pool.

38. (D) is correct. A phylogenetic tree shows the hypothetical evolutionary history of a species or a group of related species. Systematists create phylogenies in order to study the path of evolution on Earth and to better understand relationships among species that live on Earth today.

39. (A) is correct. The gametophyte is the dominant generation in bryophytes and typically is a plant that is large enough to be noticed. Bryophytes produce diploid spores that land on suitable environments and divide by mitosis to eventually grow into gametophytes. Most bryophytes lack vascular tissue, which limits their ability to grow tall. Some common bryophytes are mosses, liverworts, and hornworts.

40. (B) is correct. The flow of sap from the leaves to the other parts of the plant body is driven by hydrostatic pressure that develops inside the sieve tube, as phloem unloading creates a high solute concentration at the source end of the sieve tube—in contrast with a low solute concentration at the sink end. Water flows through the tube because the pressure is greatest at the tube's source end.

41. (D) is correct. Regulatory genes are those that code for a protein—either a repressor or an inducer—that controls the transcription of another gene or a group of genes. In this case, allolactose is an inducer. When it is present in the cell, it causes the genes that transcribe for β-galactosidase to be turned on and to produce the enzyme. This represents an example of an inducible operon.

42. (D) is correct. Scientists can use the rate of genetic recombination between two genes in order to build a genetic map, which is an ordered list of the genetic loci along the length of a chromosome. The closer two linked genes are (linked genes are those located on the same chromosome), the less likely it is that they will be separated during crossing over. The rate of recombination is proportionate to the distance between genes on a chromosome.

43. (D) is correct. The opening of the stomata in the leaves of a plant results in CO_2 being taken in from the atmosphere. This, in turn, results in photosynthesis when light energy can also be captured by the leaf. Another result is an increase in transpiration, or water loss by the leaf, by evaporation through the stomata.

44. (E) is correct. All of the answer choices are proof of the process of evolution except the existence of homologies among different species' diets. Since all living organisms require certain nutrients in order to survive—and since there are only so many consumable organic substances on Earth—the fact that two species might have similar components in their diet does not necessarily imply evolutionary relatedness.

45. (D) is correct. Since the frequency of the recessive allele is 0.35, we know that the frequency of the other allele is $1 - 0.35$, which equals 0.65. The Hardy-Weinberg equation states that if a population contains just two alleles for a given trait, and if the frequency of one of the alleles is known, the frequency of the other allele can be calculated using the equation $p + q = 1$. If you designate the frequency of the occurrence of the recessive allele as q, and use its value of 0.35, you can rearrange the equation to read $p + 0.35 = 1$. Then, $1 - 0.35 = 0.65$, which is equal to p, or the frequency of the other (in this case, the dominant) allele.

46. (C) is correct. This is an autosomal recessive trait. If it were sex-linked, it would be expressed in only one of the sexes—usually the male, since males have only one X chromosome. We know it is recessive because it does not appear in every generation; only dominant traits appear in every generation. Also, the first generation does not show the trait, but they have children who do.

47. (B) is correct. In facilitated diffusion, the transport of certain substances across a membrane is aided by transport proteins that span the membrane. These transport proteins are specialized for the solute that they transport. In the case of chemiosmosis, an H^+ gradient is established across the inner mitochondrial membrane by the electron transport chain, which moves H^+ against its concentration gradient into the mitochondrial matrix. As these ions move back to the matrix, diffusion is facilitated by ATP synthase proteins.

48. (C) is correct. When protein enters the small intestine, the enzymes that are responsible for breaking down proteins go to work. Trypsin and chymotrypsin are secreted by the pancreas in inactive form. They must be activated by an intestinal enzyme called enteropeptidase before beginning to break down the peptide bonds between the amino acids.

49. (C) is correct. This is an example of parasitic symbiosis, between the human and the tapeworm. The tapeworm gains nutrients from the human, whereas the human is harmed by the symbiotic relationship. The tapeworm causes intestinal blockage when it grows to its maximal length, and it can rob nutrients from the human to the extent that the human can develop nutritional deficiencies.

50. (D) is correct. The benthic zone in the ocean is the bottom of all aquatic zones; it is made up of sand and organic nutrients and is occupied by bacteria, fungi, seaweeds, algae, invertebrates, and some fishes. These organisms receive nutrients from the detritus that rains down from the ocean levels above it, where animals produce various metabolic wastes and die.

51. (B) is correct. Active transport involves the movement of substances across membranes against their concentration gradient. In this process, the cell must expend energy, usually in the form of ATP. One example of active transport in the cell is the sodium-potassium pump, in which three sodium ions are pumped out of the cell and two potassium ions are pumped in. In the process, ATP is hydrolyzed—it transfers one of its phosphate groups to the transport protein.

52. **(B) is correct.** An active skeletal muscle cell would be the most likely to have a high concentration of mitochondria in its cytoplasm because mitochondria are the sites of cellular respiration, and cellular respiration is the source of ATP in the cell. Muscle cells use ATP in the process of contraction, and since they store only enough ATP for a few contractions, they must have many functional mitochondria to keep up the flow of ATP production when muscle contraction is continuous.

53. **(C) is correct.** The two most closely related organisms are the lobster and the spider. Both are arthropods, which are characterized by having segmented bodies, exoskeletons, and jointed appendages. Both organisms also have an open circulatory system.

54. **(D) is correct.** The activation energy of a reaction is the initial energy investment required in order for the reaction to proceed. It is the energy required in order to break the bonds of the substrate enough for the substrate to reach the highly unstable transition state. You can tell that answer *D* is the reaction energy of the uncatalyzed reaction (because the presence of a catalyst would decrease the overall energy of the reaction), so the taller curve must be the uncatalyzed reaction.

55. **(C) is correct.** The activation energy of the catalyzed reaction is represented by answer *C*. The overall energy of this reaction is significantly lower than that of the uncatalyzed reaction. This is because enzymes speed up the course of reactions by lowering the energy of activation so that the transition state is much easier to reach.

56. **(B) is correct.** The transition state of a reaction is the highest-energy, most unstable form of the reactants in the reaction. The energy put into the reaction in order to make it "go"—also known as the activation energy—must be sufficient to enable the reactants to reach this transition state.

57. **(A) is correct.** The ovules are structures that develop in the plant ovary, and they contain the female gametophyte.

58. **(C) is correct.** The anther is the site of pollen production in the plant, and pollen grains contain the immature male gametophyte of a plant.

59. **(B) is correct.** The stigma is the sticky structure located at the end of the carpel. It is responsible for catching pollen grains.

60. **(E) is correct.** The sepals are usually green, and they are a whorl of modified leaves that enclose and protect the flower bud before it opens.

61. **(D) is correct.** The style is the stalk of the carpel of a flower; the ovary is at the base of the stalk; the stigma is at the top of the style.

62. **(B) is correct.** Growth hormone (GH) is secreted by the anterior pituitary gland and affects many different target tissues. It promotes growth and stimulates the production of growth factors.

63. **(A) is correct.** Follicle-stimulating hormone, or FSH, is secreted by the anterior pituitary. It stimulates the production of ova and sperm in the gonads.

■ **64. (D) is correct.** Androgens are the male sex hormones, and the main androgen is testosterone. Androgens are synthesized in the testes, and they stimulate the development and maintenance of the male reproductive system.

■ **65. (C) is correct.** Melatonin is secreted by the pineal gland (in the brain); it is a modified amino acid that is secreted at night. The amount of melatonin secreted depends on the length of the night.

■ **66. (C) is correct.** The tundra is characterized by having a permafrost, which is a permanently frozen subsoil, and bitterly cold temperatures. Because of the frozen subsoil, little precipitation, and high winds plants do not grow very tall.

■ **67. (D) is correct.** The chaparral is home to many dense, spiny evergreen bushes. The summers are long, hot, and dry; and the winters are mild and rainy. Plants in the chaparral are adapted for the periodic fires that ravage these biomes.

■ **68. (E) is correct.** The coniferous forest biome is characterized by frequent snowfall, harsh winters, short summers, and the presence of gymnosperms.

■ **69. (A) is correct.** Savannas are home to grazing herbivores and their predators. They contain tall grasses with sporadic clusters of trees. There is a considerable rainy season interrupted by periods of seasonal drought in areas containing savannas.

■ **70. (D) is correct.** Ribosomes are cell organelles that are constructed of rRNA and protein and function as the site of protein synthesis in the cytoplasm.

■ **71. (E) is correct.** DNA replication occurs in the nucleus of the cell. The genetic material is replicated prior to mitotic or meiotic cell division.

■ **72. (C) is correct.** Light reactions generate ATP by powering the addition of a phosphate group to ADP, a process called photophosphorylation. In the chloroplast, chlorophyll is embedded within the thylakoid membranes.

■ **73. (B) is correct.** Glycolysis takes place in the cytosol of the cell. In glycolysis, glucose is split into two molecules of pyruvate. This metabolic pathway occurs in all living cells, and it is the starting point for fermentation or cellular respiration.

■ **74. (E) is correct.** The phylum Chordata contains two groups of invertebrates plus all animals with backbones. All chordates possess a notochord, a dorsal hollow nerve cord, pharyngeal clefts, and a post-anal tail as an embryo.

■ **75. (C) is correct.** Nematodes are found in aquatic habitats and have unsegmented bodies with a tough exoskeleton called a cuticle. They have a complete digestive tract but lack a circulatory system, and they reproduce sexually.

■ **76. (A) is correct.** Rotifers have a complete digestive tract, with a separate mouth and anus, and a ring of cilia around their mouths, which draws in water. They are pseudocoelomates.

■ **77. (B) is correct.** Porifera are sponges that have a sac-like body and are suspension feeders with no nerves or muscles. They draw water into a central cavity and filter it for nutrients. Sponges are also hermaphrodites.

■ **78. (D) is correct.** Platyhelminthes are flatworms that live in marine environments and other wet habitats. They include many parasitic species, and they have a gastrovascular cavity with just one opening. They are also acoelomates.

79. (A) is correct. Telomeres are regions found at the tips of chromosomes, and they are made up not of genes but of repeating short sequences of DNA. These parts of chromosomes are copied by a special enzyme called telomerase.

80. (E) is correct. DNA ligase is an enzyme that is necessary for the replication of DNA; it catalyzes the covalent bonding of the 3′ end of the new DNA fragment to the 5′ end of the growing chain.

81. (B) is correct. DNA polymerase is another enzyme involved in DNA replication—it catalyzes the elongation of new DNA at the replication fork by adding nucleotides to the existing chain.

82. (C) is correct. Helicase is an enzyme that untwists the double helix of DNA at replication forks prior to DNA replication.

83. (D) is correct. The plant cell is probably in G_0 phase (G_0 phase is a nondividing phase). In many cells, there exists a G_1 checkpoint, and if at this checkpoint the cell is made to exit the cycle, it enters this nondividing G_0 phase. Because this plant cell has spent no time in any other phase besides the G_1, it is most likely arrested in G_0.

84. (C) is correct. The process of mitosis in the monkey liver cell took 18 minutes. The M phase of the cell cycle is the mitotic phase, and it is the phase in which the cell divides the nucleus and partitions the cytoplasm and organelles, plus the newly replicated DNA, to two new daughter cells.

85. (D) is correct. The mesoderm eventually gives rise to most organs and tissues in the body, including the kidney, heart, and inner layer of the skin (including mucous membranes). Therefore, mesoderm would have given rise to the heart, which was stained orange, and the mucous membranes, stained green in this example.

86. (B) is correct. The tissue stained blue was the brain, and the brain is derived from the ectoderm. Also arising from ectoderm is the rest of the nervous system, and the outer epidermal layer of skin.

87. (D) is correct. The genes most likely to travel together and end up in the same daughter cell are W and E. This is because they are located close together on the chromosome. The closer two genes are on the chromosome, the less likely it is that crossing over will occur between them—and that they would be recombined.

88. (D) is correct. If the rate of recombination between A and W is 5%, and the distance between gene A and gene W is 5 map units, and the distance between gene W and gene G is 15 map units, then you can calculate the rate of recombination between W and G by multiplying the 5% by 3, to get a 15% recombination rate.

89. (A) is correct. Species A is autotrophic and the primary producer of the ecosystem. This species is capable of capturing solar energy and converting it into the chemical energy contained in the bonds of organic compounds, which are used by the other organisms in this ecosystem. It also is the only one that has arrows flowing only from it, indicating that it is not a consumer.

90. (B) is correct. Species B and C represent primary consumers—they consume only the autotrophs in this ecosystem, which are the primary producers. They are presumably herbivores since their only food source is the primary producer.

91. (C) is correct. Species E consumes species A (presumably a plant), species C (presumably an animal), and species B (also presumably an animal). This means that species E is an omnivore—it eats both plants and animals. Species E is also both a primary consumer (because it consumes species A) and a secondary consumer (because it consumes species B and C).

92. (A) is correct. The most likely of these organisms to have hemolymph as its circulatory fluid are the smallest organisms, which are presumably insects. These insects have open circulatory systems, in which no distinction is made between blood and interstitial fluid, and hearts that pump the hemolymph directly into the sinuses, which are open cavities for chemical exchange.

93. (C) is correct. In all of these animals, hemoglobin (an iron-containing molecule) is used to transport oxygen. It is contained in the red blood cells, or erythrocytes, which are a component of the blood of each of these animals. Animals 1 and 2 have hemolymph, which in most circumstances does not contain hemoglobin.

94. (B) is correct. The two organisms in this chart in which gas exchange can occur without movement of some part of the animal are the small insect (#1) and the fish (#3). If insects are small enough, gas exchange simply takes place across the moist membranes of their trachea. Some fishes can sit still in water and have the water flow across their gills, with gas exchange taking place.

95. (D) is correct. The DNA fragments migrated along the gel at rates according to their size—the smaller DNA fragments migrated more quickly through the dense gel and can be found near the bottom of the gel, whereas the larger fragments migrated more slowly and can be found closer to the top.

96. (D) is correct. Sample 2 must have been cut at more restriction sites than was sample 4 because more DNA fragments of different sizes were produced. This is shown by the greater number of bands on the gel in the lane of sample 2.

97. (A) is correct. The purpose of radioactively labeling these DNA samples was to make them visible when the gel was done running. After the gel is finished, and the DNA samples have migrated to a sufficient position to be distinguishable, the radiation is detected by radiography.

98. (B) is correct. The solution on Side A is hypotonic to the solution on Side B—it is less concentrated than the solution on Side B, at the time this experiment began.

99. (B) is correct. After two hours, the amount of NaCl on side B will have increased. The membrane separating the two sides allows the passage of NaCl and not glucose, so there will be no movement of glucose—thus, no change in its concentration—but NaCl will travel down its concentration gradient to Side B.

100. (C) is correct. After two hours, the water column in Side B would be slightly higher. If the concentration of NaCl had equalized on both sides of the tube, the solution on Side B would still be hypertonic to that on Side A; thus, water would flow through the membrane in an attempt to equalize its concentration on both sides of the tube until gravitational pressures exerted an equal force to prevent it from rising farther.

1. (a) Negative feedback loops are very important in maintaining homeostasis, through factors such as hormone secretion. Negative feedback loops work much the same way as thermostats do in houses. A receptor somewhere in the body detects a change in some factor in the animal's internal environment, and it transmits this information to a control center. The control center processes the information and directs a response to an effector, which carries out the response. In negative feedback, a change in the variable triggers the control center to prevent further change in the same direction.

(b) The birth control pill blocks the secretion of GnRH by the hypothalamus, and FSH and LH by the pituitary. Together, GnRH, LH, and FSH all work in an elaborate feedback loop that synchronizes the ovarian cycle and the menstrual cycle. During the ovarian cycle, the hypothalamus releases GnRH which stimulates the pituitary to release FSH and LH. FSH stimulates the follicle to grow, and the follicle cells secrete estrogen, which in negative feedback keeps the secretions of FSH and LH relatively low. Later in the cycle, the follicles begin to secrete estrogens rapidly, which has the effect of suddenly causing the increased secretion of FSH and LH. The sudden increase in the secretion of LH is what stimulates ovulation. By blocking the release of GnRH, birth control pills also block LH, which prevents ovulation and FSH, which prevents the follicle from maturing. Ovulation is the release of the egg from the ovaries, and this is the time in the ovarian cycle when fertilization can occur if sperm are present.

This is a good free response answer because it shows working knowledge of the following terms:

negative feedback loop	*LH*
hormone	*pituitary gland*
receptor	*follicle*
control center	*estrogens*
effector	*ovarian cycle*
GnRH	*menstrual cycle*
hypothalamus	*ovulation*
FSH	*fertilization*

This response also demonstrates an understanding of the following important biological processes—a negative feedback loop and the female menstrual/ovarian cycle.

2. (a) Three ways in which gene expression in a cell is controlled are through chromatin packing, DNA methylation, and histone acetylation. In chromatin packing, when the genetic material is in heterochromatin form, it is highly condensed and proteins involved in transcription do not have access to the DNA. In DNA methylation, methyl groups are attached to specific regions of

DNA immediately after it is synthesized. In some cases, this is thought to be responsible for these genes' long-term inactivation. Finally, in histone acetylation, acetyl groups are attached to certain amino acids of histone proteins, and when the histones are acetylated, their shape alters so that they are less tightly bound to DNA; this enables the proteins involved in transcription to move in and begin work. When histones are deacetylated, DNA transcription is impossible.

(b) In order to tell how actively a certain cell is transcribing its DNA and translating its mRNA into protein, you could do a few things in the laboratory. You could monitor the rate of relaxation of the heterochromatin in the cell nucleus; the more relaxed the chromosomes are, the more DNA is being transcribed. Then you could monitor the amount of uptake of cytosine, guanine, uracil, and adenine in the cell; the rate at which these bases are taken up would be an indicator of the rate at which they are being incorporated into mRNA in transcription. Finally, you could monitor the rate at which free amino acids are being consumed in the cell. This would indicate the rate at which they are being incorporated into growing peptide chains in translation.

This is a good free-response answer because it shows working knowledge of the following key terms:

chromatin packing	*cytosine*
DNA methylation	*adenine*
histone acetylation	*uracil*
histones	*guanine*
transcription	*amino acids*
heterochromatin	*peptides*

The response also shows an understanding of the following important biological processes: control of gene expression in eukaryotes and the process of transcription and translation.

3. (a) Humans, gibbons, orangutans, and chimpanzees are primates, so all have an opposable thumb and feet that can grip. Living primates consist of three groups—the lemurs, lorises, and pottos; the tarsiers; and the anthropoids. All four of these species are anthropoids, and they are all hominoids. Evidence suggests that humans and chimps are two divergent branches of the hominoid tree that evolved from a common ancestor (that was neither human nor chimp) about 5 million years ago. The phylogenetic tree on the next page could be used to show the relationships between the primates.

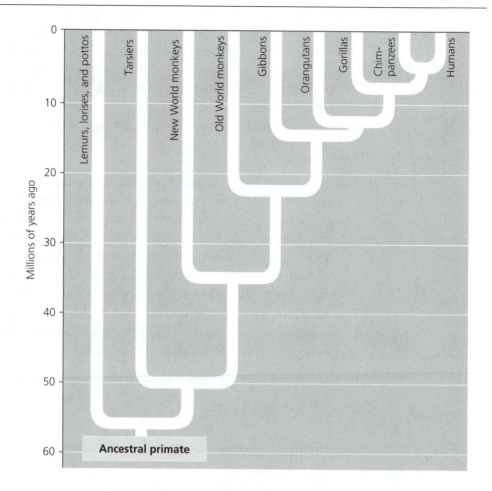

(b) One of the three kinds of evidence that was used to determine relationships among these species is fossil evidence. Fossils are impressions or parts of organisms that are preserved in rock. Most fossils are found in sedimentary rock, and the age of fossils can be determined to a certain extent by the depth of the rock layer in which they are found. Another type of evidence that is used to determine relationships among the anthropoids is the existence of structural homologies. Homologies are shared characteristics that are the result of two species having evolved from a common ancestor. The third type of evidence is molecular homologies in DNA or proteins. The likeness of two species' DNA or proteins tells how closely they are related evolutionarily.

(c) Some of the characteristics of the ancestors of *Homo sapiens* are a relatively large brain (which is associated with the use of language and with other cultural aspects), a longer jaw and certain resulting changes in the teeth, bipedal posture, and a reduced difference in the sizes of the two sexes—males and females were more nearly the same size.

This is a good free-response answer because it shows a working knowledge of the following key terms:

primates	*phylogenetic tree*
anthropoids	*fossils*
hominoids	*homologies*

The response also shows an understanding of these key biological concepts: evolution, the relatedness of primates, and how humans are similar and different from other primates.

4. (a) In the morning, the stomata in the leaves of the plant would open and allow the intake of carbon dioxide and the release of oxygen. The plant would then begin to use the light energy from the sun to convert carbon dioxide and water into sugar molecules, which it will use as food, and oxygen. The process of transpiration is a critical one. Transpiration is the loss of water through the stomata. There must be a balance between the intake of carbon dioxide and the loss of water. Therefore, the plant must keep its stomata open during the day in order to use the sun's energy for photosynthesis, while risking loss of water through transpiration. When night falls, the plant will close its stomata to prevent unnecessary water loss. Because it can no longer get energy from the sun for photosynthesis, it doesn't need carbon dioxide.

   Plants are generally bound to follow a biological clock that controls their circadian rhythms. The amounts of transpiration and enzyme synthesis fluctuate during the course of the day. Some of this is in response to changes in humidity and temperature that occur during the course of the day, but even without those external changes, the biological clock functions. Since this is a flowering plant, it will flower when the night length reaches a critical length; this is how a plant determines the time of season.

   (b) If this plant is rotated 180°, plant hormones will act to start its growth in the direction facing the sun. This is thought to be because a plant responds to light by an asymmetrical distribution of auxin going down from the tip of the plant, which causes the cells on the darker side of the plant to elongate (not divide) more than the cells on the brighter side of the plant. Growth of a plant toward a light source is known as positive phototropism.

This is a good free-response answer because it shows knowledge of the following terms:

stomata	*circadian rhythm*
transpiration	*auxin*
photosynthesis	*phototropism*
biological clock	

The response also shows an understanding of the following important biological processes: the daily metabolic cycle of plants and plant responses to light.

Practice Test 2

Biology
Section 1

Time—1 hour and 20 minutes

Directions: Each of the questions or incomplete statements below is followed by five suggested answers or completions. Select the one that is best in each case, and fill in the corresponding oval on the answer sheet.

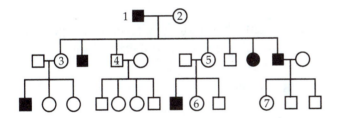

1. Which of the following patterns of inheritance best explains the transmission of the trait in the pedigree above?
   (A) Sex-linked dominant
   (B) Sex-linked recessive
   (C) Autosomal dominant
   (D) Autosomal recessive
   (E) Incompletely dominant

2. A geneticist crosses two rabbits, both of which have brown fur. In rabbits, brown fur is dominant over white fur. Six of the eight offspring produced have brown fur, and the other two have white fur. The genotypes of the parents were most likely which of the following?
   (A) $BB \times bb$
   (B) $BB \times Bb$
   (C) $Bb \times bb$
   (D) $Bb \times Bb$
   (E) $bb \times bb$

3. Which of the following is an example of simple diffusion across a membrane?
   (A) The movement of H^+ across the thylakoid membrane during photosynthesis
   (B) The uptake of neurotransmitters by the postsynaptic membrane during the transmission of a nerve impulse
   (C) The movement of oxygen in the alveoli across the epithelial membrane and into the bloodstream
   (D) The exchange of sodium and potassium across a cell membrane through the Na^+-K^+ pump
   (E) The movement of glucose across the body cell membranes and the cells of the liver, which stores it as glycogen

4. Which of the following correctly represents the order of the tissues through which water and minerals will pass on their way up from a plant's roots?
   (A) Root hair, endodermis (Casparian strip), epidermis, cortex, stele
   (B) Root hair, cortex, epidermis, endodermis (Casparian strip), stele
   (C) Root hair, stele, cortex, endodermis (Casparian strip), epidermis
   (D) Root hair, epidermis, stele, cortex, endodermis (Casparian strip)
   (E) Root hair, epidermis, cortex, endodermis (Casparian strip), stele

5. Which of the following evolved before algae?
   (A) Bacteria
   (B) Hydras
   (C) Cnidarians
   (D) Fungi
   (E) Protists

6. Insertions and deletions may cause which of the following types of mutation?
   (A) Missense mutation
   (B) Nonsense mutation
   (C) Gene substitution
   (D) Base-pair substitution
   (E) Frameshift mutation

7. The stomata—the openings on the underside of a plant leaf through which carbon dioxide is taken up and oxygen is expelled—are opened as a result of
   (A) movement of mesophylls away from the stomatal opening
   (B) increased turgidity in the guard cells
   (C) decreased turgidity in the guard cells
   (D) growth of the guard cells toward the mesophyll
   (E) elongation of the guard cells toward the mesophyll

8. Which characteristic is NOT required of a population in Hardy-Weinberg equilibrium?
   (A) The population must be very large.
   (B) There must be no migration into or out of the population.
   (C) The members of the population must be mating randomly.
   (D) There must be only two alleles present for each characteristic in the population.
   (E) Natural selection must not be operating in the population.

9. The mitotic spindle consists of microtubules and which structure?
   (A) Centromere
   (B) Centrosome
   (C) Cytoplasm
   (D) Kinetochore
   (E) Metaphase plate

10. In plants that undergo alternation of generations, the gametophyte stage is always
    (A) a large visible plant
    (B) a seed
    (C) diploid
    (D) haploid
    (E) unicellular

11. Bacteria reproduce by which of the following processes?
    (A) Mitosis
    (B) Meiosis
    (C) Binary fission
    (D) Binary division
    (E) Cleavage

12. All of the following are factors contributing to the ascent of water through the xylem in plants EXCEPT
    (A) transpiration
    (B) low water potential at one end
    (C) cohesion of water to the vessel walls
    (D) adhesion of water to the vessel walls
    (E) sources and sinks

13. In plants, the abscission, or dropping, of leaves is triggered by changes in
    (A) cytokinin
    (B) ethylene
    (C) abscisic acid
    (D) gibberellins
    (E) brassinosteroids

14. Near the lungs, a branch from the pulmonary artery would contain which of the following?
    (A) Oxygen-rich blood
    (B) Oxygen-poor blood
    (C) Dissolved nutrients from the stomach
    (D) Blood rich in carbon monoxide
    (E) Lymph

15. Gel electrophoresis can be used for which of the following laboratory procedures?
    (A) Determining the molecular weight of proteins and nucleic acids
    (B) Determining the charge of proteins and nucleic acids
    (C) Separating nucleic acids and proteins on the basis of their size
    (D) Separating nucleic acids and proteins on the basis of their charge
    (E) Breaking up proteins and nucleic acids into their monomers

16. In humans, if red hair (R) is dominant to brown hair (r), and freckles (F) are dominant to no freckles (f), what fraction of the progeny of the cross $RrFf \times RRff$ will have red hair and no freckles?
    (A) $9/16$
    (B) $1/2$
    (C) $3/8$
    (D) $3/16$
    (E) $1/16$

17. Which of the following can be observed best by using a compound light microscope?
    (A) Atoms and molecules
    (B) Proteins
    (C) Ribosomes
    (D) Bacteria
    (E) Viruses

18. The phenomenon by which plants will bend toward or away from a light source is known as
    (A) photoaffinity
    (B) taxis
    (C) phototropism
    (D) thigmotropism
    (E) photophilia

19. All of the following are functions of microtubules in the cell EXCEPT
    (A) components of cilia, used for locomotion
    (B) components of flagellum, used for locomotion
    (C) involvement in the movement of chromosomes during cell division
    (D) components of the cytoskeleton, function in cell support
    (E) part of the nuclear membrane

20. Which of the following organelles is the site of macromolecule hydrolysis in the cell?
    (A) Mitochondria
    (B) Centrosome
    (C) Lysosome
    (D) Golgi apparatus
    (E) Ribosome

GO ON TO THE NEXT PAGE

21. Which of the following describes how a dog that is prodded while asleep will respond to the touch initially, but will eventually ignore repeated prodding?
    (A) Habituation
    (B) Imprinting
    (C) Reasoning
    (D) Instinct
    (E) Trial and error

22. Which of the following is characteristic of a plant cell but not of an animal cell?
    (A) Rough endoplasmic reticulum
    (B) Cell membrane
    (C) Ribosomes
    (D) Large central vacuole
    (E) Golgi apparatus

23. When a species is split into two populations, separated by a geographic barrier that makes breeding between the populations impossible, this could eventually lead to
    (A) sympatric speciation
    (B) allopatric speciation
    (C) adaptive radiation
    (D) polyploid speciation
    (E) exaptation

24. The fact that pairs of alleles will segregate randomly during gamete formation describes which of the following laws?
    (A) The law of segregation
    (B) The law of independent segregation
    (C) The law of equal inheritance
    (D) The law of independent assortment
    (E) The law of equal segregation

25. In cows, eye color is controlled by a single gene with two alleles. When a homozygous cow with brown eyes is crossed with a homozygous cow with green eyes, cows with blue eyes are produced. If the blue-eyed cows are crossed with each other, what fraction of their offspring will have brown eyes?
    (A) 0
    (B) ¼
    (C) ½
    (D) ¾
    (E) 1

26. Which of the following is NOT an adaptation for gas exchange?
    (A) Lungs
    (B) Tracheal system
    (C) Gills
    (D) Moist epidermis
    (E) Sinuses

27. Which of the following best characterizes the reaction represented below?
    $A + B \rightarrow AB + energy$
    (A) Exergonic reaction
    (B) Endergonic reaction
    (C) Oxidation-reduction reaction
    (D) Catabolism
    (E) Hydrolysis

28. During prophase of mitosis, nuclear DNA is in which of the following forms?
    (A) Daughter chromosomes
    (B) Chromatin
    (C) Chromosomes consisting of two sister chromatids
    (D) Single sister chromatids
    (E) Single linear chromosomes

29. One way to measure the metabolic rate of a cell would be to measure the rate at which
    (A) CO_2 is consumed by the cell
    (B) O_2 is consumed by the cell
    (C) water is consumed by the cell
    (D) O_2 is produced by the cell
    (E) glucose is consumed by the cell

30. Which of the following is a site of translation in the cell?
    (A) The nucleus
    (B) The Golgi apparatus
    (C) Smooth ER
    (D) Rough ER
    (E) Mitochondria

31. In certain plant cells, the synthesis of ATP occurs in which of the following?
    (A) Ribosomes and mitochondria
    (B) Ribosomes and chloroplasts
    (C) Mitochondria and chloroplasts
    (D) Mitochondria and the cytoplasm
    (E) Chloroplasts and the cytoplasm

32. The graph above shows the rate of growth of a population of squirrels in a certain geographic area in Connecticut during the past several decades. This population is most closely exhibiting which of the following types of growth?
    (A) Logistic growth
    (B) Probable growth
    (C) r-selected growth
    (D) K-selected growth
    (E) Exponential growth

33. Genes M and N are located on different chromosomes, and the probability of their undergoing crossing over is quite low. If the probability of allele M segregating into a gamete is ⅙, and the probability of allele N segregating into a gamete is ¼, then the probability that both of them will segregate into the same gamete is
    (A) ½₁₂
    (B) ¼
    (C) ⁵⁄₁₂
    (D) ¾
    (E) 1

34. Two individuals who are carriers for cystic fibrosis (a recessively inherited disorder) have 3 children together. None of the children have cystic fibrosis. What is the probability that the couple's fourth child will be born with cystic fibrosis?
    (A) 0%
    (B) 25%
    (C) 50%
    (D) 75%
    (E) 100%

35. Which of the following groups comprise a strand of DNA?
    (A) Phosphate groups, deoxyriboses, and nitrogenous bases
    (B) Phosphate groups, riboses, and nitrogenous bases
    (C) Phosphate groups, deoxyriboses, and amino acids
    (D) Phosphate groups, riboses, and amino acids
    (E) Deoxyriboses and nitrogenous bases

GO ON TO THE NEXT PAGE

36. The statement that evolutionary changes are composed of rapid bursts of speciation that alternate with long periods in which species do not change significantly is known as
    (A) gradualism
    (B) punctuated gradualism
    (C) punctuated equilibrium
    (D) sympatric speciation
    (E) allopatric speciation

37. Which of the following vertebrates lacks an amnion during its development?
    (A) Bird
    (B) Human
    (C) Lizard
    (D) Frog
    (E) Alligator

38. Which of the following is capable of reverse transcription, with an RNA → DNA information flow?
    (A) Viruses
    (B) Retroviruses
    (C) T cells
    (D) B cells
    (E) Ciliates

39. Compared with prokaryotic cells, eukaryotic cells are generally
    (A) smaller but more complex
    (B) larger and more complex
    (C) smaller and less complex
    (D) larger but less complex
    (E) the same size but more complex

40. Which of the following plant hormones is responsible for stimulating stem elongation, root growth, and cell differentiation?
    (A) Ethylene
    (B) Abscisic acid
    (C) Cytokinin
    (D) Gibberellin
    (E) Auxin

41. In terms of evolution, which of the following is closest to fungi?
    (A) Plants
    (B) Animals
    (C) Archaea
    (D) Bacteria
    (E) Viruses

42. In humans, which of the following glands is responsible for secreting several hormones involved in reproduction?
    (A) Thyroid gland
    (B) Adrenal cortex
    (C) Adrenal medulla
    (D) Anterior pituitary
    (E) Posterior pituitary

43. Which is thought to have been the first self-replicating genetic material?
    (A) DNA
    (B) RNA
    (C) cDNA
    (D) mRNA
    (E) tRNA

44. When a break in the epidermal layer of humans occurs, which type of blood cell travels in great numbers to the break and releases clotting factors?
    (A) Leukocytes
    (B) Erythrocytes
    (C) Helper T cells
    (D) Helper B cells
    (E) Platelets

45. In photosynthesis, the functional product(s) of the light reactions
    (A) are ATP and NADPH
    (B) are ATP and NADH
    (C) is glyceraldehyde
    (D) is glucose
    (E) are carbohydrates

46. Which group is best characterized as being eukaryotic and saprophytic with hyphae?
    (A) Protista
    (B) Plantae
    (C) Archaea
    (D) Fungi
    (E) Animalia

47. In humans, color blindness is a sex-linked recessive trait. If a man and a woman have a son who is color blind, which of the following must be true?
    (A) The father is color blind.
    (B) Both parents carry the allele for color blindness.
    (C) Neither parent carries the allele for color blindness.
    (D) The father carries the allele for color blindness.
    (E) The mother carries the allele for color blindness.

48. If a horse breeds with a donkey, a mule is produced. Mules are not capable of breeding with either parental species, or each other. This is an example of what type of postzygotic barrier?
    (A) Reduced hybrid viability
    (B) Hybrid sterility
    (C) Hybrid breakdown
    (D) Mechanical isolation
    (E) Gametic isolation

49. Radioactive isotopes can be used to date fossils. The amount of time it takes for half of a radioactive isotope to decay is also known as the substance's
    (A) release rate
    (B) radioactive decay rate
    (C) half-life
    (D) time scale
    (E) decay rate

50. The female gametophytes of a plant develop in the ovaries of the plant, whereas the male gametophyte develops in which plant structure?
    (A) Stigma
    (B) Style
    (C) Carpel
    (D) Anther
    (E) Sepal

51. Which of the following is the most direct result of the presence of salivary amylase in the mouth?
    (A) The breakdown of proteins
    (B) The breakdown of polypeptides
    (C) The breakdown of lipids
    (D) The breakdown of carbohydrates
    (E) The breakdown of nucleic acids

52. The leaves of a plant appear green to us because
    (A) chlorophyll reflects green light
    (B) chlorophyll absorbs green light
    (C) chlorophyll reflects red light
    (D) chlorophyll reflects blue light
    (E) chlorophyll is green, and plants contain hundreds of chlorophyll molecules

53. What bonds are responsible for ice being less dense than liquid water and water being a good insulator?
    (A) Ionic
    (B) Covalent
    (C) Polar covalent
    (D) Hydrogen
    (E) Double

GO ON TO THE NEXT PAGE

54. Which of the following is the insulating layer wrapped around nerve cells that increases the speed of nerve impulse transmission?
(A) Axons
(B) Dendrites
(C) Synaptic terminal
(D) Myelin sheath
(E) Nodes of Ranvier

55. Which of the following is the substrate in the citric acid cycle?
(A) Carbon dioxide
(B) Acetyl CoA
(C) Citrate
(D) Oxaloacetate
(E) Glucose

56. Insects, spiders, and crustaceans are all classified in which phylum?
(A) Arthropoda
(B) Annelida
(C) Chordata
(D) Nemertea
(E) Cnidaria

Directions: Each group of questions below consists of five lettered choices followed by a list of numbered phrases or sentences. For each numbered phrase or sentence, select the one choice (or item) that is most closely related to it. Each choice may be used once, more than once, or not at all in each group.

Questions 57–59
(A) Meiosis II
(B) Meiosis I
(C) Binary fission
(D) Mitosis
(E) Interphase

57. The process during which prokaryotes reproduce

58. The process during which the diploid chromosome number is reduced by half

59. The process during which the genetic material of the cell is replicated

Questions 60–64
(A) Amphibia
(B) Reptilia
(C) Echinodermata
(D) Chordata
(E) Chondrichthyes

60. Members have cartilaginous skeletons and include sharks and sea rays.

61. Members have a water vascular system and include sea stars and sea cucumbers.

62. Members have eggs without shells, and some have a moist epithelium that participates in gas exchange.

63. Members have scales, lungs, and amniotic eggs.

64. Members have a notochord and pharyngeal clefts and include humans.

Questions 65–69
(A) Electron transport chain
(B) Chemiosmosis
(C) Glycolysis
(D) The citric acid cycle
(E) Light reactions of photosynthesis

65. Drives the synthesis of ATP through a hydrogen ion gradient

66. Occurs in all living cells and is the starting point for aerobic respiration and fermentation

67. Is part of cellular respiration and completes the breakdown of glucose into carbon dioxide

68. Shuttles electrons and releases energy that is used to make ATP

69. Photoexcited electrons pass from one photo-system to the next via an electron transport chain.

Questions 70–73
    (A) Population
    (B) Community
    (C) Species
    (D) Niche
    (E) Biome

70. Members are capable of interbreeding and are anatomically similar.

71. The biotic and abiotic resources a species uses in its environment

72. Individuals of one species that live in a discrete geographic area

73. All the organisms that live within a discrete geographic area

Questions 74–78
    (A) Antigens
    (B) Antibodies
    (C) Histamines
    (D) Eosinophils
    (E) Macrophages

74. Large phagocytotic cells that engulf microbes

75. A type of white blood cell that damages invaders with destructive enzymes

76. Proteins that bind antigens

77. Foreign molecules that elicit an immune response

78. Chemical signals released in response to injury

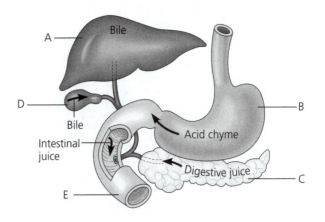

Questions 79–83
Identify the letter pointing to each organ.

79. Stomach

80. Gallbladder

81. Duodenum

82. Pancreas

83. Liver

Questions 84–86
Identify the letter that points to each description.

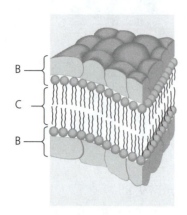

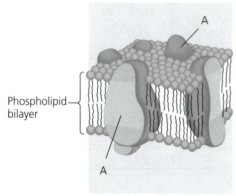

Phospholipid bilayer

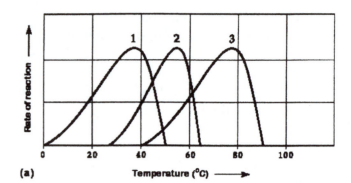

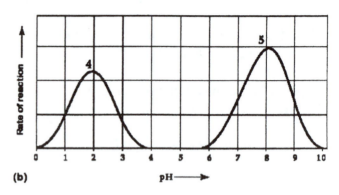

84. The hydrophilic zone of the plasma membrane

85. The hydrophobic zone of the plasma membrane

86. Allows for facilitated diffusion across the membrane

Directions: Each group of questions below concerns an experimental or laboratory situation or data. In each case, first study the description of the situation or data. Then choose the one best answer to each question following it.

Questions 87–89
The rate of reaction for 3 enzymes was calculated at different temperatures, and the rate of reaction for 2 additional enzymes was calculated at different pH levels. The results are shown in the following graphs. Assume that the *y*-axes share the same scale.

87. Which of the enzymes would most likely be able to function in the human bloodstream?
(A) 1 and 4
(B) 1, 2, and 4
(C) 1, 2, and 5
(D) 3 and 4
(E) 3 and 5

88. Which of the enzymes would be most likely to function in the geysers of Yellowstone National Park?
(A) 1
(B) 2
(C) 3
(D) 4
(E) 5

89. Which of these enzymes is most efficient—that is, has the highest rate of reaction?
(A) 1
(B) 2
(C) 3
(D) 4
(E) 5

Questions 90–92

A scientist studying the mammalian heart is experimenting on a white rat. She injects different radioactive elements into different sections of the rat's heart. The chart below lists where she injected each substance.

Heart chamber	Radioactive Isotope Used
Right ventricle	^{32}P
Right atrium	^{3}H
Left ventricle	^{14}C
Left atrium	^{238}U

90. Just after its injection, where would the radioactive isotope ^{238}U be detected first?
   (A) Left ventricle
   (B) Right ventricle
   (C) Right atrium
   (D) Systemic capillaries
   (E) Pulmonary capillaries

91. Just after its injection, the blood injected with ^{14}C would be detected performing which of the following tasks in the body?
   (A) Picking up oxygen from the systemic capillaries
   (B) Transporting oxygen to systemic capillaries
   (C) Picking up oxygen in the capillaries of the lungs
   (D) Dropping off carbon dioxide in the capillaries of the lungs
   (E) Delivering oxygen to the capillaries of the lungs

92. Just after its injection of ^{32}P, the blood injected with ^{32}P would be detected performing which of the following tasks in the body?
   (A) Picking up oxygen from the systemic capillaries
   (B) Delivering oxygen to the systemic capillaries
   (C) Picking up oxygen in the capillaries of the lungs
   (D) Dropping off carbon monoxide in the capillaries of the lungs
   (E) Delivering oxygen to the capillaries of the lungs

Questions 93–94 refer to the graph shown below.

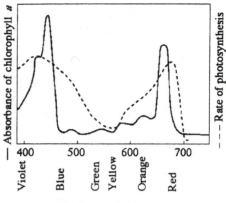

Wavelength of light (nm)

93. A biologist would use which of the following terms to refer to the solid line in the graph?
   (A) Action spectrum
   (B) Absorption spectrum
   (C) Photostimulation curve
   (D) Electromagnetic spectrum
   (E) Visible light spectrum

94. Which of the following is the best reason the curve for the absorbency of light by chlorophyll *a* does not perfectly match the rate of photosynthesis?
   (A) The rate of photosynthesis is always fractionally slower than the rate of absorbency by chlorophyll *a*.
   (B) The rate of photosynthesis is always fractionally faster than the rate of absorbency by chlorophyll *a*.
   (C) There are fewer chlorophyll *a* molecules in the cell than the other molecules involved in photosynthesis, so chlorophyll *a* is the rate-limiting reagent.
   (D) Chlorophyll *a* is not the only photosynthetically important pigment in chloroplasts.
   (E) Light of about 550 nm inhibits all photosynthesis.

GO ON TO THE NEXT PAGE

Questions 95–98
An ecologist studying a certain biogeographic area has sketched the following food web for the community that lives there. The arrows represent energy flow, the letters represent species.

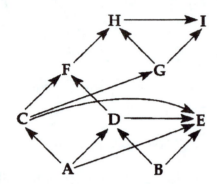

95. Which of the following is most likely to be autotrophic?
    (A) E
    (B) A
    (C) I
    (D) G
    (E) D

96. Which members of the food web are secondary consumers?
    (A) D, E, and F
    (B) A, B, and C
    (C) G, H, and I
    (D) E, F, and G
    (E) B, D, and E

97. Which of the species in the food web are exclusively carnivores?
    (A) E, F, G, H, and I
    (B) F, G, H, and I
    (C) C, D, and E
    (D) F and G
    (E) H and I

98. If this were a savanna ecosystem, what organisms are species A and B most likely to represent?
    (A) Two species of small low-growing bushes
    (B) Two species of lichen
    (C) Two species of insects
    (D) Two species of rodent
    (E) Two species of grasses

Questions 99–100 refer to the following table, which shows the temperature at which the DNA of various species has been found to denature.

Species	Temperature at Which DNA Denatures
A	25°C
B	80°C
C	72°C
D	58°C
E	57°C

99. Which of the species in this table are most likely to be related most closely evolutionarily?
    (A) A and B
    (B) B and C
    (C) C and D
    (D) D and E
    (E) A and E

100. What other experimental method besides heating could be used to denature DNA strands in order to obtain the same type of information?
    (A) Adding a buffer to the DNA samples
    (B) Adding a solvent to the DNA samples
    (C) Slowly lowering the pH of the DNA samples
    (D) Slowly adding free bases to the DNA samples
    (E) Freezing the samples

END OF SECTION I

Biology
Section II

Time—10 minutes to plan responses; 1 hour and 30 minutes for writing

Answer all questions. Number your answers as the questions are numbered below.

Answers must be in essay form. Outline form is NOT acceptable. Labeled diagrams may be used to supplement discussion, but in no case will a diagram alone suffice. It is important that you read each question completely before you begin to write.

1. The cell membrane is one of the most important parts of a cell; it allows the selective passage of materials into and out of the cell, thereby maintaining a constant, desired internal composition.
   (a) **Discuss** the components of a typical animal cell membrane, as well as the roles each of these components plays in regulating the cell's internal environment.
   (b) **Discuss** the ways in which the following can enter an animal cell, including:
      • viral DNA.
      • hormones.
      • water molecules.

2. It is thought that the terrestrial plants we see around us today evolved from aquatic algae.
   (a) **Discuss** three obstacles to the movement of plants to land.
   (b) **Discuss** three adaptations that have evolved in terrestrial plants that combat the obstacles above.

3. **Describe** the process of cell division in a typical plant cell, including:
   (a) the cell cycle of a plant cell.
   (b) the process of mitosis.
   (c) ‹cytokinesis.

4. It is theorized that glycolysis was the first metabolic pathway for the production of ATP.
   (a) **Provide** three pieces of evidence that support this point.
   (b) **Describe** how the citric acid cycle is related to chemiosmosis and oxidative phosphorylation.

END OF EXAMINATION

ANSWERS AND EXPLANATIONS

Multiple-Choice Questions

1. **(B) is correct.** The trait is sex-linked and recessive. Two sets of unaffected parents in the second generation have a child that is affected. You can tell that the trait is sex-linked because five out of the six people affected by the trait are males.

2. **(D) is correct.** Because both parents have brown fur, but they produce some white offspring, you have to conclude that both rabbits were heterozygous. Each parent must have a recessive allele to pass on.

3. **(C) is correct.** The movement of oxygen from the clusters of alveoli at the tips of the bronchioles in the lung, across the epithelial walls, and into the bloodstream is an example of passive diffusion. Carbon dioxide moves from the bloodstream back into the lungs, to be expelled during exhalation in the same way—through diffusion. All of the answers except *C* are examples of active transport.

4. **(E) is correct.** The order of tissues through which water and certain minerals will pass as they travel from the roots to the xylem (then to be transported through the entire plant) is as follows: into the root hair cell, through the epidermis, through the cortex, through the endodermis (where the Casparian strip acts as a filter), and then into the stele (which contains the xylem vessels).

5. **(A) is correct.** Bacteria are the only living organisms among the answers to evolve before algae; bacteria are prokaryotes, whereas algae are eukaryotes. The earliest eukaryotes were single-celled algae, and they are thought to have evolved about a half billion years after prokaryotes.

6. **(E) is correct.** Insertions and deletions are point mutations that occur when one nucleotide pair is added or lost in a gene. They have a detrimental effect on the protein product of the gene, because they often affect the reading frame of the gene—all of the codons downstream of the insertion or deletion will be grouped incorrectly and will be misread.

7. **(B) is correct.** The stomata open as a result of increased turgidity of the guard cells that flank them. As a result of this increased turgidity, they become turgid and buckle, causing a space to open between them. The change in turgidity in the guard cells is influenced by uptake and loss of potassium ions.

8. **(D) is correct.** All of the answers listed are conditions that must be met for a population to be in Hardy-Weinberg equilibrium—except the criteria that there must be only two alleles present for each characteristic. Any population that is not in Hardy-Weinberg equilibrium will evolve, and natural populations rarely achieve this type of equilibrium for extended periods of time.

9. **(B) is correct.** During mitosis and meiosis, the microtubules begin to be assembled at the centrosome, which is a cell organelle that organizes the microtubules. Animal cells have a pair of centrioles at the center of their centrosomes. Together, the centrosomes and microtubules form a spindle.

10. **(D) is correct.** The two forms that alternate in plants undergoing alternation of generations are the sporophyte and the gametophyte. The sporophyte is

the diploid form (cells with two sets of chromosomes), and the gametophyte is the haploid form (cells with one set of chromosomes).

▌ **11. (C) is correct.** Bacteria reproduce by a process called binary fission. Most bacteria contain a single, circular chromosome that also contains associated proteins. They start to replicate this circular chromosome. One of the copies moves toward one end of the cell. At the end of replication, the plasma membrane pinches inward and a new cell wall is formed, resulting in two identical daughter cells.

▌ **12. (E) is correct.** Sources and sinks do not contribute to the movement of water and minerals up through the xylem of the plant. The movement of sugars from sources to sinks occurs through the phloem, which distributes the products of photosynthesis from the leaves throughout the plant.

▌ **13. (B) is correct.** The plant hormone responsible for promoting leaf abscission, or the loss of leaves each fall, is ethylene. This practice prevents trees from dehydrating in winter when the ground is frozen and water is unavailable. Ethylene is also responsible for fruit ripening and for controlling the growth of roots, leaves, and flowers.

▌ **14. (B) is correct.** A branch from the pulmonary artery located near the lungs would be oxygen poor. The branches from the pulmonary artery carry oxygen-poor blood away from the heart to the lung's alveoli for oxygenation. Pulmonary veins return oxygenated blood to the heart, which then pumps the oxygenated blood to the rest of the body.

▌ **15. (C) is correct.** Gel electrophoresis is a procedure used in a laboratory to separate proteins and nucleic acids on the basis of their size—essentially, the rate at which they move through a gel when an electric field is applied. Larger segments of DNA move through gel at a slower pace than smaller segments.

▌ **16. (B) is correct.** The answer is ½. In order to deduce this, you should first determine the gametes that each of the parents could produce. The first parent could produce the gametes *RF, Rf, rF,* and *rf;* the second parent would produce only *Rf* gametes. Use a Punnett square to help you figure out the proportions of offspring based on these gametes.

▌ **17. (D) is correct.** Most bacteria can be seen with a compound light microscope, whereas all of the other structures listed are too small to be seen with this type of microscope. They must be viewed with an electron microscope.

▌ **18. (C) is correct.** Phototropism refers to a plant's growth in response to a light source. Negative phototropism occurs when a shoot grows away from a light source, and positive phototropism occurs when a shoot grows toward the light source.

▌ **19. (E) is correct.** Microtubules are involved in all of the cell functions listed except as a component of the nuclear membrane. The nuclear membrane, as all the other membranes of the cell, is composed mainly of phospholipids and associated proteins. Microtubules in the cell function in the roles of support and movement.

▌ **20. (C) is correct.** Lysosomes are digestive compartments in the cell. They are membrane-bound sacs containing hydrolytic enzymes that can digest macromolecules such as proteins, sugars, fats, and nucleic acids; the lysosomal interior has a low pH that aids in the breakdown of these large molecules.

21. (A) is correct. Habituation is one of the simplest types of learning; it is a loss of responsiveness to a stimulus that conveys limited information or no information at all. If the dog learns that the prodding is not associated with any type of outcome—soon it will learn to ignore the sensation.

22. (D) is correct. One difference between plant cells and animal cells is that plant cells have a large central vacuole. This vacuole is important to the plant because it acts as a stockroom for necessary organic compounds and is a repository for inorganic ions. The vacuole can make up about 80% of a plant's total volume.

23. (B) is correct. One of the two general types of speciation is allopatric speciation. In allopatric speciation, two populations are geographically separate, with no link between the two populations. The other type of speciation is sympatric speciation. In this case, two populations are in the same geographic area, but biological factors (such as chromosome changes and nonrandom mating) reduce gene flow.

24. (D) is correct. Mendel's law of independent assortment states that during gamete formation, each pair of alleles will segregate independently of one another; which allele travels to which gamete is independent of the actions of the other alleles.

25. (B) is correct. If you say that the brown-eyed homozygous cow has genotype $E^B E^B$ and the green-eyed homozygous cow has genotype $E^G E^G$, then all of their offspring would be genotype $E^B E^G$ (blue-eyed). With a Punnett square, you can see that crossing two individuals with genotype $E^B E^G$ would give you offspring in the following ratio: 1 $E^B E^B$:2 $E^B E^G$:1 $E^G E^G$. So ¼ of the offspring would have genotype $E^B E^B$, or brown eyes.

26. (E) is correct. The only adaptation listed that is not used for gas exchange is the sinuses, which are basically just spaces surrounding the organs of the body of animals that have open circulatory systems. Fish have gills, many insects have tracheal systems, many vertebrates have lungs, and some amphibians such as frogs also exchange gases across their moist epithelium.

27. (A) is correct. In reactions that are exergonic, energy is given off—often in the form of heat—during the course of the reaction. Conversely, endergonic reactions require the input of energy in order to proceed. In written reactions, the reactants are generally written on the left side of the arrow, and the products are written on the right side of the arrow.

28. (C) is correct. During prophase of mitosis, the DNA exists in the form of chromosomes consisting of two sister chromatids. During interphase, the DNA exists as chromatin, which is loose DNA and protein. As mitosis begins, the newly replicated DNA condenses into chromosomes composed of the two newly formed sister chromatids.

29. (B) is correct. O_2 receives electrons from the electron transport chain and forms water during the process of oxidative phosphorylation, and the cell makes ATP through the oxidative phosphorylation. So measuring the rate of consumption of O_2 by the cell is a good way to determine its metabolic rate.

30. (D) is correct. Transcription (the process by which DNA is transcribed into mRNA) takes place in the nucleus, whereas translation (the process by which mRNA is translated into the amino acid sequence of a polypeptide) takes place at ribosomes. Some ribosomes are associated with the rough endoplasmic reticulum, a series of continuous membranes in the cell. Smooth ER is not associated with ribosomes.

31. (C) is correct. In plant cells, both chloroplasts and mitochondria produce ATP by chemiosmosis. The thylakoid membrane of the chloroplasts and the inner mitochondrial membrane of the mitochondria and electron transport chain pump protons across the membrane. This energy is used to power an ATP synthase that produces ATP.

32. (E) is correct. This population is exhibiting exponential growth. Exponential growth can occur when the conditions in an environment are ideal—when there is enough of, or an excess of, required resources in an environment. The population then grows at its maximum rate until it reaches its carrying capacity.

33. (C) is correct. According to the addition rule, the probability that an event will occur in two or more different ways can be calculated by adding the separate probabilities of those two ways. In this case, you can figure out the probability of alleles M and N segregating into the same gamete by adding the probabilities that either will segregate into a gamete: $\frac{1}{4} + \frac{1}{6} = \frac{5}{12}$.

34. (B) is correct. The carrier parents would each have the genotype Aa. This means that their children would have a 25% chance of inheriting both of the recessive genes, with the genotype aa. The fact that their first three children do not have cystic fibrosis in no way affects the probability of the fourth child having cystic fibrosis—these events are unrelated.

35. (A) is correct. The groups that comprise a strand of DNA are phosphate groups, deoxyriboses, and nitrogenous bases. The four nitrogenous bases contained in DNA are adenine, thymine, guanine, and cytosine.

36. (C) is correct. Punctuated equilibrium is the term used for the idea that evolutionary change in a species occurs in rapid bursts alternating with long periods of little or no change. Gradualism is the model of evolution in which species evolve gradually and diverge more and more as time passes.

37. (D) is correct. The frog is the only animal listed that does not have an amnion at some stage in its development. An amnion is the innermost of the four extraembryonic membranes; it contains a fluid-filled sac in which the embryo is suspended. Frog eggs can develop without this protective sac because they are laid in aquatic environments.

38. (B) is correct. Retroviruses are viruses that are capable of reverse transcription—they use an enzyme called reverse transcriptase to transcribe DNA from an RNA template. This newly made cDNA integrates into the chromosome of an animal cell and is copied along with the animal cells' DNA.

39. (B) is correct. Eukaryotic cells are generally larger than prokaryotic cells, and they are more complex. Unlike eukaryotes, they have no nucleus (their genetic material is concentrated in a nucleoid region); they also lack many of the cell organelles that eukaryotes have. They are very simple cells.

40. (E) is correct. Auxins are plant hormones that are responsible for stimulating stem elongation (when they are present in low concentration), root growth, cell differentiation, and shoot branching. They also regulate the development of fruits, and they function in gravitropism and phototropism.

■ **41. (B) is correct.** In several important characteristics, such as nutritional mode, structural organization, growth, and reproductive technique, the fungi are more similar to animals than to plants. Molecular studies have also supported this finding.

■ **42. (D) is correct.** The anterior pituitary is responsible for the secretion of some hormones that are involved in the human reproductive cycle, such as follicle-stimulating hormone (which stimulates production of sperm and ova) and luteinizing hormone (which stimulates the ovaries and testes).

■ **43. (B) is correct.** The first genetic material may have been short pieces of RNA that served as templates for aligning amino acids in polypeptide synthesis and for aligning nucleotides in a primitive form of self-replication. Early protobionts with self-replicating, catalytic RNA would have been more effective at using resources and would have increased in number through natural selection.

■ **44. (E) is correct.** Platelets are small, enucleated blood cell fragments that are derived from bone marrow. They travel to the site of a break in the skin and release clotting factors, which through a complex set of reactions transform fibrinogen to fibrin, which in turn aggregates into threads that form a framework for a clot, sealing the break.

■ **45. (A) is correct.** The light reactions of photosynthesis convert solar energy to chemical energy in the form of ATP and NADPH. They do this when light is absorbed by various pigments in the thylakoid membrane of the chloroplasts; the pigments pass the energy down a chain of electron acceptors, and in the process, ATP and NADPH are produced.

■ **46. (D) is correct.** Fungi are eukaryotes that are decomposers—also known as saprobes. They absorb nutrients from nonliving organic material such as decomposing plants, dead animals, or wastes from living animals. The bodies of fungi are composed of hyphae—tiny filaments that form a mat called a mycelium.

■ **47. (E) is correct.** If a sex-linked trait is recessive, the female will express it only if she is homozygous for it, whereas a male needs only to receive the affected allele from his mother in order to be affected. If this couple produces a color-blind son, then the mother must carry the allele for color blindness. If the father is color blind, he cannot pass the trait on to a son because all sons inherit a Y chromosome from their father, rather than an X.

■ **48. (B) is correct.** This is an example of hybrid sterility—if two species mate and produce offspring, they can still be reproductively isolated if their offspring cannot reproduce.

■ **49. (C) is correct.** The half-life of a radioactive isotope is the amount of time it takes for half of the original sample to decay. The half-life is useful because it is unaffected by temperature, pressure, or any other changes in environment.

■ **50. (D) is correct.** Pollen grains are the male gametophytes of flowering plants. In the anthers of the plant are microspores, which divide by mitosis to produce a generative cell nucleus and a pollen tube cell nucleus. A pollen grain consists of these two nuclei enclosed in a thick wall.

■ **51. (D) is correct.** Salivary amylase is an enzyme that is found in human saliva and secreted into the oral cavity. It is capable of hydrolyzing starch (a glucose polymer found in plants) and glycogen (a glucose polymer in animal tissues). Salivary amylase breaks down these carbohydrates into maltose and other disaccharides.

■ **52. (A) is correct.** We perceive the leaves of plants to be green because chlorophyll absorbs blue and red light while reflecting and transmitting green light.

■ **53. (D) is correct.** Because water molecules can form relatively strong hydrogen bonds between them, many unique characteristics of water result. These characteristics include good insulation properties with a high specific heat, greater density in liquid form than solid form, and high surface tension.

■ **54. (D) is correct.** In the nervous system, some nerve cells (neurons) are covered by Schwann cells, which are wrapped in myelin. As the nerve impulse travels the length of the nerve cell, it jumps along the gaps between Schwann cells. These gaps are called nodes of Ranvier.

■ **55. (B) is correct.** In cellular respiration, the pyruvate that is produced in glycolysis is converted to acetyl CoA, which then enters the citric acid cycle. In each "turn" of the citric acid cycle, acetyl CoA is oxidized, CO_2 is reduced, and the following are produced: 1 ATP, 3 NADH, and 1 $FADH_2$.

■ **56. (A) is correct.** All of the listed animals are part of the phylum Arthropoda. Arthropods are characterized by their segmentation, hard exoskeleton, and jointed appendages. The exoskeleton of arthropods is composed of protein and chitin—and in order to grow, arthropods must shed their hard exoskeleton in a process called molting.

■ **57. (C) is correct.** Binary fission is the process by which bacteria reproduce. The bacterium replicates its DNA, and the DNA migrates to opposite ends of the cell. Then the plasma membrane infolds, and a cell wall begins to divide the two daughter cells.

■ **58. (A) is correct.** In meiosis II, the chromosome number of the cell is reduced by half. The parent cell is diploid, with a chromosome number of $2n$; the daughter cells are haploid, or n.

■ **59. (E) is correct.** Interphase of the cell cycle is the phase during which the cell grows and replicates its genetic material. Mitosis, which is divided into several phases, is the part of the cell cycle during which the cell divides.

■ **60. (E) is correct.** Sharks and rays are in the class Chondrichthyes; they have cartilaginous skeletons as well as jaws and paired fins.

■ **61. (C) is correct.** Sea stars and sea cucumbers are both part of the phylum Echinodermata. These animals are slow moving and have a radial body plan. Echinoderms have a water vascular system, which is a network of canals that function in movement, feeding, and gas exchange.

■ **62. (A) is correct.** These are characteristics of the class Amphibia, which includes frogs and salamanders. Amphibians are characterized by being both aquatic and terrestrial, as well as by laying eggs that have no exterior shell (which would dehydrate quickly if not laid in water).

■ **63. (B) is correct.** Reptiles such as snakes and lizards have several adaptations for land that amphibians don't have, such as scales and lungs. Reptiles are ectotherms; they absorb external heat instead of regulating their internal body temperature.

■ **64. (D) is correct.** Chordates such as humans are characterized by having a notochord; pharyngeal clefts (at some point in their development); a post-anal tail (again at some point in their development); and a dorsal, hollow nerve cord.

■ **65. (B) is correct.** Chemiosmosis is an energy-coupling reaction. The energy created by a hydrogen gradient formed across a membrane is used to drive the synthesis of ATP.

■ **66. (C) is correct.** Glycolysis is the process by which glucose is split into two molecules of pyruvate. It is a metabolic pathway that occurs in all living cells and is the first part of cellular respiration and fermentation.

■ **67. (D) is correct.** The citric acid cycle completes the breakdown of glucose started in glycolysis. In the citric acid cycle, acetyl CoA is broken down completely into carbon dioxide. It occurs in the mitochondria and is the second part of cellular respiration.

■ **68. (A) is correct.** An electron transport chain is composed of a series of electron carriers (which are proteins) embedded in a membrane. They shuttle electrons and, in the process, release energy that is used to make ATP.

■ **69. (E) is correct.** In the light reactions of photosynthesis, solar energy is converted to the chemical energy of ATP. One step of this occurs when photoexcited electrons are passed from one photosystem to the next in an electron transport chain that functions similarly to the one in cellular respiration.

■ **70. (C) is correct.** A species is defined as a population or group of populations whose members can interbreed with one another to produce viable, fertile offspring.

■ **71. (D) is correct.** An organism's ecological niche is the sum total of its use of biotic and abiotic resources in its environment.

■ **72. (A) is correct.** A population is defined as a group of individuals of one species that live together in a certain geographic area.

■ **73. (B) is correct.** A community consists of all of the populations of organisms that live in a particular geographic area.

■ **74. (E) is correct.** Monocytes develop into macrophages—phagocytic cells found in many tissues that function in innate immunity by destroying microbes and in acquired immunity as antigen-presenting cells.

■ **75. (D) is correct.** Eosinophils are leukocytes that act against large parasitic invaders; they position themselves against the wall of a parasite and inject destructive enzymes into the invader.

■ **76. (B) is correct.** Antibodies are secreted by B cells. They are proteins that are specific to antigens. They represent the effectors in an immune response.

■ **77. (A) is correct.** Antigens are foreign particles in the body that elicit a response from the immune system.

■ **78. (C) is correct.** Histamines are chemical signals that are released by cells of the body in response to an injury; they are stored in and released by mast cells.

■ **79. (B) is correct.** The stomach is an elastic muscular sac that stores and starts to digest food. It secretes gastric juice, which contains enzymes that start the initial digestion of proteins. The churning motion of the stomach walls also facilitates digestion.

80. (D) is correct. The gallbladder is an organ found near the liver that stores bile and releases it into the small intestine when needed to emulsify fats.

81. (E) is correct. The duodenum is the first section of the small intestine, where acid chyme from the stomach mixes together with digestive juices from the pancreas, liver, gallbladder, and walls of the small intestine.

82. (C) is correct. The pancreas secretes digestive enzymes—as well as an alkaline solution—into the small intestine.

83. (A) is correct. The liver produces bile, and bile is stored in the gallbladder. It acts as a detergent, aiding in the digestion of fats.

84. (B) is correct. The phospholipid bilayer is made up of phospholipids, which have a hydrophilic head group and two hydrophobic, fatty acid chains. The head group points inward to the cytoplasm and outward to the interstitial fluid.

85. (C) is correct. The hydrophobic fatty acid tails of the phospholipids point toward each other, avoiding contact with the cytoplasm and the watery environment outside of the cell.

86. (A) is correct. Proteins are both embedded in the phospholipid bilayer (these are integral proteins) and associated with the cytosol face of the cell membrane (these are called peripheral proteins).

87. (C) is correct. The enzymes that would be able to function in the bloodstream of humans according to this data are 1, 2, and 5. The temperature of the human body is about 36–38 degrees Celsius, so enzymes 1 and 2 would be somewhat functional; the pH of the bloodstream is about 7.4. Only enzyme 5 would function at that pH.

88. (C) is correct. The enzyme most likely to function in the hot springs of Yellowstone National Park is enzyme 3, which has an optimal activity at a temperature of about 80 degrees Celsius. Enzymes of thermophilic (heat-loving) bacteria work best at very high temperatures.

89. (E) is correct. Enzyme 5 is the most efficient enzyme in this example—it has the highest rate of reaction. This means that its optimal rate is faster than that of any of the other enzymes depicted, assuming a standard measurement for the y-axes of both graphs.

90. (A) is correct. Just after the ^{238}U was injected into the left atrium, it would travel to the left ventricle. Blood enters the heart through the right and left atria; next it travels to the right or left ventricle.

91. (B) is correct. Just after it was injected, the blood carrying the radioactive carbon would be detected dropping off oxygen in the systemic capillaries. Blood leaves the left ventricle, enters the aorta, and then travels through the body, dropping off oxygen to active metabolic tissues.

92. (C) is correct. Just after it was injected, the blood carrying ^{32}P would be found in the lung capillaries, where it would be picking up oxygen and dropping off carbon dioxide before returning to the heart through the left atrium. Oxygen-depleted blood is pumped through the right side of the heart on its way to the lungs.

93. (B) is correct. An absorption spectrum is a graph plotting a particular pigment's light absorption—that is, the fraction of light not reflected or transmitted versus the wavelength of light. This is a plot of the absorbency of the chlorophyll a pigment versus light wavelength.

▍**94. (D) is correct.** The reason the action spectrum for photosynthesis doesn't match the absorption spectrum for chlorophyll *a* is because chlorophyll *a* is not the only photosynthetically important pigment in the chloroplast. Two other photosynthetically important pigments are chlorophyll *b* and carotenoids.

▍**95. (B) is correct.** Species A is most likely to be autotrophic—autotrophs convert light energy from the sun into chemical energy of organic molecules. As you can see, species A does not consume any other species (likewise with B), so it must produce its own food by obtaining energy from the sun.

▍**96. (D) is correct.** Species E, F, and G are all secondary consumers in this food web. Secondary consumers consume primary consumers, and primary consumers consume primary producers. In this case, for example, species F eats species C and D, which in turn consume primary producers A and B.

▍**97. (B) is correct.** The carnivores in this food web are represented by species F, G, H, and I. Carnivores are generally secondary, tertiary, and quaternary consumers, and they are distinct from omnivores, which eat both plants and animals. Species E is an example of an omnivore.

▍**98. (E) is correct.** If this were a food web drawn to represent the ecosystem of a savanna, species A and B most likely would represent two species of grasses, since grasses are the predominant primary producers of the savanna biome. Insects are also predominant as primary consumers, so species C, D, and E could represent different species of insects.

▍**99. (D) is correct.** The two species that are likely to be most closely related evolutionarily are species D and E. Because their DNA denatures at about the same temperature (57°C and 58°C, respectively), one could hypothesize that the composition of their DNA is similar.

▍**100. (C) is correct.** Another way to denature DNA is to lower the pH of its environment—DNA denatures at low pH, and keeping track of the pH at which each of the DNA samples degraded would give you the same kind of data as the table above. DNA that degraded at relatively higher pH would be less tightly bound than DNA that degraded at lower pH.

Free-Response Questions

1. (a) The cell membrane of a typical animal cell is composed of three main components: phospholipids, which are two fatty acids joined to two glycerol hydroxyl groups and a phosphate group connected to the third glycerol hydroxyl group; proteins, both integral (embedded in the cell membrane) and peripheral (associated with the outside of the membrane); and membrane carbohydrates and glycolipids, which are small carbohydrates associated with the outside of the membrane.

Phospholipids form a semisolid foundation for the rest of the molecules in the cell membrane. Some very small molecules and ions can pass through the lipid membrane unaided. This type of movement across the membrane is called passive diffusion, because energy is not needed to move the substance across the membrane.

The function of the proteins is multifold, but one important function is to facilitate the passive transport of water and certain other solutes across the membrane. The proteins that serve this function are called transport proteins. Proteins can also participate in the active transport of certain substances across the membrane; they can act as pumps that use ATP energy to transport substances against their concentration gradient. Proteins can also act as important cell-surface receptors.

Membrane carbohydrates and glycolipids on the cytosolic surface of the cell membrane are important in cell-cell recognition; these carbohydrates and glycolipids vary from species to species, and from cell type to cell type, so they function in cell-cell recognition.

(b) Glycoproteins found on viral envelopes bind to specific receptor molecules on the surface of a host cell. This promotes entry of the capsid and viral genome into the cytoplasm where cellular enzymes digest the capsid and release the genetic material. Viral reproduction follows, but the specific steps differ depending on the type of virus.

Hormones are the chemical messengers of the body. They travel through the bloodstream to their target cells. There are two ways by which hormones can gain entry into a target cell. Lipid-soluble hormones diffuse through the cell membrane, then through the cytoplasm, and then bind to a specific receptor protein in the nucleus. Water-soluble hormones bind to a specific receptor protein found on the target cell membrane. This binding signals a series of biochemical signal transduction pathways.

Water enters the cell through a process called facilitated diffusion. This means that water crosses the cell membrane down its concentration gradient, but with the help of specific transport proteins. Transport proteins are specific for the molecules they assist across the membrane, but they do not require the input of energy.

This is a good free-response answer because it shows knowledge of the following important biological terms:

phospholipid	*passive diffusion*
fatty acid	*facilitated diffusion*
glycerol	*active transport*
integral protein	*hormones*
peripheral protein	*target cell*
carbohydrate	*signal transduction pathway*
glycolipid	*receptor*
capsid	

The response also shows knowledge of the following important biological processes: the importance and function of the cell membrane; how molecules get across the cell membrane, how viruses infect cells; and mechanisms of hormonal signaling.

2. (a) Three major obstacles the plants faced for living on land were dehydration, reproduction, and support. Aquatic plants do not require adaptations for *conserving* water (because water is all around them and available at all times), but land plants are in danger of losing water through evaporation. Plants faced a reproductive obstacle—fertilization—as well. In water, the gametes could float from one plant to another, allowing for fertilization. On land, however, there is no watery environment for these gametes to float in. Also even after fertilization occurs, what will stop the zygote from desiccating? A third obstacle to plants' living on land was structural. In aqueous environments, plants are supported by water. On land, they would need to develop mechanisms for support.

(b) The problem of dehydration was solved by the development of a waxy epidermal layer on the outside of leaves called the cuticle. Plants also have stomata, which allow for gas exchange. Stomata can open and close to control excessive transpiration. Vascular tissues (xylem and phloem) also evolved for the transport of water and other nutrients around the plant body. Xylem transports water from the roots up to the leaves, and phloem carries sugar from the leaves (source) to other parts of the plant (sink).

The problem of how to reproduce on land was solved in part by the evolution of the seed. A seed consists of a plant embryo combined with a food supply and encased in a protective coat. This protective coat prevents the embryo from dehydrating even if the seed sits on dry ground for a relatively long period of time. Another adaptation to aid in plant reproduction on land was the spore. Spores are reproductive cells that can develop into a mature plant without fusing with another cell. They are generally lightweight and can travel significantly far from the parent through the air and grow into a new plant.

Plants adapted to living on land without the structural support of water. They developed hard, stiff shoots that enabled them to grow to great heights. The shoots of plants are made up of several types of plant tissues, including the vascular tissues xylem and phloem. Both xylem and phloem are made up of dead water-conducting cells joined together to form long, stiff tubes that help support the plant in its growth.

This is a good free-response answer because it shows a working knowledge of the following important terms:

fertilization	*seed*
gametes	*xylem*
epidermis	*phloem*
cuticle	*spores*
stomata	

The response also shows a working knowledge of the important biological concept of the evolutionary adaptations of plants that enabled them to colonize terrestrial environments.

3. The cell cycle of a plant cell, much like the cell cycle of most other types of cells, includes two main phases: interphase and mitosis. Interphase is divided into three phases: the G_1 (gap 1), S (synthesis), and G_2 (gap 2). The cell undergoes a tremendous amount of biochemical activity during the G_1 phase, in which the plant cell grows and produces new organelles. In the S phase, synthesis of new DNA material takes place. During the G_2 phase, there is continued growth, and organelles needed for cell division or mitosis are replicated. There are several checkpoints in the cell cycle of a plant cell, the most crucial of which is the G_1 checkpoint. If the plant cell gets the go-ahead signal at this checkpoint, it will be committed to divide. If not, it will enter the G_0 phase, which is a nondividing phase.

In late interphase, just before the mitotic phase, the nucleus is intact and the chromosomes are not well defined in the nucleus. When the plant enters the first mitotic phase (prophase), the chromatin fibers become more condensed into visible chromosomes, each of which has two sister chromatids. In the cytoplasm, the mitotic spindle forms, and the centrosomes move away from each other and toward the opposite poles of the cell. The next phase that the plant cell would enter is prometaphase, in which the nuclear envelope would fragment and the microtubules would attach to the kinetochores of the chromatids. In metaphase, all of the chromosomes would line up on the metaphase plate at the equator of the cell, and microtubules would be attached to the kinetochores of every sister chromatid. When the plant cell enters anaphase, the sister chromatids are "dragged" apart from each other to opposite ends of the cell by the retracting microtubules.

The final stages of plant cell mitosis are telophase and cytokinesis. In telophase, two new nuclei begin to form around the groups of sister chromatids that have now migrated to opposite ends of the cell. The chromosomes becomes less condensed, and the cell plate—which divides the cytoplasm in two—starts to grow at the center of the cell, eventually growing all the way to the perimeter of the parent cell and effectively dividing the cell in two.

Following the outline above, this response deals with the following important terms:

cell cycle	*chromatin fibers*
G_1 phase	*sister chromatids*
S phase	*mitotic spindle*
G_2 phase	*prometaphase*
G_1 checkpoint	*metaphase*
G_0 phase	*nuclear envelope*
nondividing phase	*microtubules*
interphase	*kinetochores*
mitotic phase	*telophase*
centrosomes	*cytokinesis*
prophase	*cell plate*

The response also describes thoroughly the following processes: the cell cycle including its checkpoints, plant cell mitosis, and cytokinesis.

4. (a) There are three convincing reasons the theory that glycolysis was the first ATP-producing metabolic pathway to evolve is probably true. The first reason is that long ago, Earth's atmosphere contained almost no oxygen, and only relatively recently have the current atmospheric levels of gases come to be what they are. Glycolysis does not require oxygen, so it is possible that prokaryotes (which evolved before eukaryotes) used this method for making ATP.

The second substantiating clue is that glycolysis is a very common method for making ATP; in fact, almost all living organisms use it. This commonality implies that it originated very early in the evolution of metabolic pathways.

The final reason has to do with the site of glycolysis—that is, it takes place in the cytosol, and not in an organelle. Prokaryotic cells, which evolved first, are much simpler than eukaryotic cells, and they contain no membrane-bound organelles (not even a nucleus). Therefore, if glycolysis were to take place in an early prokaryotic cell, it would have to evolve such that it was capable of taking place in the cytosol—for instance, it would have to evolve such that it did not rely on a specialized membrane in order to function.

(b) In the course of the citric acid cycle, acetyl CoA is first joined to oxaloacetate to form citrate, and then the molecule is manipulated extensively to finally re-form a molecule of oxaloacetate. In the course of these reactions, the citric acid cycle produces 1 ATP molecule, 3 NADH molecules, and 1 FADH$_2$ molecule (per turn).

The way that the citric acid cycle is related to oxidative phosphorylation is that the NADH and FADH$_2$ molecules produced during the citric acid cycle donate electrons to the electron transport chain, which is embedded in the wall of the inner mitochondrial membrane. This electron transport chain shuttles the electrons down its length, in an exergonic reaction. The energy produced in this series of electron transfers is used to power an enzyme called ATP synthase, which is also embedded in the wall of the inner mitochondrial membrane; ATP synthase catalyzes the phosphorylation of ADP to form ATP.

This is a good free-response answer because it knowledgeably uses the following key terms:

glycolysis	*oxaloacetate*
prokaryotes	*citrate*
eukaryotes	*citric acid cycle*
ATP	*NADH*
cytosol	*FADH$_2$*
organelle	*inner mitochondrial membrane*
nucleus	*phosphorylation*
acetyl CoA	

The response also shows a working knowledge of the following important biological concepts: the origin of life and ancient Earth, relationships among living organisms, glycolysis, the citric acid cycle, and the electron transport chain.

Index

A

A band, 273
A site, 126
Abiotic components of environment, 291
Abscisic acid, 221
Absorption, 234
Absorption of nutrients, 185
Absorption spectrum, 83
Acetylcholine, 267, 273
Acid chyme, 235
Acid precipitation, 304
Acids, 30
Acoelomates, 188, 191
Acquired immunity, 244–247
Acrosomal reaction, 261
Acrosome, 261
Actin, 47, 273
Action potential, 266
Action spectrum, 83
Activation energy, 54
Activator proteins, 130
Active immunity, 246
Active site, 54
Active transport, 51
Adaptations, 152
Adaptive radiation, 160, 163
Addition rule in calculating probability, 106
Adenine (A), 36, 118, 126
Adenosine diphosphate (ADP), 54, 82
Adenosine triphosphate. See ATP.
ADH (antidiuretic hormone), 251, 252
Adhesion, 29, 346
Adipose cells, 34
ADP (adenosine diphosphate), 54, 82
Aerobic respiration, 75
Afferent arterioles, 250
Agonistic behaviors, 277
Agricultural applications of DNA
    technology, 139
Albinism, 107
Alcohol fermentation, 80
Aldosterone, 251
Alimentary canals, 235
Allantois, 196
Alleles, 104
    multiple, 106
Allergies, 247
Allopatric speciation, 160
Allosteric site on enzyme, 57
Alpha helix, 34
Alternation of generations in plants, 178
Altruism, 277
Alveoli, 242

Amino acids, 34, 125
    essential, 233
Aminopeptidase, 236
Ammonia, 248
Amniocentesis, 108
Amnion, 196
Amniotes, 196, 263
Amniotic egg, 196, 197
Amylase, 235, 236
Anabolic pathways, 53, 80
Anaerobic conditions, 80
Analogous likenesses, 153, 172
Anaphase, 62, 63
Anaphase I, 100, 102
Anaphase II, 101, 102
Anchorage dependency, 64
Aneuploidy, 111
Angiosperm, 182
    features of life cycle, 217
    reproduction and biotechnology,
        216–219
Angiotensin, 251
Angiotensin II, 251
Animal behavior, 275–277
    lab, 354–357
Animal development, 261
Animal diversity, 186–190
Animal form and function, basic
        principles, 231–233
Animal nutrition, 233–237
Animal reproduction, 255–260
Annual plants, 209
Anterior pituitary, 252
Antheridia, 179
Anthophyta, phylum, 182
Antibiotics, 176
Antibodies, 244
Anticodon, 125
Antidiuretic hormone (ADH), 251, 252
Antigen-presenting cells (APCs), 244
Antigens, 244
Antimicrobial proteins, 243, 244
Antiparellel strands in DNA, 118
Aorta, 239, 240
AP Scholar Awards, 3
APCs (antigen-presenting cells), 244
Aphotic zone, 292
Apical meristems, 209
Apoplastic route, 213
Apoptosis, 60, 132, 221
Aposematic coloration, 298
Appendix, 236
Aquaporins, 50, 212

Aquatic biomes, 292
Aquatic primary productivity lab, 357–363
Aqueous humor, 272
Archaea, 43, 184–176
Archegonia, 179
Archenteron, 186
Arteries, 238
Arterioles, 238
Artificial selection, 152
    of crops, 219
Ascending loop of Henle, 250
Ascomycota fungi, 185
Asexual reproduction, 97, 255
    of plants, 219
Associative learning, 276
Astrocytes, 269
Atomic number, 27
Atoms, 27
ATP (adenosine triphosphate), 54, 82, 83
    in Calvin cycle, 87–88
    pump, 51
    synthase, 79, 80, 86
ATP-consuming phase of glycolysis, 77
ATP-producing phase of glycolysis, 77
Atria, 238
Atrioventricular (AV) valve, 240
Auditory canal, 271
Auditory signals, 276
Autonomic nervous system, 269, 270
Autopolyploid plants, 160
Autosomes, 98
Autotrophs, 81, 301
Auxin, 220
    hormone group, 220
    role in gravitropism, 222
AV node, 240
Axillary buds, 208
Axons, 264

B

B cells, 244, 245, 246
B lymphocytes (B cells), 244, 245, 246
Bacteria, 43, 174–176
Bacterial chromosome, 120
Bacterial transformation, 329
Bacteriophages, 117, 132
Bark, 210
Barr body, 110
Barrier defenses, 243
Basement membrane, 231
Base-pair substitution of nucleotide, 128
Bases, 30
Basidiomycota, 185

Basophils, 244
Batesian mimicry, 299
B-cell activation, 244
Behavior, 275
Behavioral isolation, 159
Benign tumor, 64
Benthic zone, 292
Beta pleated sheet, 34
Bicarbonate fluid, 235
Biennial plants, 209
Big-bang reproduction, 295
Bilateral symmetry of animal body, 186
Bilaterally symmetrical animals, 191
Bile, 235
Binary fission, 174
Binomial nomenclature, 151, 171
Biofuels, 219
Biogeochemical cycles, 302
Biogeographic factors in community, 300
Biogeography, 154
Biological magnification, 304
Biological species concept, 159
Biomass, 299
Biomes, 292
Bioremediation, 176
Biosynthesis, 80
Biotechnology, 134
Biotic influences, 291
Birds, 197
Birth rates, factors reducing, 297
Blastocoel, 262
Blastocyst, 263
Blastomeres, 262
Blastopore, 186
Blastula, 186, 187, 262
Blind spot, 272
Blood, 237
Blood components, 240–241
Blood pressure, 240
Blood types, 246
Blue-light photoreceptors, 221
Bohr shift, 243
Bolus, 235
Bottleneck effect, 157
Bottleneck event, 341
Bowman's capsule, 248
Brainstem, 270
Breathing, 242
Broadleaf forest, temperate, 293
Bronchi, 242
Bronchioles, 242
Bryophytes, 179, 180
Budding, 255
Buffers, 30
Bulbourethral glands, 257
Bulk flow, 212
Bundle-sheath cells, 89

C

Calcitonin, 254
Calcium ion release, 262
Calcium ions, 273
Calvin cycle, 82, 83, 84
    converting CO_2 to sugar, 87–88
CAM photosynthesis, 89
Cambium, 210
Cancer cells, 64
Cancer from genetic changes, 131–132
Canopy, 293
Capillaries, 238
Capillary beds, 239
Capsid, 132
Carbohydrates, 32, 49
Carbon atoms, 31
Carbon cycle, 302, 303
Carbon fixation, 82
Carbon skeleton, 31
Carbonic acid, 30
Carbonic anhydrase, 243
Carboxypeptidase, 236
Cardiac cycle, 239
Cardiac muscle, contraction of, 240
Carnivorous plants, 216
Carotenoid molecules, 84
Carpels, 183
Carrying capacity, 296
Casparian strip, 213
Catabolic pathway, 53, 75–76
Catalysts, 54
Catastrophism, 151
CD8 protein, 245
cDNA library, 136
Cecum, 236
Cell body of neuron, 264
Cell-cell signals, 130
Cell communication, 57–60
Cell cycle, 60–64
    control system, 63
Cell differentiation, 130
Cell division, 130
    zone of, 210
Cell-mediated immune response, 245
Cell membranes 34
Cell plate, 63, 102
Cell respiration lab, 326–329
Cell suicide, 60
Cell-surface receptors, 252
Cell wall, 45, 48
Cells, types of differentiated in plants,
    208–209
Cellular innate defenses, 243
Cellular membranes, 48–49
Cellular respiration, 46, 75–80
    chemiosmosis in, 86
Cellulose, 33

Central nervous system (CNS), 266
Central vacuole, 45, 48
Centrioles, 47
Centromere, 61
Centrosome, 44, 47
Cephalization, 188, 268
Cerebellum, 270
Cerebral cortex, 270
Cerebrospinal fluid, 269
Cerebrum, 270
Cervix, 256
Chaparral, 293
Chaperonins, 35
Charophytes, 178
Chase, Martha, 117
Checkpoints in control system, 63
Chemical bonds, 28
Chemical cycling process, 301
Chemical energy, 53
    harvesting, 75–80
Chemical synapse, 267
Chemiosmosis, 78–80, 85, 86
Chemoautotrophs, 175
Chemoheterotrophs, 175
Chemoreceptors, 271
Chiasmata, 101
Chi-square data analysis, 336–338
Chitin, 33, 185
Chlorophyll, 81, 83, 84, 322
Chlorophyll *a*, 84
Chloroplast, 45, 48, 81, 82
    generating ATP, 85
Choanocytes, 191
Cholecystokinin (CCK), 236
Cholesterol, 34
Chordates, characteristics of, 194
Chorion, 196
Chorionic villus sampling, 108
Choroid, 272
Chromatin, 44, 45, 46, 63, 120
Chromosomal mutations, 155
Chromosome theory of inheritance, 109
Chromosomes, 43
    independent assortment of, 103
Chymotrypsin, 236
Cilia, 47, 242
Circadian rhythms, 214, 221, 276
Circulation and gas exchange, 237–243
Circulatory system, 237–241
    components of, 237
    physiology lab of, 351–353
Cisternae, 46
Citric acid cycle, 78, 80
Clade, 172
Cladogram, 172
Class Amphibia, 196
Class Chondrichthyes, 196

Class I MHCs, 245
Class II MHCs, 245
Class of organism, 171
Class Osteichthyes, 196
Classical conditioning, 276
Cleavage, 186, 262
Cleavage furrow, 63, 102
Climate, 291
Clitoris, 256
Clonal selection, 244
Clone, 97
Cloning genes, steps in, 135–136
Cloning vector, 135
Closed circulatory systems, 238
Club fungi, 185
CNS (central nervous system), 266, 269
Cochlea, 271
Codominance in inheritance patterns, 106
Codon recognition, 127
Codons, 123, 125, 126
Coelomates, 188, 191
Coenocytic fungi, 185
Coenzymes, 56
Cofactors, 56
Cognition, 276
Cognitive map, 276
Cohesion, 29, 346
Collecting duct, 250, 251
Collenchyma cells, 208, 350
Colon, 236
Commensalism, 175, 299
Community, 298
Community ecology, 298–300
Companion cells, 209
Competent cells, 330
Competitive exclusion principle, 298
Competitive inhibitors, 56
Complement system, 244
Complete digestive tracts, 235
Complete dominance in inheritance
    patterns, 106
Complete metamorphosis, 192
Compound, 27
Compound eyes, 272
Concentration gradient, 50
Conceptual-thematic questions, 9
Condensation reactions, 32
Cones for color vision, 272
Conformation, 32
Coniferous forest, 294
Conjugation of genes, 174
Connective tissue, 231
Continental drift, 154, 162–163
Controlled experiment, 355
Convergent evolution, 153, 172
Coral reef, 292
Corepressor, 129

Cork cambium, 210
Cornea, 272
Corpus callosum, 270
Corpus luteum, 256
Cortical reaction, 261
Cost of exam, 4
Cotransport, 51, 211
Cotyledons, 217
Countercurrent exchange, 233, 242
Countercurrent system, 251
Covalent bond, 28
    of carbon, 31
Craniates, 195
Creutzfeldt-Jakob, 134
Crick, Francis, 117, 118
Cristae, 46
Critical values table, chi-square, 337
Crossing over, 103, 110
    of alleles, 155
    of DNA, 101
Cryptic coloration, 298
Cuticle of leaves, 208
Cuvier, Georges, 151
Cyclic electron flow, 85
Cyclin-dependent kinases (Cdk), 64
Cyclin proteins, 64
Cystic fibrosis, 108
Cytochrome complex, 86
Cytokines, 245
Cytokinesis, 61, 62, 63, 100, 101, 102
Cytokinins, 220
Cytoplasm, 44
Cytoplasmic activities, 60
Cytoplasmic determinants, 130, 264
Cytosine (C) , 36, 118, 126
Cytoskeleton, 44, 45, 46
Cytosol, 43, 46, 77
Cytotoxic T cells, 245, 246

D

Darwin, Charles, 152
Darwinian view of life, 151–154
Daughter cells, 99
Day-neutral plants, 222
Day of exam, tips for, 17–18
Death rates, factors increasing, 297
Decomposer prokaryotes, 175
Decomposers, 301
Dehydration reactions, 32
Dehydration synthesis, 32, 34
Deletion of chromosomal fragment, 111
Deletions of nucleotide pairs, 128
Demographic transition, 297
Demography, 295
Denaturation, 35
Denaturing, 32
Dendrites, 264

Denitrification, 340
Density, population, 294
Density-dependent factors, 297
Density-dependent inhibition, 64
Density independent population, 297
Dentition, 237
Deoxyribonucleic acid. *See* DNA.
Dermal tissue, 208
Descending loop of Henle, 250
Desert, 293
Desmosomes, 48
Determination of cells, 131
Detritivores, 301
Detritus, 292
Deuterostomes, 189
Deuterostomia, 192
Diaphragm, 242
Diastole, 239, 351
Diastolic pressure, 240
Differential gene expression, 130
Diffusion lab, 312–314
Digestion, 234
Dihybrid, 336
Dihybrid cross, 105
Dipeptidases, 236
Diploid, 99
Diploid cells, 60
Diploidy, 158
Directional selection, 157
Directive words, 19
Dispersion of population, 294
Disruptive selection, 157
Dissolved oxygen lab, 357–363
Distal tubule, 250
Disturbance, 300
Disulfide bridges, 34
DNA (deoxyribonucleic acid), 35, 45, 46,
    117–122, 329–335
    chip, 136
    ligase, 119, 134
    marker, 332
    methylation, 130
    microarray assays, 136
    polymerases, 119
    technology, uses for, 139
    technology and genomics, 134–139
Domain, 171
Dominant allele, 104, 105
Dominant species, 299
Dormancy of seed, 218
Dormant seed, 326
Dorsal nerve cord, 194
Double fertilization, 217
Double helix, 118
Down syndrome, 111
DPIP, 322, 323
Drought, 222

Duchenne muscular dystrophy, 110
Duodenum, 235, 236
Duplication of chromosome fragment, 111
Dynamic equilibrium, 312

E

E. coli (Escherichia coli), 237
E site, 126
Eardrum, 271
Echinodermata, 192
Ecological footprint, 297
Ecological niche, 298
Ecological succession, 300
Ecology, 291
    community, 298–300
Ecosystems, 301–304
Ectoderm, 186
    cell layer, 262
Ectotherm, 232, 352
Effector cells, 244
Efferent arterioles, 250
Electrical synapse, 267
Electrochemical gradient, 51
Electrogenic pump, 51
Electromagnet receptors, 271
Electron shells, 27
Electronegativity, 28, 29
Electrons, 27
Element, 27
Elimination, 234
Elongation
    of DNA, 124
    stage of translation, 127
    zone of, 210
Embryonic development, 261
Embryonic homologies, 153
Embryophytes, 179
Endemic species, 154
Endergonic process, 82
Endergonic reaction, 54
Endocrine glands, 252
Endocrine system, 232, 251–254
Endocytosis, 52
Endoderm, 186
    cell layer, 262
Endodermis, 213
Endometrium, 256
Endoplasmic reticulum (ER), 44, 46
Endoskeletons, 275
Endosperm, 217
Endosymbiosis, 177
Endosymbiotic hypothesis, 162
Endothelium, 238
Endotherms, 197, 232
Energy, 53
Energy coupling, 54
Energy glow process, 301
Energy storage polysaccharides, 33

Enhancer regions, 130
Enterogastrone, 236
Entropy, 53
Environmental cleanup through DNA
    technology, 139
Enzyme catalysis lab, 315–318
Enzymes, 34, 54
Enzyme-substrate complex, 54, 55
Eosinophils, 244
Epicotyl, 217
Epidermal cells, 350
Epididymis, 257
Epiglottis, 235
Epinephrine, 240
Epiphytes, 216, 294
Epistasis, 107
Epithelial tissue, 231
Erythrocytes, 240
Escherichia coli (E. coli), 237
Esophagus, 235
Essential amino acids, 233
Essential fatty acids, 233
Essential nutrients, 233
Estradiol, 260
Estrogen, 34, 259
Estrous cycles, 258
Estuaries, 292
Ethology, 275
Ethylene, 221
Euchromatin, 122
Eudicots, 183
Eudocots, 220
Eukarya, 43, 174
Eukaryotes, multicellular, 162
Eukaryotic cells, 43, 44
Eukaryotic chromosomes, 120
Eukaryotic organisms, 60
Eustachian tube, 271
Eutherians, 197
Eutrophic lakes, 292, 302
Evaluating AP grades, 3
Evapotranspiration, 302
Evo-devo, 163
Evolution
    of populations, 155–158
    scientific evidence for, 153
Evolution and population genetics lab,
    341–345
Exam fees, 4
Exaptations, 163
Excitatory neurotransmitters, 267
Excretion, 247
    osmoregulation and, 247–251
Excretion process, 248, 251
Excretory systems, 248
Exergonic reaction, 54
Exergonic release of energy, 76
Exocytosis, 52

Exons, 124
Exoskeletons, 275
Exponential population growth, 295
External fertilization, 256
Extinction, 300
    mass, 163
Extracellular digestion, 234
Extracellular matrix (ECM), 47
Extraembryonic membranes, 196
Extreme halophiles, 175
Extreme thermophiles, 175
Extremophiles, 175

F

F_1 (first filial) generation, 103
F_2 (second filial) generation, 103
Facilitated diffusion, 50
Factual multiple-choice questions, 7–8
Facultative anaerobes, 80, 175
Family of organism, 171
Fats, 33, 236
Fatty acids, 33
    essential, 233
Feathers, 197
Feedback
    inhibition, 57
    loops, 254
    systems, 232
Female
    anatomy, 256
    hormones, 256
Fermentation, 75, 80
Fertilization, 97–99, 155, 256
Fibrin, 241
Fibrinogen, 241
Fibrous roots, 207
Filtration process, 248, 251
Fission, 255
5' cap addition to RNA, 124
Fixed action pattern (FAP), 275
Fixed allele, 155
Flagella, 47, 174
Flagellum, 44
Flower, 183
Follicles, 256
Follicle-stimulating hormone (FSH),
    252, 260
Food chain, 299
Food webs, 299
Foraging behavior, 276
Forensic applications of DNA
    technology, 139
Fossil records, 153, 162
Founder effect, 157
Four-chambered heart, 197
Fragmentation, 255
Frameshift mutation, 128
Franklin, Rosaline, 117

Free energy, 54
Free-response questions, 11–12
  example of, 20–21
  grading procedures for, 13–14
  strategies for, 19–21
Freshwater biomes, 292
  types of, 292
Fruit, 183, 217
FSH (follicle-stimulating hormone),
  252, 260
Functional groups, 31
Fundamental niche, 298
Fungi, 184–185
Fungus roots, 207

G

G protein, 58
G_0 phase, 64
G_1 phase, 61
  checkpoint, 63, 64
G_2 phase, 62
checkpoint, 63
Gametangia, 179
Gametes, 97, 98, 256
  human, 60
Gametic isolation, 159
Gametophytes, reduced, 181
Gametophyte stage of plant, 178
Gap junctions, 48
Gas exchange, 241
  circulation and, 237–243
Gastric juice, 235
Gastrin, 236
Gastrovascular cavity, 234, 237
Gastrula, 186, 187
Gastrulation, 262
Gel electrophoresis, 136, 138, 329, 331
Gene cloning, 134
Gene expression, 122
  regulation of, 128–132
Gene flow, 157, 159
Gene of interest, 330
Gene pool, 155
Gene therapy, 139
Gene to protein, 122–128
Generative nucleus, 217
Genes, 97
  Mendel's ideas about, 103–108
  portion of operon, 129
Genetic crosses, predicting, 106
Genetic disorders from altered
  chromosomes, 111
Genetic drift, 157, 341
Genetic engineering, 134
Genetic recombination, 110
Genetic testing, 108
Genetic transformation, 329
Genetic variation, 101

Genetically modified organism (GM
  organism), 139, 219
Genetics of organisms lab, 335–340
Genome, 60, 132
Genomic library, 136
Genomics, DNA technology and, 134–139
Genotype, 105
Genus, 171
  of organism, 171
Geologic events, speciation by, 160
Gestation, 260
Gibberellins, 220–221
Gills, 242
Glia, 269
Globular proteins, 34, 35
Glomerulus, 248
Glyceraldehyde-3-phosphate (G3P), 88
Glycerol, 33
Glycogen, 33
Glycolysis, 77, 80
Glycoproteins, 47
Gnathostomes, 196
Goals of AP course, 5
Golgi apparatus, 44, 45, 46
Gonadotropic hormones, 260
Gonads, 256
GPP (gross primary production),
  301–302
G-protein-linked receptor, 58
Grading procedures, 13–14
Gradualism, 160
Gram-negative cells, 174
Gram-positive bacteria, 174
Graphing data, reviewing, 16
Grassland, temperate, 293
Gravitropism, 222
Gray matter, 269
Green algae, land plants evolved from,
  178–181
Geenhouse effect, 304
Gross primary production (GPP),
  301–302
Gross productivity, 358, 359
Guanine (G) , 36, 118, 126
Guard cells, 210
Guard cells, 213–214, 347
Gymnosperms, 182

H

Habitat isolation, 159
Habituation, 276
Haldane, 161
Half-life of isotope, 162
Haploid cells, 60, 98
Haploid gametes, 255
Haploid nuclei, 217
Haploid number of chromosomes, 98
Haploid spores, 180

Hardy-Weinberg
  equation, 155–156
  equilibrium, 156
  theorem, 156
Hardy-Weinberg Law of Genetic
  Equilibrium, 341
  conditions for remaining at, 341, 342
hCG (human chorionic gonadotropin), 260
Heart, 237
Heart rate, 240
  regulation of, 240
Heat, denaturation of protein through, 35
Helper T cells, 245, 246, 247
Hemoglobin, 240, 243
Hemolymph, 237
Hemophilia, 110
Hepatic portal vessel, 236
Herbivores, 237
Herbivory, 299
Hermaphroditism, 255
Hershey, Alfred, 117
Heterochromatin, 122
Heterochrony, 163
Heterospory, 181
Heterotrophs, 81, 301
  multicellular, 185
Heterozygote advantage, 158
Heterozygous organism, 105
Histamines, 244
Histone acetylation, 130
Histones, 122, 130
HIV, 247
Hollow dorsal nerve cord, 268
Homeostasis, 232
Homeotic genes, 163
Homologous chromosomes, 98
Homologous structures, 153, 172
Homology, 153
Homozygous organisms, 105
Hormones, 220, 232, 236, 251–254
Host range of viruses, 133
Hox genes, 163
Human chorionic gonadotropin (hCG), 260
Human digestive system, 234
Human evolution, features of, 197
Humoral immune response, 245
Huntington's disease, 108
Hybrid breakdown, 160
Hydrochloric acid, 235
Hydrogen bonding, 346
Hydrogen bonds, 29, 34
Hydrolysis, 32
Hydrolytic enzymes, 47
Hydrophilic molecules, 49, 50
Hydrophilic substances, 30
Hydrophobic, 33
  barrier, 49
  interactions, 34

molecules, 49, 50
substances, 30
Hydrostatic skeletons, 275
Hypertonic solution, 50, 313
Hyphae, 185
Hypothalamus, 252, 270
Hypothesis, 355
Hypotonic solution, 50, 313

I

I band, 273
Immigration rates, 300
Immune system, 243–247
Imprinting, 276
Inclusive fitness, 277
Incomplete dominance in inheritance
    patterns, 106
Incomplete metamorphosis, 192
Incus, 271
Independent assortment
    of chromosomes, 103, 155
    Mendel's law of, 104
Indoleacetic acid, 220
Inducer, 129
Inducible operon, 129
Induction of cells, 130
Inductive cell signals, 264
Inflammatory response, 244
Ingestion, 234
Inheritance
    of acquired characteristics, 152
    chromosome theory of, 109
    laws of probability in Mendelian, 106
Inhibitory neurotransmitters, 267
Initiation stage
    of transcription, 123
    of translation, 126
Innate behaviors, 275
Innate immune responses, 243
Innate immunity, 243–244
Inner cell mass, 263
Inner ear, 271
Insertions of nucleotide pairs, 128
Integral proteins, 49
Intercellular junctions, 48
Interferon, 244
Intermediate disturbance hypothesis, 300
Intermediate filaments, 44, 45, 47
Internal fertilization, 256
Interneurons, 266
Interphase, 61
    of meiosis, 100
Interspecific interactions, 298
Interstitial fluid, 238
Intertidal zone, 292
Intracellular digestion, 234
Intracellular receptors, 57, 252
Introns, 124
Invagination, 262

Inversion of chromosome fragment, 111
Invertebrates, 190–193
Ion, 28, 50
    channels, 261, 266
    pumps, 266
Ionic bonds, 28–29
Iris in eyeball, 272
Island biogeography, 300
Isomers, 31
Isotonic solution, 50, 313
Isotopes, 27

K

Karyotype, 98
Keystone species, 299
Kidneys, 248
Kin selection, 277
Kinases, 64
Kinesis, 275
Kinetic energy, 53
Kingdom, 171
Klinefelter syndrome, 111
K-selection, 296, 297

L

Lab-based multiple-choice questions, 10
Lab review for essay, 15–16
Labia, 256
Laboratory
    animal behavior, 354–357
    cell respiration, 326–329
    diffusion and osmosis, 312–314
    dissolved oxygen and aquatic primary
        productivity, 357–363
    enzyme catalysis, 315–318
    genetics of organisms, 335–340
    mitosis and meiosis, 318–321
    molecular biology, 329–335
    physiology of circulatory system,
        351–353
    plant pigments and photosynthesis,
        321–325
    population genetics and evolution,
        341–345
    transpiration, 346–351
Lacteal, 236
Lactic acid fermentation, 80
Lagging strand of DNA, 119
Lamarck, Jean-Baptiste de, 152
Lampreys, 195
Large intestine, 236
Larynx, 242
Lateral meristems, 209
Law of independent assortment,
    Mendel's, 104
Law of segregation, Mendel's, 104
Laws of probability in Mendelian
    inheritance, 106
Laws of thermodynamics, 53

Leading strand of DNA, 119
Leaf-to-root translocation, 215
Learning, 276
Leaves, 208
Left atrium, 239
Left ventricle, 239
Lethal dominant alleles, 108
Leukocytes, 240
Leydig cells, 257
LH, 260
Lichens, 185
Life cycle, 97
Life history of organism, 295
Ligand, 57
Ligand-gated ion channels, 58
Light-harvesting complex, 84
Light reactions, 82
Limnetic zone, 292
Linear (noncyclic) electron flow, 84
Linkage map, 110
Linked genes, 110
Linnaeus, Carolus, 151, 171
Lipase enzyme, 236
Lipids, 33–34, 48
Littoral zone, 292
Liver, 236
Locus, 97
Long-day plants, 222
Looped domains of 30nm fiber, 121, 122
Loops of Henle, 242, 250
Lorenz, Konrad, 276
Lungs, 242
Lupus, 247
Luteinizing hormone, 252
Lyell, Charles, 152
Lymph, 237, 240
Lymph nodes, 240
Lymphatic system, 240
Lymphocytes, 244
Lysis, 246
Lysogenic cycle, 133
Lysosome, 44, 47
Lysozyme, antimicrobial, 243
Lytic cycle, 133

M

Macroclimate patterns, 291
Macroevolution, 159
Macromolecules, 32–36
Macronutrients for plants, 215
Macrophages, 244
Mad cow disease, 134
Major histocompatibility complex
    molecules (MHCs), 244, 245, 247
Malignant tumor, 64
Malleus, 271
Malpighian tubules, 248
Mammals, 197
Mammary glands, 197

Map unit, 110, 319
Marine biomes, 292
Marker, DNA, 332
Marsupials, 197
Mass number, 27
Mast cells, 244
Matching questions, 9–10
Mating systems, 277
Matter, 27
Maturation, zone of, 210
Mechanical isolation, 159
Mechanoreceptors, 271
Mediator proteins, 130
Medulla oblongata, 270
Megaspores, 181
Meiosis, 60, 99, 217
    differences from mitosis, 102–103
    in mammals, 258
    and sexual life cycles, 97–103
    simulation, 318
Meiosis I, 100
Meiosis II, 101, 102
Meiosis lab, 318–321
Membrane potential, 51, 266
Membrane structure and function, 48–52
Memory cells, 244
Mendel, 103–108
Mendelian patterns of inheritance, 106–107
Menstrual cycles, 258
Menstrual flow phase, 258
Meristems, 209
Mesoderm, 262
Mesoderm cell layer, 262
Mesophyll cells, 89
Mesophyll tissue, 81
Messenger RNA (mRNA), 45, 122
Metabolism, 53–57
    and enzyme activity, 57
Metamorphosis, 186, 192
Metanephridia, 248
Metaphase, 62, 63
Metaphase chromosome, 121, 122
Metaphase I, 100, 101
Metaphase II, 101, 102
Metastasis, 64
Methanogens, 175
Methylation, 110
MHCs (major histocompatibility complex
    molecules), 244, 245, 247
Micro RNAs (miRNA), 130
Microclimates, 291
Microevolution, 155, 159
Microfilaments, 44, 45, 47
Micronutrients for plants, 215
Microsporangia, 217
Microspores, 181
Microtubules, 44, 45, 46–47, 63
Microvilli, 44, 236
Midbrain, 270

Middle ear, 271
Migration, 275
Miller, 161
Mineral absorption by roots, 212
Minerals, 233
Mismatch repair, 119
Missense mutations, 128
Mitochondria, 46
    generating ATP, 85
Mitochondrion, 44, 45
Mitosis, 60, 61, 217
    differences from, 102–103
    observing, 318
Mitosis lab, 318–321
Mitotic phase, 61–63
Mitotic spindle, 63
Molecular basis of inheritance, 117–122
Molecular biology lab, 329–335
Molecular clocks, 173
Molecular homologies, 153
Molecular systematics, 172
Molecules, formation and function, 28–29
Monocots, 183, 220
Monocytes, 244
Monogamous mating systems, 277
Monohybrid, 335
Monohybrid cross, 105
Monomers, 32
Monophyletic group, 176
Monosaccharides, 32
Monosomic eggs, 111
Monotremes, 197
Morgan, Thomas Hunt, 109
Morphogenesis, 130, 210
Morphogens, 131
Morula, 262
Motile prokaryotes, 174
Motor mechanisms, 271–275
Motor molecules, 47
Motor nervous system, 269
Motor neurons, 266, 273
Mouth, 235
MPF molecules, 64
mRNA (messenger RNA), 45
Mucus produced in stomach, 235
Mucus-producing cells in trachea, 242
Müllerian mimicry, 299
Multicellular heterotrophs, 185
Multiple-choice questions, 7–10
    grading procedures for, 13
    strategies for, 18–19
Multiple sclerosis, 247
Multiplication rule in calculating
    probability, 106
Muscle contraction, 273
Muscle tissue, 231
Mutagens, 128
Mutations, 155, 174
Mutualism, 175, 299

Mycelium, 185
Mycorrhizae, 207, 212, 216
Mycorrhizal fungi, 185
Myelin sheath, 264
Myofibrils, 273
Myofilaments, 273
Myosin, 47, 274

N

NAD^+, 76, 80
NADH, 76, 77, 80
$NADP^+$, 82, 85, 322
$NADP^+$ reductase, 86
NADPH, 82, 83, 85, 322
    in Calvin cycle, 87–88
Natural killer (NK) cells, 244
Natural selection, 152, 156
Negative feedback, 252, 254
Negative feedback systems, 232
Negative gravitropism, 222
Negative phototropism, 220
Nephridia, 192
Nephrons, 248, 250–251
Neritic zone, 292
Nerve cell, 232
Nerve impulses, 266
Nerve net, 268
Nerves, 266
Nervous system, 232, 268–270
Nervous tissue, 232
Net primary production (NPP), 302
Net productivity, 358, 359
Neural plate, 263
Neuromuscular junction, 273
Neurons, 232, 264–267, 268
Neurotransmitter release, 267
Neurotransmitters, 265, 267
Neurulation, 263
Neutralization, 246
Neutrons, 27
Neutrophils, 244
Nitrification, 304
Nitrogen cycle, 302
Nitrogen fixation, 175, 304
Nitrogenous wastes, 247
    types of, 248
Nodes of Ranvier, 266
Nomograph, 358
Noncoding RNAs, 130
Noncompetitive inhibitors, 56
Nondisjunction, 111
Nonpituitary hormones, 254
Nonpolar covalent bonds, 28
Nonpolar molecules, 50
Nonsense mutations, 128
Nonvascular plants, 179
Notochord, 194, 262
NPP (net primary production), 302
Nuclear envelope, 44, 45, 63

Nuclear pores, 45
Nuclear transplantation, 138
Nucleases, 119
Nucleic acid hybridization, 136
Nucleic acids, 35–36, 236
Nucleolus, 44, 45, 46
Nucleosomes, 121, 122
Nucleotide excision repair, 119
Nucleotides, 35, 126
Nucleus of cell, 44, 45
Null hypothesis, 337, 338
Nutrients, classes of, 233

O

Obligate aerobes, 175
Obligate anaerobes, 175
Oils, 33
Okazaki fragments of DNA, 119
Old Testament, 151
Oligodendrocytes, 269
Oligotrophic lakes, 292
Ommatidia, 272
Oncogenes, 131
One gene–one polypeptide hypothesis, 122
Oocyte, 256
Oogenesis, 258
Oogonia, 258
Oparin, 161
Open circulatory systems, 237
Operant conditioning, 276
Operator portion of operon, 129
Operon, 129
Operon genes, 129
Opsonization, 246
Optimal foraging model, 276
Oral cavity, 235
Order of organism, 171
Organ of Corti, 271
Organ systems, 231
Organogenesis, 262
Organs, 231
Origin of species, 159–160
Origins of replication, 119
Osculum, 191
Osmoregulation and excretion, 247–251
Osmosis, 50, 211, 312
Osmosis lab, 312–314
Outer ear, 271
Ova, 98
Oval window of ear, 271
Ovarian cycle, 258
Ovaries, 256
Oviduct, 256
Ovulation, 255, 256
Ovules in seed plants, 181
Ovum, 255
Oxaloacetate, 89
Oxidation, 76

Oxidation-reduction (redox) reactions, 76
Oxidative phosphorylation, 78–80
Oxygen, 82
Oxytocin, 252
Ozone layer, 304

P

P (parental) generation, 103
P site, 126
p53 gene, 132
Pain receptors, 271
Paleontologists, 162
Paleontology, 153
Pancreatic amylases, 236
Paper chromatography, 322
Paraphyletic group, 176
Parasitic plants, 216
Parasitism, 175, 299
Parasympathetic division of nervous
    system, 270
Parasympathetic nerves, 240
Parenchyma, 348
Parenchyma cells, 208, 210
Parental types, 110
Parthenogenesis, 255
Partial pressure, 241
Parturition, 260
Passive diffusion, 50
Passive immunity, 246
Passive transport, 50–51, 211
Pathogenic prokaryotes, 175
Pattern formation, 131, 210
Pedigree, 107
Pelagic biome, 292
Penis, 256
Pentose, 35
PEP carboxylase, 89
Pepsin, 235
Pepsinogen, 235
Peptide bond formation, 127
Peptide bonds, 34
Peptidoglycans, 174
Perception, 271
Perennial plants, 209
Peripheral nervous system (PNS), 266, 269
Peripheral proteins, 49
Peristalsis, 235
Peritubular capillaries, 250
Permafrost, 294
Peroxisome, 44, 45, 46
pH
    denaturation of protein due to change
        in, 35
    and protein enzymes, 56
    scale, 30
Phages, 132
Phagocytic white blood cells, 243, 244
Phagocytosis, 52, 244

Pharmaceuticals, DNA technology in
    production of, 139
Pharyngeal clefts, 194
Pharynx, 235
Phenotype, 105
Pheromones, 276
Phloem, 208, 214, 350
Phloem cells, 208
Phosphate group, 35
Phosphoenolpyruvate (PEP), 89
Phospholipids, 34, 49
    bilayer, 49, 50
Phosphorylate ADP to ATP, 86
Phosphorylation cascade, 59
Photic zone, 292
Photoautotrophs, 175
Photoheterotrophs, 175
Photons, 83
Photoperiodism, 221
Photophosphorylation, 82
Photorespiration, 88
Photosynthesis, 81–89
    chemiosmosis in, 86
    overall reaction of, 82
    and plant pigments lab, 321–325
Photosystem I (PS I), 84, 85, 86
Photosystem II (PS II), 84, 85, 86
Photosystems, 84
Phototropism, 220
Phylogenetic trees, 172
Phylogeny, 171–173
Phylum, 171
Phylum Annelida, 192
Phylum Arthropoda, 192
Phylum Chordata, 192, 194
Phylum Cnidaria, 191
Phylum Mollusca, 191
Phylum Nematoda, 191
Phylum Platyhelminthes, 191
Physical defenses of plants, 222
Phytochromes, 221
Pigments, 83
Pili, 174
Pinna, 271
Pinocytosis, 52
Placental mammals, 197
Plant nutrition, 216
Plant pigments and photosynthesis lab,
    321–325
Plant structure, growth, and development,
    207–210
Plasma, 240
    membrane, 44, 45, 48
    membrane receptors, 57
Plasmids, 174, 329
Plasmodesmata, 45, 48, 215
Platelets, 241
Pleiotropy, 107

PNS (peripheral nervous system), 266, 269
Point mutations, 128, 155
Polar covalent bonds, 28
Polar molecules, 29, 50
Polar nuclei, 217
Polarity, 210
Pollen grain, 217
Pollen in seed plants, 181
Pollen tube, 217
Pollination, 217
Poly-A tail addition to RNA, 124
Polygamous mating systems, 277
Polygenic inheritance, 107
Polymerase chain reaction (PCR), 136
Polymers, 32
Polypeptide chains, 35
Polyphyletic group, 176
Polyploid speciation, 160
Polyploidy, 111
Polysaccharides, 32
Polyspermy, 261, 262
Pons, 270
Population, 155, 294
   ecology, 294–297
   genetics, 155
Population genetics and evolution lab,
   341–345
Positional information, 210
Positive feedback, 252, 254
Positive feedback systems, 232
Positive gravitropism, 222
Positive phototropism, 220
Posterior pituitary, 252
Posterior vena cava, 239
Postzygotic barriers, 159
   examples of, 160
Potential energy, 53
Potometer, 348
Predation, 298
Predator, 298
Pregnancy, 260
Pressure flow, 214
Pressure potential, 212
Prey, 298
Prezygotic barriers, 159
Primary consumers, 301
Primary electron acceptor, 84
Primary immune response, 245
Primary oocytes, 258
Primary producers in ecosystem, 301
Primary production, 301
Primary productivity, 358
Primary structure, 34
Primary succession, 300
Primates, 197
Prions, 134
Products from substrate, 54
Progesterone, 260

Prokaryotes
   as first living organisms, 162
   uses by humans, 176
Prokaryotic cells, 43
Proliferative phase, 258
Prometaphase, 62, 63
Promiscuous mating systems, 277
Promoter portion of operon, 129
Promoter sequence of DNA, 123
Prophage, 133
Prophase, 62, 63
Prophase I, 100
Prophase II, 101, 102
Prostate gland, 257
Protein enzymes
   and changes in pH, 56
   and changes in temperature, 56
Protein kinases, 59
Protein kinetochores, 63
Proteins, 34–35, 48, 236
Protists, 176–177
   examples of, 177
Protobionts, 161
Proton motive force, 79
Proton pump, 211, 215
Protonephridia/flame-bulb system, 248
Protons, 27
Proto-oncogenes, 131
Protostomes, 189
Proximal tubule, 250
Proximate causes of behavior, 275
Pseudocoelomates, 188, 191
Pulmonary arteries, 239
Pulmonary veins, 239
Pulse, 238, 352
Punctuated equilibrium, 160
Pupil of eyeball, 272
Pyloric sphincter, 235
Pyramids of energy, 302
Pyruvate, 80
Pyruvate molecules, 77, 78

Q

Quaternary structure, 35

R

Radial symmetry of animal body, 186
Radially symmetrical animals, 191
Radicle, 217
Radiometric dating of rocks or fossils, 162
Random fertilization, 103
Random mating, 342
Reabsorption process, 248, 251
Reaction center of photosystem, 84
Realized niche of species, 298
Reception, 271
   in cell signaling, 57–58
   of signals , 220

Receptor-mediated endocytosis, 52
Receptor protein, 57
Receptor tyrosine kinase, 58
Recessive allele, 104, 105
Recessively inherited disorders, 108
Recognition sites, 330
Recombinant DNA, 134
Recombinants, 110
Red blood cells (RBCs), 240
Reduced hybrid fertility, 160
Reduction, 76
Redundant genetic code, 123
Reflex, 169
Regeneration, 255
Regulatory genes, 129
Relative dating of fossils, 162
Relative fitness of organism, 157
Renal artery, 248
Renal vein, 248
Renin, 251
Repeated reproduction, 295
Replication bubble, 119
Replication of DNA, 118, 119
Reporter gene, 331
Repressible operon, 129
Repressor protein, 129
Reproduction in animals, 255–260
Reproductive anatomy
   female, 257
   male, 257
Reproductive isolation, 159
Reptiles, 196
Respiration, 241, 359
Respiratory medium, 241
Respiratory surface, 241
Respirometer, 326
Response
   in cell signaling, 57, 60
   cellular, 220
Resting potential of nerve cell, 266
Restriction enzymes, 134, 330
Restriction fragment length
    polymorphisms, 139
Restriction fragments of DNA, 134
Restriction sites, 134
Retina, 272
Retroviruses, 133
Reverse multiple-choice questions, 8–9
Reverse transcriptase, 133, 136
Rheumatoid arthritis, 247
Rhizobacteria, 216
Rhizobium, 304
Rhizobium bacteria, 216
Rhizosphere, 216
Rhodopsin, 272
Ribonucleic acid (RNA), 35, 36, 45
   See also RNA.
Ribosomal RNA (rRNA), 125, 126

Ribosomes, 44, 45, 46, 122
Ribozyme, 124
Ribulose bisphosphate (RuBP), 88
Right atrium, 239
Right ventricle, 239
Rivers, current in, 292
RNA (ribonucleic acid), 35, 45
    polymerase, 123
    processing, 122
    splicing, 124
Rods in eye, 272
Root cap, 210
Root hairs, 207
Root pressure, 213
Root system of plants, 207
Root tip, 210
Rough endoplasmic reticulum (ER), 45, 46
Round tissue, 208
r-selection, 297
Rubisco, 88
Rule of addition in probability, 106
Rule of multiplication in probability, 106

S

S phase, 61
    checkpoint, 63
SA node, 240
Sac fungi, 185
Saliva, 235
Salivary amylase, 236
Saltatory conduction, 266
Sample questions and answers
    animal form and function, 277–289
    cell, 65–73
    chemistry of life, 36–41
    ecology, 304–310
    evolution, 164–170
    evolutionary history, 198–206
    Mendelian genetics, 112–116
    molecular genetics, 139–150
    plant form and function, 223–229
    respiration and photosynthesis, 90–96
Sample tests, taking, 16–17
San Andreas Fault, 162
Sarcomere, 273
Sarcoplasmic reticulum, 273
Saturated fatty acids, 33
Savannas, 293
Scala naturae, 151
Scales, 196, 197
Schwann cells, 269
Sclera, 272
Sclerenchyma cells, 208, 350
Scores on exams, 4
Scrotum, 256
Second messenger ions, 59
Secondary consumers, 301
Secondary immune response, 245

Secondary oocyte, 258
Secondary structure, 34
Secondary succession, 300
Secretin, 236
Secretion process, 248, 251
Secretory phase, 258
Section I of AP exam, 7–10
    grading procedures for, 13
    strategies for, 18–19
Section II of AP exam, 11–12
    grading procedures for, 13–14
    strategies for, 19–21
Seed coat, 217
Seed plants, evolution of, 181–183
Seeds, 181, 326
Segregation, Mendel's law of, 104
Selection gene, 331
Selectively permeable cell, 48
Self-replicating RNA as first genetic
    material, 161
Semicircular canals, 271
Semiconservative replication, 118
Semilog paper, 332
Semilunar valves, 240
Seminal vesicles, 257
Seminiferous tubules, 256, 257, 258
Sensory mechanisms, 271–275
Sensory neurons, 266
Sensory receptors, 266
Septa, 185
Set point in homeostatic control
    systems, 232
Sex-linked characteristic, 336
Sex-linked gene, 109
Sex reversal, 255
Sexual life cycles, meiosis and, 97–103
Sexual reproduction, 97, 255
Shoot system, 207
Short-day plants, 222
Sickle-cell disease, 108
Sieve-tube elements, 208
Sieve tubes, 214
Sign stimuli, 275
Signal, 275
Signal peptide, 128
Signal transduction, 219
Signal transduction pathway, 252
Signaling, 264–267
Single-lens eyes, 272
Sinoatrial (SA) node, 240
Sinuses, 237
Sister chromatids, 61, 102
Skeletal muscle, 272, 273
Sliding-filament model, 273
Small interfering RNAs (siRNAs), 130
Small intestine, 235, 236
Smooth endoplasmic reticulum (ER), 45, 46
Sodium-potassium pump, 51

Solute, 30
Solute potential, 212
Solution, 30
Solvent, 30
Somatic cells, 60, 98
Somatic nervous system, 269
Somites, 263
Sordaria, meiosis in, 319
Southern blotting, 136
Speciation, 159, 160
Species, 171
Species diversity, 299
    of organism, 171
Specific heat, 30
Sperm, 98, 255, 256
    production in seed plants, 181
Spermatogenesis, 258, 260
Spermatogonia, 258
Sphincters, 235
Sphygmomanometer, 240, 351
Spicules, 191
Spindle, 63
Spliceosome, 124
Sponges, 191
Spongocoel, 191
Sporangia, 180
Spores, 180
Sporophyte stage of plants, 178, 180
Stabilizing selection, 157
Stamens, 183
Stapes, 271
Starch, 33
Stem cells, 138, 241
Stems, 207
Steroids, 33, 34
Sticky end of DNA, 134, 330
Stomata, 81, 210, 346
    light-induced opening of, 221
Streams, current in, 292
Stroke volume, 240
Stroma, 81
Structural support polysaccharides, 33
Structure of water, 29
Subkingdom Eumetazoa, 191
Subkingdom parazoa, 191
Substrate, 54
Sucrose loading, 215
Sugar sink, 214, 215
Sugar source, 214
Symbiosis, 299
Symbiotic prokaryotes, 175
Sympathetic division of nervous system, 270
Sympathetic nerves, 240
Sympatric speciation, 160
Symplasm, 215
Symplastic route, 213
Synapses, 264–267
Synapsis, 100

Synaptic terminals, 265
Synthetic auxins, 220
Systematics, 171
Systemic inflammatory responses, 244
Systole, 239, 351
Systolic pressure, 240

T

T cells, 245
T lymphocytes (T cells), 244
T tubules (transverse tubules), 273
Taproots, 207
Taste buds, 272
Taxon, 171
Taxonomy, 151, 171
Tay-Sachs disease, 108
T-cell activation, 244
Telomeres, 120
Telophase, 62, 63
Telophase I, 100, 102
Telophase II, 101, 102
Temperate broadleaf forest, 293
Temperate grassland, 293
Temperature changes and protein
    enzymes, 56
Template strand, 123
Temporal isolation, 159
Terminal bud, 207–208
Termination of DNA sequence, 124
Termination stage of translation, 127
Terminator sequence of DNA, 123
Terrestrial nitrogen cycle, 303
Tertiary consumers, 301
Tertiary structure, 34
Testcross, 105
Testes, 2567
Testosterone, 34
Test-taking strategies, 15–21
Tetrapods, 196
Thalamus, 270
Themes in biology, 5
Thermoclines, 292
Thermodynamics, 53
Thermoreceptors, 271
Thermoregulation, 232
Thick filaments, 273
Thigmotropism, 222
Thin filaments, 273
30nm fiber, 121, 122
3-bisphosphoglycerate, 88
Three-domain system of life, 173
3-phosphoglycerate, 88
Threshold of depolarization, 266
Thylakoid membrane, 84
Thylakoid space, 81
Thylakoids, 81
Thymine (T) , 36, 118, 126
Tight junctions, 48

Tinbergen, Nicholas, 275
Tissues, 231
Topic correlation chart, 22–23
Topics in biology, 5
    outline of, 5–6
Totipotent cells, 264
Trace elements, 27
Trachea, 242
Tracheal systems, 242
Tracheids, 208
Transcription, 122
Transcription factors, 123
Transcription initiation complex, 123, 130
Transcription unit of DNA, 123
Transduction, 59
    in cell signaling, 57, 59
    of genes, 174
    of signal, 220
Transfer RNA (tRNA), 125
Transformation
    of blood filtrate to urine, steps in,
        250–251
    of cell, 64
    efficiency, 331
    genetic, 329
    in prokaryote, 174
Translation, 123
    three stages of, 126–128
Translocation, 214
    of chromosome fragment, 111
Translocution, 127
Transmembrane protein, 79
Transpiration, 29, 213, 346
Transpiration-cohesion-tension
    mechanism, 213
Transpiration lab, 346–351
Transport epithelia, 247
Transport proteins, 50–51, 211
Transport vesicles, 46
Transverse tubules (T tubules), 273
Triglycerides, 33
Triplet code, 123
Trisomic eggs, 111
Trophic levels, 299
Trophic structure, 299
Trophoblast, 263
Tropic hormones, 252
Tropical forest, 294
Tropism, 220
Troponin, 273
True coelom, 188
Trypsin, 236
Tube nucleus, 217
Tumor, 64
Tumor-suppressor genes, 131, 132
Tundra, 294
Turner syndrome, 111
Tympanic membrane, 271

U

Ultimate causes of behavior, 275
Uniformitarianism, 152
Unsaturated fatty acids, 34
Urea, 248
Ureters, 248
Urethra, 248
Urey, 161
Uric acid, 248
Urine, 248
Uses and disuse theory, 152

V

Vacuoles, 48
Vagina, 256
Valence electrons in carbon, 31
Van der Waals interactions, 29, 34
Vas deferens, 257
Vascular cambium, 210
Vascular plants, resource acquisition and
    transport in, 211–215
Vascular tissue, 208
Vegetative reproduction, 219
Veins, 238
Ventral nerve cord, 268
Ventricles, 238
Venules, 239
Vertebrae, 194
Vertebrate brain, 270
Vertebrates, 194–197
Vesicles, 265
Vessels, 237
Vessels cells, 208
Vestigial organs, 153
Villi in small intestine, 236
Viral envelopes, 132
Viral reproduction, 133
Viroids, 134
Viruses, 132–134
Visual signals, 276
Vitamins, 233
Vitreous humor, 272
Voice box, 242
Von Frisch, Karl, 276

W

Waggle dance, 276
Warning coloration, 298
Water, 29–30
    emergent properties of, 29
    structure of, 29
Water absorption by roots, 212
Water potential, 211, 313
    calculating, 314
    equation, 211
    factors affecting, 314
Watson, James, 117, 118

Waxes, 33
White blood cells (WBCs), 240
White matter, 269
Wild type form of trait, 335
Wilkins, Maurice, 117
Windpipe, 242
Wobble, 126

X

X chromosome, 98
X-ray crystallography, 117

Xylem, 208
  cells, 208, 350
  sap, 213

Y

Y chromosome, 98
Yolk sac, 196

Z

Z lines, 273

Zone
  of cell division, 210
  of elongation, 210
  of maturation, 210
Zygomycota fungi, 185
Zygote, 99, 186, 187, 217, 255
Zygote fungi, 185